THE ROUTLEDGE HANDBOOK OF PHILOSOPHY OF BIODIVERSITY

W0113370

Biological diversity – or "biodiversity" – is the degree of variation of life within an ecosystem. It is a relatively new topic of study, but has grown enormously in recent years. Because of its interdisciplinary nature, the very concept of biodiversity is the subject of debate among philosophers, biologists, geographers, and environmentalists.

The Routledge Handbook of Philosophy of Biodiversity is an outstanding reference source to the key topics and debates in this exciting subject. Comprising twenty-three chapters by a team of international contributors, the *Handbook* is divided into six parts:

- **Historical and sociological contexts:** focusing on the emergence of the term and early attempts to measure biodiversity.
- **What is biodiversity?** How should biodiversity be defined? How can biodiversity include entities at the edge of its boundaries, including microbial diversity and genetically engineered organisms?
- **Why protect biodiversity?** What can traditional environmental ethics contribute to biodiversity? Topics covered include anthropocentrism, intrinsic value, and ethical controversies surrounding the economics of biodiversity.
- **Measurement and methodology:** including decision theory and conservation, the use of indicators for biodiversity, and the changing use of genetics in biodiversity conservation.
- **Social contexts and global justice:** including conservation and community conflicts and biodiversity and cultural values.
- **Biodiversity and other environmental values:** how does biodiversity relate to other values like ecological restoration or ecological sustainability?

Essential reading for students and researchers in philosophy, environmental science and environmental studies, and conservation management, it will also be extremely useful to those studying biodiversity in subjects such as biology and geography.

Justin Garson is Associate Professor in the Department of Philosophy at Hunter College of the City University of New York, USA.

Anya Plutynski is Associate Professor in the Department of Philosophy at Washington University in St Louis, USA.

Sahotra Sarkar is Professor in the Departments of Integrative Biology and Philosophy at the University of Texas at Austin, USA.

Routledge Handbooks in Philosophy

Routledge Handbooks in Philosophy are state-of-the-art surveys of emerging, newly refreshed, and important fields in philosophy, providing accessible yet thorough assessments of key problems, themes, thinkers, and recent developments in research.

All chapters for each volume are specially commissioned, and written by leading scholars in the field. Carefully edited and organized, *Routledge Handbooks in Philosophy* provide indispensable reference tools for students and researchers seeking a comprehensive overview of new and exciting topics in philosophy. They are also valuable teaching resources as accompaniments to textbooks, anthologies, and research-orientated publications.

Also available:

The Routledge Handbook of Embodied Cognition
Edited by Lawrence Shapiro

The Routledge Handbook of Neoplatonism
Edited by Pauliina Remes and Svetla Slaveva-Griffin

The Routledge Handbook of Contemporary Philosophy of Religion
Edited by Graham Oppy

The Routledge Handbook of Philosophy of Well-Being
Edited by Guy Fletcher

The Routledge Handbook of Philosophy of Imagination
Edited by Amy Kind

The Routledge Handbook of the Stoic Tradition
Edited by John Sellars

The Routledge Handbook of Philosophy of Information
Edited by Luciano Floridi

The Routledge Handbook of Philosophy of the Social Mind
Edited by Julian Kiverstein

THE ROUTLEDGE HANDBOOK OF PHILOSOPHY OF BIODIVERSITY

Edited by

Justin Garson, Anya Plutynski, and Sahotra Sarkar

LONDON AND NEW YORK

First published 2017 by Routledge

2 Park Square, Milton Park, Abingdon, Oxfordshire OX14 4RN
52 Vanderbilt Avenue, New York, NY 10017

Routledge is an imprint of the Taylor & Francis Group, an informa business

First issued in paperback 2019

British Library Cataloguing in Publication Data
A catalogue record for this book is available from the British Library

Library of Congress Cataloguing in Publication Data
Names: Garson, Justin, editor. Title: The Routledge handbook of philosophy of biodiversity /
edited by Justin Garson, Anya Plutynski, and Sahotra Sarkar. Description: 1 [edition]. |
New York : Routledge, 2016. | Series: Routledge handbooks in philosophy |
Includes bibliographical references and index. Identifiers: LCCN 2016011130|
ISBN 9781138827738 (hardback : alk. paper) | ISBN 9781315530215 (e-book)
Subjects: LCSH: Biodiversity–Philosophy.
Classification: LCC QH541.15.B56 R686 2016 | DDC 333.95–dc23
LC record available at https://lccn.loc.gov/2016011130

ISBN 13: 978-1-138-82773-8 (hbk)
ISBN 13: 978-0-367-37049-7 (pbk)

Typeset in Bembo
by Out of House Publishing

CONTENTS

Contents

LIST OF FIGURES

LIST OF TABLES

NOTES ON CONTRIBUTORS

Jeremy Brooks is an Assistant Professor in the School of Environment and Natural Resources at The Ohio State University, USA. He is an environmental social scientist whose research areas are community-based conservation, sustainable development, and sustainable consumption.

J. Baird Callicott is co-Editor-in-Chief of the *Encyclopedia of Environmental Ethics and Philosophy*; he is author or editor of a score of books, latest among them *Thinking Like a Planet: The Land Ethic and the Earth Ethic*; and author of dozens of journal articles and book chapters.

Shaun Cunningham is a Research Fellow at the School of Biological Sciences, Deakin University, Melbourne, Australia. His research interests include predicting vegetation condition and extent across regions, restoring vegetation and ecological processes in agricultural landscapes, and physiological explanations for plant distribution.

Jacqueline England is a Research Scientist at the Land and Water Unit of the Commonwealth Scientific and Industrial Research Organisation, Australia. Her main research interest is the ecology of natural and planted forests, including assessing the co-benefits of carbon plantings in agricultural landscapes.

Daniel P. Faith is a Senior Principal Research Scientist at the Australian Museum, Sydney, Australia. His research is concerned with theory and applications of quantitative biodiversity assessment. Special emphasis is on the best possible use of museum collections in regional biodiversity assessment, and on the links from biodiversity assessment to sustainability.

Timothy Farnham is Director of the Miller Worley Center for the Environment and Associate Professor of Environmental Studies at Mount Holyoke College in South Hadley, MA, USA. His research interests lie in environmental values and the history of environmental thought. He is the author of *Saving Nature's Legacy: Origins of the Idea of Biological Diversity* (2007).

David M. Frank is currently a Lecturer in the Department of Philosophy and Department of Ecology and Evolutionary Biology, University of Tennessee, Knoxville, USA. His research focuses on topics at the intersection of ethics and philosophy of science.

Yayoi Fujita-Lagerqvist is a Lecturer of Human Geography in the School of Geosciences at the University of Sydney, Australia. Her research interests are in natural resource management and rural livelihood in mainland Southeast Asia.

Justin Garson is an Associate Professor of Philosophy at Hunter College of the City University of New York, USA. His main research areas are the philosophy of biology, environmental philosophy, and the history of neuroscience. He is the author of *The Biological Mind: A Philosophical Introduction* (Routledge, 2015).

Yrjö Haila is Professor of Environmental Policy (Emeritus) at the University of Tampere, Finland. Educated as an ecologist, his main research interests have centered on the nature–society interface. He has published *Humanity and Nature* together with Richard Levins (1992), and *How Nature Speaks,* co-edited with Chuck Dyke (2006).

Lisa Heinzerling is the Justice William J. Brennan, Jr, Professor of Law at Georgetown University Law Center, USA. In 2009 and 2010, she was the senior climate advisor to the administrator of the US Environmental Protection Agency and then the Associate Administrator of EPA's Office of Policy.

Christopher Lean is a PhD student at the Australian National University. His research is within philosophy of biology with a primary focus on debates in ecology and conservation science. He recently published a defense of the use of phylogenetic measures as proxies for "biodiversity" with James Maclaurin and on cancer and levels of selection with Anya Plutynski.

James Maclaurin is an Associate Professor in the Department of Philosophy at the University of Otago, New Zealand. He is a philosopher of biology with particular interests in biodiversity and in the application of evolutionary principles outside biology, in domains such as computer science and economics.

Katie McShane is an Assistant Professor of Philosophy at Colorado State University, USA, whose research focuses on environmental ethics and ethical theory. Her work has been published in journals such as *Philosophical Studies, Environmental Ethics, Environmental Values,* and *Ethics & the Environment.*

Lynn A. Maguire is Professor of the Practice of Environmental Decision Analysis in the Nicholas School of the Environment, Duke University, Durham, NC, USA. Her interests combine decision analysis, environmental dispute resolution, and collaborative planning.

Christophe Malaterre is Assistant Professor of Philosophy, and Canada Research Chair in Philosophy of the Life Sciences at the Université du Québec à Montréal (UQAM), Canada. His main research is in the philosophy of biology and focuses on the epistemic questions that arise at interface between the chemical and the biological worlds.

Ben A. Minteer holds the Arizona Zoological Society Endowed Chair at Arizona State University, USA and is a Professor of Environmental Ethics and Conservation in ASU's School of Life Sciences. His most recent book is *After Preservation: Saving American Nature in the Age of Humans* (2015).

Jay Odenbaugh is an Associate Professor of Philosophy at Lewis & Clark College in Portland, OR, USA. His main research area is the philosophy of science focusing on biology and psychology. He is currently writing a book tentatively titled *In a Sentimental Mood: Emotion, Evolution, and Expression.*

Anya Plutynski is Associate Professor in the Department of Philosophy at Washington University in St Louis, MO, USA. She is a historian and philosopher of biology and medicine. She co-edited Blackwell's *Companion to Philosophy of Biology*, and her forthcoming book is *Explaining Cancer: Philosophical Issues in Cancer Causation, Evidence & Explanation.*

Carlos Santana is an Assistant Professor of Philosophy at the University of Utah, USA. His research is in the areas of environmental philosophy, the philosophy of biology, and the philosophy of linguistics.

Sahotra Sarkar is Professor in the Departments of Integrative Biology and Philosophy at the University of Texas at Austin, USA. He is a philosopher of science and a conservation biologist and disease ecologist with a PhD from the University of Chicago.

David Sepkoski is a Senior Research Scholar at the Max Planck Institute for the History of Science, Germany. His most recent book is *Rereading the Fossil Record: The Growth of Paleobiology as an Evolutionary Discipline* (2012). He is currently writing a book on the history of extinction.

Helena Siipi works as a collegium researcher at the Turku Institute for Advanced Studies (TIAS) and the Philosophy Unit of the University of Turku, Finland. Her main research areas are applied ethics, environmental philosophy, ethics of new biotechnologies, and philosophy of food.

Kim Sterelny is a Professor of Philosophy at the ANU, though he maintains his long association with Victoria University of Wellington as well. His central interests are in philosophy of biology, recently with a special focus on the evolution of human social behaviour and social complexity.

Alan R. Templeton is the Charles Rebstock Professor Emeritus of Biology at Washington University in St Louis, MO, USA, and a part-time Professor of Evolutionary and Environmental Biology at the University of Haifa, Israel. Trained in human genetics, he also applies genetics to conservation programs throughout the world.

Kerrie Wilson is an Associate Professor and ARC Future Fellow at The University of Queensland, Australia and an Affiliated Professor in Conservation Science at The University of Copenhagen, Denmark. She has authored over 100 publications in the field of environmental decision-making.

ACKNOWLEDGMENTS

We are grateful to Tony Bruce, our editor at Routledge, for suggesting the idea of this volume and inviting us to edit it. We would also like to thank two anonymous referees at Routledge for their valuable comments and feedback on the initial proposal. Finally, we wish to thank Neil Sentance for his thorough copyediting.

INTRODUCTION

Justin Garson, Anya Plutynski, and Sahotra Sarkar

The term "biodiversity" was coined in 1986 as shorthand for "biological diversity," and soon became the primary focus of the emerging discipline of conservation biology. Despite its prominence, the concept of biodiversity sits uncomfortably at the intersection of different discourses, and is often invoked to perform different, and perhaps incongruous, roles. Partly as a consequence of these competing roles, the nature of biodiversity, its measurement, and its value, have been the subjects of heated debate among philosophers, biologists, environmental planners, and social scientists. *The Routledge Handbook of Philosophy of Biodiversity* brings the diverse facets of these debates together in a single volume, with the hopes of fostering further discussion. One feature of the volume is its interdisciplinary character: it includes contributions from philosophers, biologists, lawyers, environmental planners, and social scientists.

There are at least four general areas of discourse in which the concept of biodiversity is embedded: conceptual, ethical, methodological, and sociopolitical. We will briefly summarize significant problems in each of these four domains. The first is *conceptual*: what is biodiversity? How should "biodiversity" be defined? Conservationists commonly define "biodiversity" as diversity at all levels of biological structural, taxonomic, and functional organization, from alleles to ecosystems. Yet the problem with such a definition is apparent immediately, as it makes "biodiversity" synonymous with the entire biological world, which is fruitless for the purposes of conservation planning.

More generally, if biodiversity is a legitimate theoretical concept in biology, then we should expect to be able to identify law-like generalizations in which it plays a role. But such law-like generalizations are hard to find. For example, ecologists have argued for decades over whether there is a significant correlation between the diversity of an ecosystem and its stability. The answer has remained elusive, and the debate continues to this day. The fact that the debate has stretched on for so long suggests that the disputants do not entirely agree on the meaning of the term. Another possibility, one that would explain the relative paucity of law-like generalizations involving biodiversity, is that "biodiversity" is a value-laden term but not a scientific one. If "biodiversity" means, in part, whatever aspect of the biological world the speaker finds valuable, then we would expect its extension to be fluid.

When we reflect on the nature of biodiversity, we are quickly brought to questions about its value, and to ethical questions. Why is it worth preserving? Do we have an obligation to preserve it? This is not an idle question: successful conservation projects consume money, labor, time, expertise, and other valuable resources – resources that may have gone to other worthwhile ends such as reducing poverty or disease. One problem stems from the following consideration. A standard sort of ethical justification for protecting some non-human animals is that those animals possess *intrinsic value* (though theorists differ on what exactly intrinsic value is and which creatures have it). But can collective entities, such as populations, or even biodiversity, possess intrinsic value? Is biodiversity valuable in its own right, or merely because of the way it serves the needs and goals of individuals?

The third area of discussion is the domain of *methodology* and measurement. How should biodiversity be measured in the field? What data must one collect to demonstrate that one region has more biodiversity than some other? One might think that once the conceptual question is solved – what is biodiversity? – the measurement problem would be fairly straightforward. But the situation is not so simple. In fact, arguably, these two roles are in tension with one another, that is, the role that biodiversity plays in biological theory and the role it plays as a tool of measurement. For the purpose of measurement, environmental planners require an instrument that permits them to prioritize areas for conservation action *rapidly* and *reliably*. Certain concepts of biodiversity that are reasonable in the context of biological theory, such as phylogenetic concepts of biodiversity, may be difficult to employ for practitioners in the field. Reconciling these two roles is no easy task.

Finally, discussions about the meaning, value, and measurement of biodiversity take place in a social and political setting, which is our fourth area of concern. Social realities infiltrate discussions about biodiversity in at least two ways. First, if biodiversity is a normative concept, then what counts as "biodiversity" on any given occasion is determined, in part, by the goals and values of the community that uses the term. In North America, for example, large, charismatic fauna fit the prototype of biodiversity far better than, say, insect diversity or even microbial diversity. But not all cultures share these preferences. Hence, understanding what biodiversity *means* for a community reveals the character of the culture and must be taken into account when conservation strategies are being designed and implemented.

Social realities enter conservation practices through a second route. Conservation projects, as a rule, take place in the context of competing land-use preferences. Many of these competing preferences may have *prima facie* importance, such as a preference for human habitation, or the production of biomass for human consumption. This raises the problem of negotiation among stakeholders. This problem is particularly salient when the needs and desires of conservationists to protect biodiversity run up against a community's legitimate claim to exercise sovereignty over its own land. The design and coordination of international biodiversity conservation projects must recognize and respect the constraints of social and global justice. But what does justice demand of us? And who (e.g. conservationists, social scientists, local communities) is best able to prescribe and manage the way that people relate to their land?

We have two main goals with this volume as a whole. Our first goal is simply to foster more discussion among philosophers, biologists, conservationists, and social scientists, regarding these important questions. We believe that the time is ripe for biodiversity to assume a more prominent role in philosophical discussion. Our second goal, however, is a more general one. Over the last two decades, the field of *environmental philosophy* has started to emerge as a well-defined area of philosophical inquiry in contrast to *environmental ethics*. If environmental ethics is thought of as preoccupied with the nature and foundation of our moral responsibilities to the non-human world, then environmental philosophy encompasses environmental ethics, but places it within a more comprehensive framework for thinking philosophically about the environment. As the contributions in this volume will testify, environmental philosophy includes the kind of rigorous conceptual analysis associated with other areas of philosophy, as well as themes that have been more strongly associated with the philosophy of science (including the philosophy of the social sciences and the philosophy of economics). Though this volume focuses primarily on biodiversity, one role of the book is to promote the field of environmental philosophy as a project that is distinct from, but closely related to, traditional environmental ethics.

The outstanding contributions in this volume speak for themselves. However, an overview of the structure of the book, and a brief introduction to the contents of the chapters, is warranted. This will provide the reader with an initial orientation to a complex topic.

The contributions that make up the book fall into six main sections. Part I, "Historical and sociological contexts," characterizes the social concerns and forces that propelled biodiversity into the foreground of current international attention. It also looks at early attempts in ecology to understand and define biological diversity (that is, prior to the solidification of conservation biology as a self-identified scientific discipline in the 1980s). Understanding the historical problem-context that motivated a concern with biodiversity provides, we believe, the proper orientation for assessing its contemporary meaning and value.

Timothy Farnham, in his contribution, explores the question of why interest in biodiversity exploded in the 1980s. He shows how the recent concern for biodiversity grew out of three distinctive sorts of concerns raised by biologists in the course of the twentieth century. The first was the concern for species loss, the second was the concern for the diminution of genetic variety, and the third was the concern for the disappearance of ecosystem types. In Farnham's view, our current preoccupation with biodiversity protection represents the convergence or "confluence" of these streams.

David Sepkoski explores one particular strand of this confluence at greater length. Here, he advances the thesis that our current preoccupation with biodiversity stemmed specifically from a widespread concern for species loss, which concern started in the mid-twentieth century. This concern is expressed, among other ways, in the view that the Earth is in the throes of a sixth major extinction crisis. Sepkoski traces this concern with species loss to changing conceptions about the very nature of extinction. Throughout the early twentieth century, many leading biologists regarded extinction in the rosy context of the balance of nature; if one species was driven to extinction, another would probably emerge to fill its place. After the mid-twentieth century documentation of mass extinctions, as well as broader social and cultural changes, it became impossible to think of extinction in such optimistic terms.

Theorists have offered several different, and even competing, characterizations of what biodiversity is. Part II, "What is biodiversity?," explores some of these characterizations. The contributors examine various proposals for understanding biodiversity. Should we even retain the concept of biodiversity in light of the multiplicity of characterizations? Another method for thinking about biodiversity is to explore what happens when we apply the concept to traditionally marginal or borderline cases. How do existing concepts of biodiversity apply to the diversity of microbial life? How do they apply to synthetic life?

Sahotra Sarkar reviews different approaches to defining biodiversity, scientistic, normative, deflationary, and eliminative. His contribution notes how biodiversity has been operationalized within systematic conservation planning for the selection of representative conservation area networks. But it also notes the problems with such scientific approaches to the question, including a lack of consensus among practitioners. Finally, it reviews the other major approaches to the problem of defining and determining biodiversity.

James Maclaurin begins by noting the explosion of "biodiversity" measures and concepts. In light of this multiplicity, should biodiversity be thought of as a natural kind (analogous to "selection pressure" or "memory")? Or is it a loosely related hodgepodge of different properties? Whether or not biodiversity is a natural kind (or natural quality) depends on what we think natural kinds are, and it depends on what we think biodiversity is. Maclaurin provides an overview of the recent literature on natural kinds and discusses how biodiversity can plausibly constitute a natural kind.

Daniel P. Faith takes a broader look at the concept of biodiversity and its use than most other contributions in the volume. He argues that the conceptual apparatus of systematic conservation planning, which assumes that planning is about spatial areas, can be seamlessly extended to a broader (e.g. phylogenetic) context where the units can include taxonomic and trait groups. This broadening captures a lot of what made biodiversity an attractive concept in biological subdisciplines beyond conservation biology in which it originated.

Carlos Santana argues that, in general, "biodiversity" should be eliminated from scientific contexts (eliminativism about biodiversity). This is not just because of the confusing proliferation of definitions and measurements of biodiversity. More problematically, he thinks that different definitions of biodiversity embody value judgments in ways that are not entirely transparent. By doing away with the term, not only will our scientific language be more precise, but our underlying values can be rendered more perspicuous.

David M. Frank develops the notion of *definitional risk*. When we develop a specific operational definition of "biodiversity," we risk neglecting, inadvertently, certain aspects of the biological world that we take to be valuable or significant. As a consequence, he develops a view called *contextual pluralism*, which holds that different definitions of biodiversity are appropriate in different contexts, subject to some minimal constraints. He distinguishes his view from eliminitivism, the position endorsed by Santana, and provides some arguments against eliminitivism.

Christopher Lean and Kim Sterelny note that biodiversity protection often focuses on species diversity, as if species were the only significant units of biodiversity. But some ecologists have suggested that we target the diversity of larger ecological units, and in particular, ecological assemblages or landscapes. Nonetheless, Lean and Sterelny urge that biodiversity assessments *not* include units above the species level. In their engaging and challenging contribution, they draw heavily upon the work of philosopher of biology William Wimsatt to show that ecological assemblages and landscapes are generally not "system-like" enough – that is, they are too ephemeral – to constitute units of biodiversity in their own right.

Helena Siipi observes that in many standard discussions, biodiversity is explicitly, or implicitly, restricted to natural entities, that is, those that have not undergone extensive human modification (such as domesticated animals or genetically engineered organisms). But what is it for something to be unnatural, and should such entities be excluded from the scope of biodiversity? Siipi surveys a number of different senses of "natural" and "unnatural" and concludes that there is probably no single idea of naturalness that can capture everything we want in a definition of "biodiversity."

Christophe Malaterre points out that when we think about defining and protecting biodiversity, we typically think neither of microbial diversity, nor of the diversity of viruses. But why not? He identifies several fascinating philosophical, ethical, and strategic questions that arise when we consider the diversity of microbes. First, assessing the diversity of microbial species raises thorny questions about what species are. Second, it raises ethical problems: clearly, protecting microbial diversity can sometimes be justified in terms of its potential benefit to people and ecosystems, but do microbes have intrinsic value the way that some macroorganisms (presumably) do? And what constitutes an ethically acceptable way of preserving them?

Part III, "Why protect biodiversity?," turns attention to traditional environmental ethics, specifically insofar as it concerns the nature and foundation of human obligations toward the non-human world. The justification for the protection of biodiversity, per se, does not appear to be straightforward, because existing justifications for the protection of individual organisms, such as those offered by animal rights or biocentrism, do not necessarily explain why we should focus on the diversity of life itself. Here, the contributors will discuss different approaches to thinking about the value of biodiversity, including those that center upon intrinsic value, economic value, and moral psychology.

Katie McShane pursues the question what exactly does it *mean* to attribute intrinsic value to biodiversity as such? One problem here is that intrinsic value has been defined in several different, non-overlapping ways. She surveys at least eight different ways in which philosophers have tried to make sense of this perplexing idea. Although she refrains from rendering a final verdict on the question, she lays out the specific challenges that proponents of intrinsic value must face, particularly those who wish to attribute intrinsic value to biodiversity as such.

J. Baird Callicott elaborates a theory of intrinsic value that permits biodiversity, as well as other parts of the natural world including non-human animals, to possess intrinsic value. The key move is to think of *value* as primarily an activity of cognitive agents (a verb), and *intrinsic* as a feature of this activity (an adverb). Hence, valuing something intrinsically stands in contrast to valuing something instrumentally, and such values are amenable to rational deliberation. Crucially, in his view, intrinsic value assessments can be ranked (e.g. the intrinsic value of a moth can be less than that of a mountain gorilla) and one function of law is to codify these intrinsic value rankings.

Lisa Heinzerling assesses attempts to place economic value on biodiversity. Drawing on previous work, she brings one particular method of economic valuation, cost–benefit analysis (CBA), under critical scrutiny. CBA, she argues, is riddled with problems, some of which are methodological (for example, methodological problems with willingness-to-pay assessments) and some of which are ethical (for example, the process of discounting future benefits, which has morally troubling implications when applied to biodiversity). Even when cost–benefit analysts are willing to acknowledge the existence of non-quantifiable values, she argues, such considerations are frequently ignored in the real-world political deliberations that are informed by CBA.

Jay Odenbaugh argues that philosophical discussions about the value of biodiversity often miss the broader question, namely, how should we get people to care about biodiversity? This is as much a question of moral psychology as it is of ethics. Odenbaugh begins by criticizing the idea that biodiversity, per se, has intrinsic value. Even if it does, however, it is not clear whether intrinsic value attributions actually motivate people to act. He closes by arguing that we should shift our attention to the factors that motivate people to act, including the ecosystem services provided by endangered species, and considers the Pacific salmon as an example of this approach.

Ben A. Minteer tackles the ethical dimensions of assisted colonization, that is, the practice of artificially relocating species to places that are more favorable to their long-term persistence. While some commentators think of assisted colonization as a morally problematic human meddling in nature, others see it as a necessary means to saving species. Minteer strikes a thoughtful balance between these two positions by appealing to the writings of Aldo Leopold. He shows how a careful analysis of Leopold's ethical philosophy should lead us to cautiously accept assisted colonization, when we have good reason to believe that the species cannot be protected by traditional *in situ* techniques.

Part IV, "Measurement and methodology," considers the *how* of biodiversity conservation. Conservationists are divided not only by conceptual and ethical disagreements, but by methodological ones. There are several topics to be considered here. Given that we must rely on (often imperfect) proxies, or indicators, for biodiversity, what sorts of biodiversity indicators should we rely on? Corresponding to this, how do we accurately assess the rate of biodiversity loss? How do we make rational decisions in the face of risk and uncertainty? Moreover, although conservationists wish to better incorporate ecological and evolutionary theory into conservation practice, it is unclear how to best do this. This section as a whole will demonstrate potential points of tension between the role that the concept of biodiversity plays in biological theory, the role it plays in sociopolitical negotiation, and the role it plays in providing accurate and rapid field measurements.

Kerrie Wilson, Jacqueline England, and Shaun Cunningham provide a masterful assessment of the problem of biodiversity indicators. What, precisely, must we measure in the field when we measure biodiversity, and how do we know that the thing we are measuring adequately represents biodiversity itself? The authors consider several standard biodiversity indicators as well as the criteria by which one judges the goodness of these indicators. They argue that indicators must be "fit for purpose," that is, that they should be sensitive to the specific objectives at hand, and that these objectives can be shaped by social values. They also discuss points of tension between the ways that indicators are used by ecologists and by policy-makers.

Lynn A. Maguire considers a variety of different approaches to structured decision-making in the context of conservation. She argues that the trade-offs among multiple goals and uncertainty about outcomes suggest that it is a wise policy to break such complex decisions down into simpler parts so as to best articulate and prioritize values, make these values transparent to stakeholders, and adapt to new information. She considers a variety of contexts where such multi-criteria structured decision-making is relevant to conservation: public land management, conservation organizations' allocations of their limited resources, and individuals' attempts to align their values with their actions, and particularly their consumer choices. She considers the assumptions such approaches make, and the critiques of such assumptions from advocates of intrinsic value, and argues for a middle ground.

Alan R. Templeton discusses the role of genetic measurement in monitoring biodiversity and in conservation, due to developments in the fields of genetics and genomics. The newer techniques of molecular genetics and genomics have revolutionized the study and management of rare and elusive species, and have revealed whole realms of biodiversity that were previously hidden from our view.

It is often said that we are in the midst of a sixth mass extinction. Clearly, one motive of this volume is to contribute to the effort to halt biodiversity loss. **Yrjö Haila**, however, discusses how difficult it is to construct reliable measures of extinction rates. Part of the problem stems from the inherent ambiguity of "biodiversity" itself. Another problem is that, when we measure extinction rates, we must rely on indirect indicators, or proxies (such as the destruction of habitat), and scientists disagree about the correct measures. As an alternative, he suggests that we focus more on preventing the factors that cause biodiversity loss, instead of trying to quantify biodiversity loss per se.

Part V, "Social contexts and global justice," explores the delicate relationship between biodiversity conservation and the sovereignty of local communities over their land. Traditionally, local communities have had little say over conservation decisions, which have largely been designed and implemented by foreign non-governmental organizations in collaboration with government agents. Yet ethical considerations require conservationists to approach projects in a way that fosters local autonomy. Not only might one think that local communities should have greater control over the ways in which biodiversity plans are designed and managed, but also over what *constitutes* biodiversity in the planning context. A second question, closely connected, is the issue of disciplinary authority: who (e.g. conservationists, social scientists, local communities) is best qualified to prescribe and manage the patterns of interaction between people and land?

Anya Plutynski and Yayoi Fujita-Lagerqvist discuss how conservation science has become a more interdisciplinary field, integrating insights from social science as well as conservation biology, traditionally conceived. They argue that as a practical matter, as well as a matter of justice, successful conservation policies need to include all relevant stakeholders in the negotiation of systematic conservation planning from the outset. Attention to local context is essential as well as ongoing adaptation and negotiation. In sum, putting conservation into practice requires a fine balance of attention to both local and indigenous peoples' interests in development, historical rights to use and management of land, the wishes of international conservation organizations, the state's interests in and control of resources, and larger economic interests in natural resources.

Jeremy Brooks explores the topic of trade-offs and synergies in conservation planning. Conservation takes place in a context of conflicting goals, values, and needs. For example, there may be conflicts between conserving a rare species and developing land for human habitation, or even between protecting two different species with different habitat needs. It is particularly important to assess potential trade-offs when we are considering how to best conserve while respecting the values and goals of local communities. Yet conservationists have not analyzed the problem of trade-offs in any explicit way. Brooks provides a theoretical framework for approaching the problem, particularly in the context of community-based conservation, and emphasizes the need for monitoring multiple outcomes.

Part VI, "Biodiversity and other environmental values," provides two overviews of other major concepts in conservation planning. How does biodiversity relate to ecological sustainability? And how does it relate to ecological restoration? One problem here is that, just as there are multiple characterizations of biodiversity, so too are there multiple, and sometimes competing, characterizations of sustainability and restoration.

J. Baird Callicott provides a critical assessment of attempts to define the notion of sustainability. He offers a criticism of the notion of sustainable development as it appears in the UN's Brundtland Report and he also criticizes philosopher Bryan Norton's place-based account of sustainability. As an alternative, he recommends that we pursue *ecological sustainability*, which calls for a more radical transformation of human economic systems than the other approaches. The claim is that human economic systems should be modeled on what he calls the "economy of nature." He sets out several principles of the economy of nature and considers ethical implications.

Justin Garson considers the nature and value of ecological restoration. He considers multiple definitions of "restoration" and introduces the controversial idea of historical fidelity. He then goes on to examine three major arguments against restoration, namely, that it is artificial rather than natural, that the choice of a baseline is arbitrary, and that restoration always has an instrumental character. Garson attempts to defend the value of restoration from each of these, and then considers the relation between restoration and biodiversity conservation.

In short, we hope that this volume will do several things. First, we hope it will invigorate existing philosophical discussions about the nature and value of biodiversity. Second, we hope it will demonstrate, for a new audience, why biodiversity is such a rich philosophical subject. Above all, we hope that it will help environmental scientists, social scientists, philosophers, and policy-makers reflect well on how we can better protect it.

PART I

Historical and sociological contexts

1

A CONFLUENCE OF VALUES

Historical roots of concern
for biological diversity

Timothy Farnham

Biological diversity was originally crafted as an all-encompassing concept for conservation that could unite many different interests under one umbrella goal: the protection of life on Earth. The term itself, in its modern usage, has been around for less than fifty years. While several prominent individuals and groups in the American environmental movement used "biological diversity" sporadically in the literature in the late 1960s and through the 1970s, the first published definition did not appear until 1980, in the second chapter of the *Annual Report* of the Council for Environmental Quality (CEQ), a White House advisory body that helps to coordinate US federal environmental policy (Farnham 2007). The chapter was entitled "Ecology and Living Resources: Biological Diversity" and it was written by Elliott Norse and Roger McManus. At the time the CEQ staff was planning the report, the destruction of tropical forests and the subsequent threat to endangered species were drawing a great deal of public attention. Norse and McManus were asked by senior staff members to research and write "on an unprecedented subject: the status of life on Earth" (Norse 1996: 6). Norse, however, also felt that such a broad topic deserved a larger scope than simply focusing on species extinctions. Indeed, there was a strong concern in certain conservation circles over the loss of genetic diversity – often referred to as "germplasm resources" – and at the other end of the spectrum were groups who primarily focused on protecting the natural world at the ecosystem level. As Norse describes in an interview, "we were talking about the loss of diversity at all stages" (Norse 1999). Certainly, the plight of endangered species was popular, but the protection of genes and ecosystems was deeply interconnected to the successful conservation of the variety of plants and animals. Combining these different levels – genes, species, and ecosystems – would become the practice for those who sought to draw attention to the threats to life on Earth. As Norse wrote, "Knowing no existing term that encompassed all that was being lost, we called it 'biological diversity'" (Norse 1996: 6).

While Norse and McManus's original 1980 definition did not explicitly identify the three levels of genetic, species, and ecosystem diversity, their work served as the foundation for the concept as public and private groups started to use and write about the term. One of the more popular definitions often cited in later literature was published in a report by the US Congress's Office of Technology and Assessment (OTA) entitled *Technologies to Maintain Biological Diversity*:

> Biological diversity refers to the variety and variability among living organisms and the ecological complexes in which they occur. Diversity can be defined as the number

of different items and their relative frequency. For biological diversity, these items are organized at many levels, ranging from complete ecosystems to the chemical structures that are the molecular basis of heredity. Thus, the term encompasses different ecosystems, species, genes, and their relative abundance.

(OTA 1987: 3)

The contracted form "biodiversity" was first introduced in 1986 at the National Academy of Science's *National Forum on Biodiversity*, where E. O. Wilson became the most prominent spokesperson for the cause and edited the papers presented at the conference in the book *Biodiversity* (1988). Thus, the late 1980s marked the years when the concern for biological diversity rose to prominence, fueled by attention from the government, numerous private environmental groups, and an increasingly aware general public.

For those interested in the evolution of the environmental movement (both in the United States and internationally), one question worth exploring is why the term "biological diversity" became so popular in such a short amount of time. A likely reason is the initial careful construction of the concept by Norse and McManus in the 1980 CEQ *Annual Report*. By bringing together various levels of diversity under "biological diversity," they were able to combine the foci of different conservation concerns under one banner. The reason that the term gained a foothold in the greater conversation is because it had been introduced into an environmental community that needed unification. Norse was keenly aware of this situation: "The concept has real value … It looked at conservation differently. It had the potential to bring people together who had not been together before …" (Norse 1999). By the time the OTA definition was published, the three tiers of diversity – genes, species, and ecosystems – were firmly partnered together and framed as the three levels at which research and conservation of biological diversity should occur. These tiers were not new topics of inquiry or concern, but it was novel that they were so explicitly linked in a single term. Tying them together affirmed the interconnectedness of the natural world while at the same time taking advantage of already established conservation traditions. Thus, the rise of popularity of the biological diversity cause was not necessarily a paradigm shift, but it was a confluence of values and concern that had been fostered over time, coming together in one concept that represented the protection of the living components of the natural world.

In trying to trace the historical roots of such a confluence, difficulties arise in identifying boundaries. Any evidence of concern for living nature could reasonably be seen as a precursor for the rise of concern for biological diversity. However, if we confine ourselves to the categories of the three tiers, we find stories of human interest in protecting genes, species, and ecosystems that provide a compelling background for understanding where the concern for biological diversity came from. In addition, it is useful to overlay the more recent rise of concern for "diversity" with the stories of those three categories. This variable helps to limit what precursor concerns we examine. For example, we could discuss seventeenth-century hunting laws in colonial America as an early example of the modern concern for species conservation. But such laws were not cultural expressions affirming the value of species diversity; they were more specifically to protect a food – and later, a recreational – resource. Similarly, the great land preserves set aside by the United States federal government in the late nineteenth century that formed the basis of the national parks, forests, and wildlife refuges could be seen as early evidence of concern for ecosystems. But while numerous factors were at play in this conservation effort, including aesthetic, cultural, and economic values, the specific goal of protecting a diversity of ecosystems was not originally articulated in these monumental efforts to protect the land and its resources.

In general, the concern for a "diversity" of genes, species, and ecosystems is a twentieth-century phenomenon that came about because of the swift encroachment on the natural world by unprecedented human population growth and the consequent extraction of resources. By the 1970s, the desire to protect all of the natural variety present on Earth was most often expressed in conjunction with a reminder of all the benefits humans would lose should the diversity of nature be reduced. It is these instances that are arguably most useful for illustrating the history of concern for biological diversity. By examining related examples of concern for living diversity at the three different levels of genes, species, and ecosystems, we can begin to see the accretion of societal and cultural values that would become associated with the conservation of biological diversity.

Concern for species diversity

Of genes, species, and ecosystems, the species level is arguably the most recognizable and most commonly understood component of biological diversity. Interactions with plants and animals are as old as humanity itself, and our close relationship to the living things that provide us with food and material resources is not surprising. The classification of different species is how humans have made sense of nature for millennia.

In the United States, efforts to protect the buffalo and the passenger pigeon in the nineteenth century were well publicized and are often referred to as the precursors to the modern sentiment of protecting endangered species. The Lacey Act of 1900 was the first federal action that moved beyond protecting game species, and sought to conserve the populations of wild birds that were being decimated for their feathers, used in making stylish women's hats (Cart 1971). Further federal action in the early to mid-twentieth century would protect individual species, such as the bald eagle and whooping crane, and increasing numbers of states adopted laws that limited hunting of certain valuable game species. But these examples of societal concern were not specifically directed at preserving the *diversity* of species. The first expressions of the imperative to protect the full range of variety came from the scientific community, with individuals voicing the worry that if we were to lose some of the different parts of nature, we would never completely understand how nature worked.

In many ways, this initial impulse for the conservation of species diversity was an outgrowth of the nineteenth-century tradition of the naturalist/explorer who collected specimens of life from the far reaches of the globe. By the early twentieth century, scientists were becoming well aware of the human impact on wild populations; this was most apparent on islands where unique species had small population numbers and were easily exterminated. An early example of this concern in the scientific literature is Willard Van Name's article published in *Science* in 1919 entitled "Zoological Aims and Opportunities." Van Name wanted to draw attention to the "protective work which is very important to science … This is the protection of what remains of the unique and peculiar forms of animal and plant life that inhabit many of the remote islands … in various parts of the world" (Van Name 1919: 83). While Van Name's call to protect the variety of living forms in the name of science was one that likely found a sympathetic audience among his fellow zoologists and biologists, he also recognized that non-scientists may not find this argument compelling; as he wrote, the importance of preserving species for study was not something that "the general public [could] be expected to appreciate" (Van Name 1919: 83). Other scientists began to express similar values in arguments to protect diversity, even when the threatened species were not popular with the public. For example, in 1915 the federal government had established the Predator and Rodent Control (PARC) program and by the 1920s, 35,000 coyotes were being killed each year in the American West (Dunlap 1987: 51).

13

This eradication was supported by a commonly held belief that predators no longer had a place in a world where humans could manage nature. To make certain that livestock would survive and to guarantee that there would be enough game for humans to hunt, it was reasoned that the predators should be removed. In response, some members of the scientific community claimed that exterminating any species would be a significant loss for understanding how the natural world functioned. They also argued that because predators played such a significant ecological role, their removal could have many unforeseen consequences. This viewpoint was best represented by the words of Lee R. Dice, from his 1925 article "The Scientific Value of Predatory Mammals." Dice wrote that:

> The lives of all species of animals living in one locality are closely interrelated; especially close are the relations between carnivores and the forms on which they prey … with the predatory mammals eliminated, it will become more difficult to explain the origin of many adaptive structures and habits in the remaining species.
>
> (Dice 1925: 27)

But Dice also went one step further and made an early plea for preservation of diversity for science's sake: "every kind of mammal, as well as every other kind of organic being, is of great scientific significance, and the world can ill-afford to permit the extermination of any species or subspecies" (Dice 1925: 25). In the face of losing animals that many people disliked and claimed no longer had any useful function, an argument for protecting species diversity arose, largely for the sake of scientific understanding of the natural world.

This kind of argument in the scientific world served to chip away at an older, well-established, notion that there were "good" and "bad" animals and plants in nature. A newer ecological understanding, related to the "balance of nature," supported protecting all living components of any natural community. As E. L. Scovell wrote in 1938 in an essay entitled "Overlook No Living Thing":

> Man cannot escape his dependence on all forms of life … Who knows what is good or bad, friend or foe? Under certain conditions, a plant, animal, bird, insect, fish or reptile may be an enemy. Under other conditions it may be a real friend … We cannot overlook any species of living thing.
>
> (Scovell 1938: 295–296)

This was an extension of the ecological argument articulated by Dice, claiming that we knew little about the interconnections between the living components of the world, and that we should take care not to remove any important working parts of ecosystems.

While support for preserving variety for scientific and ecological reasons continued to grow, others began to argue that we should protect the diversity of species for largely aesthetic reasons. Aldo Leopold, in his seminal text *Game Management* (1933), offered a pointed observation about non-game species near the end of his book. He wrote:

> The objective of a conservation program for non-game wild life should be … to retain for the average citizen the opportunity to see, admire and enjoy, and the challenge to understand, the varied forms of birds and mammals indigenous to his state. It implies not only that these forms be kept in existence, *but that the greatest possible variety of them exist in each community.*
>
> (Leopold 1933: 403; emphasis in original)

This pointed declaration for diversity is especially noteworthy as it occurred in a book that focused largely on a single class of valued species: game. Leopold's sentiments were shared by colleagues, and the call to preserve variety for aesthetic reasons would become more common in texts and articles in wildlife magazines (see Rush 1937, Lehmann 1938, Gabrielson 1942, Devoe 1944, and Skutch 1948).

Certainly, one of the most important precursors to efforts to preserve biological diversity was the interest in protecting endangered species. On a global level, the International Union for the Conservation of Nature and Natural Resources (IUCN) began earnest work on gathering data by establishing the Survival Service Commission in 1956, whose mission was to "collect data on, and maintain lists of, all wild animals and plants that may be in danger of extinction, and to initiate action to prevent it" (Fisher *et al.* 1969: 10). These lists would evolve into the Red Data Books, which by the 1960s were viewed as the authoritative source for all information on endangered species. In partnership with the IUCN, and in direct response to the imminent loss of species in Africa, the World Wildlife Fund (WWF) was established in 1961, with the first American chapter of the WWF forming in 1962. By the late 1960s, WWF had raised millions of dollars, a testament to how concerned the public had become about losing species (Crowe 1970).

Arguably, concern for endangerment is not the same as concern for diversity, and many donors likely responded to the WWF campaigns because people cared specifically for the charismatic mega-vertebrates (e.g. tigers, elephants, panda bears) whose photos are often found on fundraising materials. However, it seems evident that these endangered animal "ambassadors" opened the door for laypeople to understand that all species deserve protection. In this regard, the evolution of the wording of the United States Endangered Species Act is revealing. There were three different Endangered Species Acts, in 1966, 1969, and 1973. The first served to establish and coordinate the National Wildlife Refuges and delineated criteria for whether a species qualified for federal protection. But the law only applied to "selected species of native fish and wildlife," and it was apparent from the national list that only vertebrates would be protected. The 1969 Endangered Species Conservation Act was passed amidst the growing concern for global species loss and included the prohibition of importing any endangered species. The 1969 act also amended the 1966 definition of eligible species to include "any mammal, fish, wild bird, amphibian, reptile, mollusk, or crustacean" (Bean and Rowland 1997: 197). But the most dramatic change came with the 1973 Endangered Species Act, which not only made it illegal to "harass, harm, pursue, hunt, shoot, wound, kill, trap, capture, or collect" any threatened or endangered species, it also expanded the category of protected species to include *all phyla* of animals and plants.

The law had become a *de facto* expression of the importance of protecting the full diversity of living things. The federal government had decided that all species were worth preserving, not just ones that provided us with food, or recreational opportunities, or scientific study, or aesthetic pleasure. In an important way, diversity itself had become the value that served as the umbrella to protect all of the benefits humans experience from the animal and plant species of the world.

The concern for genetic diversity

Like the concern for protecting species diversity, the concern for preserving natural genetic variety was first most forcefully expressed by the scientific research community. At the beginning of the twentieth century, there was a great deal of excitement surrounding the growth of knowledge about the genetic mechanisms of evolution. Mendel's seminal work on pea plants had been "rediscovered" and many scientists were applying this new understanding of inheritance to their own experiments, exploring how crossbreeding could produce numerous combinations of traits. The agricultural community quickly picked up on the implications of Mendel's work. In 1906, R. H. Biffen published "Mendel's Laws of Inheritance in Wheat Breeding," illustrating how resistance to a specific disease was dictated by individual "factors of inheritance" and passed on in wheat, following the rules Mendel had established (Pistorius and van Wijk 1999: 36). For plant breeders, this new understanding would revolutionize how crop species could be improved. In the past, breeders would look for plants that had better overall qualities, but now they could simply look for plants with particular individual traits and use newly developed hybridizing techniques to produce novel varieties with reproducible characteristics. This application of genetic understanding was, according to one historian, "a watershed in the development of plant breeding in the United States. No longer was the breeder's task to adapt elite germplasm from other countries to American conditions, it was now to improve established varieties by incorporating particular exotic characters" (Kloppenburg 1988: 80). This change is significant because scientists began to focus sharply on the variety of plant material at the genetic level.

In the United States, there was already an established tradition of collecting plants from around the world, and bringing them back to test in North American soils. Any who travelled abroad – consuls, diplomats, naval personnel – were encouraged "to collect seeds and send them to Washington for distribution to farmers and anyone else who wanted them" (Witt 1985: 29). The United States Department of Agriculture even hired explorers to travel the globe searching for plants that might add to and improve America's food crops for the future. After the development of the new agronomic techniques for breeding, it was no longer simply "useful" plants that were the target of these explorers; it was now any plant that might have a "useful" characteristic. Ancient landraces (old cultivars) and wild ancestors of crop species suddenly took on new value. These plants might exhibit particular traits that could be isolated and bred into modern crops. Scientists began to develop theories about which sectors of the globe might house the species with the highest variety of valuable genetic traits. One Russian scientist, Nikolai Vavilov, proposed a map that illustrated "centers of diversity," based on his collecting expeditions to over fifty countries. It is estimated that he gathered more than 50,000 seed samples of crop plants and their relatives (Plucknett *et al.* 1987: 62). Vavilov's work was inspirational to the plant collectors of the early twentieth century, and served as an example of how the strategy had changed: "Plant breeders looked not so much for new introductions as for breeding material, not so much for a superior variety that might be adapted to American conditions but for a plant with perhaps only one superior characteristic. Hence they collected a much broader range of germplasm" (Kloppenburg 1988: 80). The emphasis was now on collecting as much variety as possible, on the off chance that it would have a useful characteristic for the future. Most importantly, the focus of value was no longer on individual organisms – it was on the diversity of genetic material.

Crop scientists also began to issue warnings that the variety of genetic material was in danger of being lost, threatened by the encroachment of burgeoning human numbers and the conversion of land to industrial monocultures. One major worry was the replacement of old cultivars and wild relatives of modern crops with new, improved crop varieties. Harlan and Martini were among the first to call attention to the loss of germplasm resources in a 1936 article on barley breeding:

> In the hinterlands of Asia there were probably barley fields when man was young. The progenies of these fields with all their surviving variations constitute the world's

priceless supply of germplasm. It has waited through long centuries. Unfortunately, from the breeder's standpoint, it is now being imperiled. When new barleys replace those grown by farmers of Ethiopia or Tibet, the world will have lost something irreplaceable.

(Harlan and Martini 1936, 303, quoted in Pistorius and van Wijk 1999: 65)

This became known as the problem of genetic erosion (Wilkes 1983), and it is an issue still recognized today as hectares of historic farmland continue to be converted into modern monocultures. Fortunately in the 1930s and 1940s, the global effort to collect a representative cross-section of the world's genetic diversity was well underway, and it had been so successful that the established gene banks were close to reaching capacity. Over the next several decades, the United States built four regional Plant Introduction Stations and a National Seed Storage Laboratory (Brown 1984: 32). Also, on the international stage, the Food and Agriculture Organization (FAO), established in 1945 by the United Nations, would become a coordinating point for global efforts to protect and distribute plant genetic resources.

Efforts to draw attention to the loss of genetic diversity received a large boost in the 1960s, when the FAO partnered with the International Biological Programme (IBP) on a long-term project to evaluate the conservation of plant genetic resources. The IBP was a massive multinational effort established by the International Council of Scientific Unions to coordinate ecosystem studies across national boundaries, and its research network provided the FAO with significant global connections. The organizations chose Otto Frankel, an Austrian-born plant geneticist, to chair several committees and coordinate conferences and program reviews. Frankel's dynamic personality, prolific publishing, and fervent pleas to conserve genetic diversity brought awareness of the issue to new heights. His prominent voice at the 1967 FAO/IBP Technical Conference on the Exploration, Utilization, and Conservation of Plant Genetic Resources lent an air of urgency to the issue of genetic erosion. In the introduction to the volume of conference papers, Frankel and his co-author Erna Bennett acknowledge the benefits of the recent advances in crop production made possible by modern breeding techniques. But the impact of monocultures in areas where diversity once existed was the take-home message of the article. The success of modern crops, they wrote, "represents a very real and immediate threat that the treasuries of variation in the centres of genetic diversity will disappear without a trace" (Frankel and Bennett 1970: 9). Their conclusion recalls Harlan and Martini's words from thirty years before: "the extinction of the natural sources of adaptation and productivity represented by primitive varieties may turn out to be an irreparable loss to future generations" (Frankel and Bennett 1970: 12).

The momentum for protecting genetic diversity carried into the 1970s and was propelled forward by further recognition at the 1972 United Nations Conference on the Human Environment in Stockholm. Actions to conserve genetic resources were spelled out in conference recommendations 39–45, and though it is unlikely that governments fully followed these recommendations, the inclusion of the plight of genetic resources was significant (Stone 1973: 165). Frankel, who played no small part in the UN Conference, continued to chair FAO committees and publish papers. In one article entitled "Genetic Resources: The Past Ten Years and the Next," Frankel and co-author John Hawkes wrote:

Perhaps the biggest transformation of the last decade has been the development of widespread awareness and concern. Ten years ago … the problem of "genetic resources" was unknown. In the last few years, it has sunk into the consciousness of scientists

17

and administrators and of the many people who have become concerned about the resources of the earth.

(Frankel and Hawkes 1975: 4)

This observation certainly seemed accurate. Numerous conferences and symposia focused on genetic resources. In 1978, the National Research Council (NRC) published *Conservation of Germplasm Resources: An Imperative*, and essentially declared a national emergency:

Saving the rich diversity of genetic material that has been provided by natural muta- tion and evolution can be achieved and is worth whatever effort may be required. It is critically important that the people of the United States recognize the long-term dangers inherent in the loss of specific genes and of genetic material, recognize that diversity in germplasm is an essential national treasure, and treat it as such.

(NRC 1978: 4)

This passage perhaps best illustrates the height to which concern for genetic diversity had risen. With many different constituencies and with evident links to protecting species and ecosystems, the protection of genetic diversity was set up to be a prominent part of any comprehensive conservation strategy.

Concern for ecosystem diversity

As mentioned earlier, early land protection efforts could be considered precursors to a more directly articulated concern for preserving a diversity of ecosystems. But the land first set aside for national parks, forests, and wildlife refuges were not recognized as valuable because they were distinct *ecosystems*, nor because they conserved a range of *diversity*. Mostly, these protected areas contained monumental examples of the American landscape, or huge swaths of com- mercially valuable forests, or habitats for important game species. Certainly, in protecting parks, forests, and wildlife refuges, the United States did protect intact ecosystems for the use, study, and enjoyment of future generations, but the goal of protecting a diversity of ecosystems was not part of the conservation mission until later in the twentieth century.

The word *ecosystem* was first introduced in 1935 by British ecologist Arthur Tansley. The ecosystem concept was meant to combine the inorganic and organic components of a defined area of nature into one object of study – not in a philosophically holistic way, but in a more mechanistic, systems-based model in the spirit of physics (Tansley 1935). The term would become popular, especially in the 1950s, through Eugene Odum's 1953 textbook *Fundamentals of Ecology*. Inevitably, the term would become firmly embedded in the conservation language of the 1960s and 1970s.

Like the history of concern for species and genetic diversity, the first calls for protecting ecosystem diversity came from the scientific community in the form of concern for preserv- ing different "types" of landscapes. In 1915, scientists studying ecology were numerous enough to form the Ecological Society of America (ESA). Led by Victor Shelford, an animal ecologist from the University of Chicago, the group's mission was "to promote the scientific study of organisms in relation to the environment and to facilitate an exchange of ideas among ecolo- gists" (Behlen 1981: 7). But under Shelford's influence, a protectionist agenda for the ESA soon developed with the formation of the Committee on the Preservation of Natural Conditions, which worked to identify nearly 600 areas in North America that were "preserved or worth preserving." The committee's letterhead slogan stated their goal: "An undisturbed area in every national park and public forest" (Croker 1991: 124). Shelford and his colleagues understood that if scientists wanted to have examples of the original landscapes to study, they would have to actively work for the preservation of those landscapes. As Shelford wrote in 1926, because ecology "obtains its inspiration in the natural order in original habitats [we] must depend upon the preservation of natural areas for the solutions of many problems" (Shelford 1926: 3).

In the 1930s and 1940s, other scientists began to call for protecting a "full sample" of undisturbed nature as standards for scientific comparisons to manipulated landscapes. Foresters wrote about protecting "a study area in undisturbed state in each great biotic region of the country" to serve as "natural yardsticks to measure man's land management by" (quoted in Hanson 1939: 132). The idea of protecting a representative sample of nature was a popular refrain. Henry Baldwin in 1941 wrote of the need for "adequate examples of all major types of vegetation." Willard Van Name published an article entitled "Need for Preservation of Natural Areas Exemplifying Vegetation Types" (1941). The challenge in implementing this agenda was that there was no set category for a preservation unit. "Ecosystem," though introduced in 1935, was not yet recognized as a proper object of study. Thus, there were many suggestions including "biotic provinces," "cover types," and "ecological associations" (Clements and Shelford 1939, Dice 1943). The fact that the scientific community was looking to protect the complete variety of these categories represents a significant precursor to the interest in protecting ecosystem diversity.

The cause of setting aside valuable examples of land in the United States took a significant leap forward with the establishment of The Nature Conservancy (TNC). The roots of TNC lie in the original Committee on the Preservation of Natural Conditions started by Shelford. Over the years, the ESA had decided that Shelford's protectionist activism was not an appropriate activity for a scientific organization, and in 1945 the Society voted to take away the Committee's lobbying power. Shelford then founded the Ecologists Union, which in 1950 changed its name to The Nature Conservancy. The new organization was helped immensely by a feature article in the *New York Times*, which conveyed a convincing sense of urgency about preserving natural areas. Specifically, there was a forceful emphasis on protecting diversity:

> In addition to their obvious use in the study of basic biological research, these virgin tracts of forest, desert, prairie, mountain, swamp … will one day be of incomparable interest as the sole survivors of the America that was. The important thing is to save as many of the diverse types as possible – now.
>
> (In Behlen 1981: 10)

TNC's membership numbers and funds grew quickly, and in 1955 the group acquired its first preserve. TNC would become what many would consider one of the most popular and effective land protection organizations in the United States.

At the international level, the first group to focus on preserving large areas of land was the same organization that led early efforts to protect species at the global level, the International Union of the Conservation of Nature and Natural Resources (IUCN). Originally founded as the International Union for the Protection of Nature in 1948 by the United Nations Educational, Scientific, and Cultural Organization, the name was changed to the IUCN in 1956 to recognize the importance of resource use to the modern economy (hence "conservation" instead of "protection"), but there was still a significant emphasis on the preservation of nature in the group's activities. The IUCN carried on its work through appointed "commissions;" the most powerful of these focused on species survival, national parks and protected areas, and the promotion of ecological study. The Union, and its extensive network of experts, also played a large role in several international initiatives that would have a significant impact on global conservation. As mentioned earlier, one was the establishment of the World Wildlife Fund in 1961. Another was the IUCN's participation in the International Biological Programme (IBP), a joint long-term venture between numerous countries that was first proposed in 1960 and sought to collect data from around the globe on the biological processes of the planet. The IBP research areas were organized into seven different "sections," one of which was called *Conservation Terrestrial* (CT). Because the IUCN had already laid the groundwork for research on conservation areas with its own Commission on National Parks and Protected Areas, their expertise was called upon for the CT section of the IBP.

The IUCN hosted the First World Conference on Parks in 1962, and with the IBP partners in attendance, one recommendation to come out of the conference directed the IUCN to "work closely with the IBP to bring into existence a series of Natural Reserves providing permanent examples of the many diverse types of habitats, both natural and seminatural, so as to preserve them permanently for world science" (Coolidge 1963: vii). "Representativeness" continued to be a common theme in the language of conservation at the time, reminiscent of American ecologists earlier in the century. As the IBP initiative began in earnest in 1964, the goal of protecting examples of "diverse types" became more distinctly articulated. For example, the CT section of the IBP identified five reasons that supported their objective of preserving natural and seminatural areas:

1. The maintenance of large, heterogeneous gene pools;
2. The perpetuation of samples of the full diversity of the world's plant and animal communities in outdoor laboratories for a wide variety of research;
3. The protection of samples of natural and seminatural ecosystems for comparisons with managed, utilized, and artificial ecosystems;
4. Outdoor museums and areas for study, especially in ecology;
5. Education in the understanding and enjoyment of the natural environment … (Worthington 1975: 30).

The first three objectives are most relevant to this discussion. Not only is the protection of diversity highlighted, but also the three levels (genes, species, and ecosystems) that would come to define the concept of biological diversity are purposefully linked together.

The IBP also contributed to the rise of the term "ecosystem" in the language of conservation, particularly in the United States branch of the CT section. Led by several prominent ecologists, including Eugene Odum, the Americans decided to rename their contribution to the IBP, calling it the "Conservation of Ecosystems" program. While they were dedicated to studying a full range of biome types, the CE program also declared one singular conservation objective: "The establishment within the United States and its possessions of a comprehensive system of protected research reserves" (Darnell 1976: 105). The language was remarkably similar to that of Shelford's Committee of the Preservation of Natural Conditions – the practice of scientists working to protect the objects of their study was now connected to a broader international agenda.

As the IBP wound down, UNESCO sponsored a conference in 1968 whose purpose was to outline a plan that would build on the work of the international network established by the IBP. The meeting was commonly referred to as the Biosphere Conference, and one of the major recommendations (with the title "Research on Ecosystems") laid out an agenda for continuing the systematic scientific study of all natural systems, both natural and modified by humans. Another interesting and perhaps more significant recommendation was entitled "Utilization and Preservation of Genetic Resources." In order to "preserve the rich genetic resources that have evolved over millions of years and are now being irretrievably lost as a result of human actions," the conference participants called for the "preservation of representative and adequate samples of all significant ecosystems in order to preserve the habitats and ecosystems necessary for the survival of populations of species" (UNESCO 1970: 216). As in the original goals of the CT section of the IBP, scientists were expressing the importance of linking all levels of life together (genes, species, and ecosystems) and safeguarding the diversity that still exists on the planet.

Out of this conference came the concept of UNESCO's Man and the Biosphere Programme and the eventual development of the concept of "biosphere reserves." Serving much like a system of international parks, the primary objective of the biosphere reserve network was "to conserve for present and future human use the diversity and integrity of biotic communities of plants and animals within natural ecosystems, and to safeguard the genetic diversity of species on which their continuing evolution depends" (UNESCO 1974: 6). Thus, the administrative tool to carry out the objective was specifically one that merged the three different levels of biological diversity. The roots of the biosphere reserve concept are arguably in efforts to protect ecosystems, but the recognition of genetic and species diversity as objectives along with the protection of ecosystems was a significant precursor concern for what would become known as biological diversity.

The importance of diversity

In addition to becoming a central tenet in discussions about the conservation of genes, species, and ecosystems, the importance of protecting diversity and variety in general was a significant topic for scientists and conservationists in the decades before the 1980s. Charles Elton, an animal ecologist, published an influential book entitled *Ecology of Invasions by Plants and Animals* in 1958 in which he explored how the diversity of an ecosystem might be impacted by introductions of new species. His final chapter "The Conservation of Variety" is frequently cited as a seminal discussion of the importance of diversity in general. Elton identified what he believed to be the ultimate goal of nature protection: "Keeping or creating sufficiently rich plant and animal communities in our changing landscape – that is … conserving ecological variety" (Elton 1958: 153).

One discussion that arose among ecologists in the 1950s and 1960s was whether or not higher diversity contributed to the "stability" of an ecosystem. In 1955, Robert MacArthur published "Fluctuations of Animal Population and a Measure of Community Stability" in *Ecology* and proposed an equation that illustrated the relationship between trophic diversity and community stability. Although MacArthur's equation did not "prove" any measurable connection between diversity and stability, he did give voice to a relationship that many scientists believed existed. Other articles would follow (see Hutchinson 1959, Margalef 1963, MacArthur and Wilson 1967) articulating some facet of the relationship between diversity and stability, and in fact the entire 1969 Brookhaven Symposium ("Diversity and Stability in Ecological Systems") was dedicated to the topic. While papers in the early 1970s by Robert May (1971, 1972, 1973a, 1973b) would call into question the ability to predict the stability of ecosystems through diversity, the common belief of the scientific community was best stated by Eugene Odum: "it is now generally assumed, but without much real scientific evidence, that the 'advantage' of a diversity of species … lies in increased stability" (Odum 1963: 34). Certainly, even with some dissention in the scientific world, the notion that decreasing diversity might "unravel" an ecosystem was firmly entrenched in popular opinion.

The conservation community followed the scientific debate with interest, and the implications were more fully explored in the 1965 Conservation Foundation conference entitled Future Environments of North America. One organizer summed up the central messages of the landmark meeting:

> One of the significant truths to emerge from this conference was the value of diversity. Perhaps the most appalling aspect of modern man's insensitive degradation of the environment has been the mounting destruction of earth's natural diversity and the creation of monotonous, uniform human habitats. Ecology has shown us that varied ecosystems are healthy, relatively stable environments better able to withstand stresses; seen in this context, the contemporary trend toward creating an artificial, bland, standardized biosphere is a fundamental threat to the quality of human existence, if not man's very survival. With each loss of variety, our potential for human choice, freedom, and change narrows.
>
> (Milton 1966: xvi)

This conference provides a revealing benchmark for how the conservation community was envisioning its future mission. Various participants had discussed the plights of gene pools, endangered species, and significant ecosystems, and placed them in the context of the value of diversity. The meeting represents one of the first single conference events that brought together all of the interests that would become the driving forces behind the rise in popularity of the concept of biological diversity.

A final, significant illustration of how the conservation community was implementing the protection of diversity is the campaign of The Nature Conservancy (TNC) in the 1970s. The scientific interest in protecting ecosystems at both the national and international level served as the platform for recognizing the imperative of protecting the component parts of ecosystems. Numerous other interested parties had separately been voicing concern for the conservation of species and genetic variety. "Diversity" had become a term that tied these parts together and was growing in popularity as an objective of conservation efforts. TNC published several reports and initiated a series of programs in the 1970s that explored strategies for preserving diversity. Robert Jenkins, the director of science for TNC, wrote several articles on protecting the full range of ecosystem types, such that preserved:

> tracts of land [are] set aside as living space for the full diversity of biological species and types, as well as their physical environment with its own innate properties. This diversity represents genetic and other information, and each datum in the system is a potential or actual contributor to human well-being and/or the health and stability of the biosphere on which we depend.
>
> (Jenkins 1972: 18)

In 1975, when TNC released a report entitled *The Preservation of Natural Diversity*, it was apparent that the land protection organization was hinging its conservation strategy on the concept of diversity:

> We need to set aside, in viable units, adequate examples of extant ecosystems, biological communities, endangered species habitats, and endangered physico-chemical environmental features. Only in this way can we maintain the full diversity of genetic variability, ecological relationships, and special processes and elements.
>
> (TNC 1975: 10)

In 1977, TNC even was successful in having a bill introduced in the US House of Representatives that proposed a "Natural Diversity Act," which would protect "different types of terrestrial and aquatic communities and ecosystems … [and the] lifeforms [which] constitute a vast genetic reservoir that supplies the material for the continuing evolution of planetary life" (Metcalf and Sebalius 1978: 7–8). Although the bill would never become law, this concern for "natural diversity" was yet another example of conservationists trying to link the protection of genes, species, and ecosystems.

It is evident from the evolution of language in the scientific and conservation community through the twentieth century that by the 1970s, the cultural, social, and political landscapes were well-prepared for the introduction of a term that would successfully tie conservation interests together. With so many similarities in content, it is curious that "natural diversity" did not take off in the same way that "biological diversity" would a few years later. Timing, apparently, is everything. With the help of several spotlight conferences, support from federal government agencies in partnership with prominent environmental groups, and promotion by dynamic spokespeople like E. O. Wilson, biological diversity was propelled to the forefront of environmentalism in the 1980s. The different constituencies necessary to support the cause were all in place, and their connected concerns were coming together under one umbrella goal. Simply put, biological diversity was a term in the right place at the right moment in history – as Elliott Norse observed, "it was an idea whose time had come" (Norse 1999).

References

Baldwin, Henry I. 1941. "An Inventory of the Natural Vegetation Types and the Need for Their Preservation." *Science* 93(2404): 81–82.

Bean, Michael J., and Melanie J. Rowland. 1997. *The Evolution of National Wildlife Law*, 3rd edn. Westport, CT: Praeger.

Behlen, Dorothy. 1981. "Taking Root." *Nature Conservancy News* 31(4): 7–11.

Biffen, R. H. 1906. "Mendel's Laws of Inheritance in Wheat Breeding." *Journal of Agricultural Sciences* 1: 4–48.

Brown, William L. 1984. "Conservation of Gene Resources in the United States." In *Plant Genetic Resources: A Conservation Imperative*, ed. Christopher W. Yeatman, David Kafton, and Garrison Wilkes. Boulder, CO: Westview.

Cart, Theodore Whaley. 1971. "The Struggle for Wildlife Protection in the United States, 1870–1900: Attitudes and Events Leading to the Lacey Act." PhD dissertation, University of North Carolina, Chapel Hill.

Clements, Frederic E., and Victor E. Shelford. 1939. *Bio-ecology*. New York: John Wiley and Sons.

Coolidge, Harold. 1963. Introduction to *Scientific Use of Natural Areas Symposium*, ed. Julia Field and Henry Field. Washington, DC: International Congress of Zoology.

Croker, Robert A. 1991. *Pioneer Ecologist: The Life and Work of Victor Ernest Shelford, 1877–1968*. Washington, DC: Smithsonian Institution Press.

Crowe, Philip K. 1970. *World Wildlife: The Last Stand*. New York: Charles Scribner's Sons.

Darnell, Rezneat M. 1976. "Natural Areas Preservation: The US/IBP Conservation of Ecosystems Program." *BioScience* 26(2): 105–108.

Devoe, Alan. 1944. "On Salvaging Nature." *American Mercury* 59(249): 366–369.

Dice, Lee R. 1925. "The Scientific Value of Predatory Mammals." *Journal of Mammalogy* 6(1): 25–27.

Dice, Lee R. 1943. *The Biotic Provinces of North America*. Ann Arbor: University of Michigan Press.

Dunlap, Thomas R. 1987. *Saving America's Wildlife*. Princeton, NJ: Princeton University Press.

Elton, Charles S. 1958. *The Ecology of Invasions by Animals and Plants*. New York: John Wiley and Sons.

Farnham, Timothy J. 2007. *Saving Nature's Legacy: Origins of the Idea of Biological Diversity*. New Haven, CT: Yale University Press.

Fisher, James, Noel Simon, and Jack Vincent. 1969. *Wildlife in Danger*. New York: Viking.

Frankel, Otto H., and E. Bennett. 1970. "Genetic Resources." In *Genetic Resources in Plants: Their Exploration and Conservation*, ed. Otto H. Frankel and E. Bennett. Philadelphia, PA: F. A. Davis.

Frankel, Otto H., and J. G. Hawkes. 1975. "Genetic Resources: The Past Ten Years and the Next." In *Crop Genetic Resources for Today and Tomorrow*, ed. Otto H. Frankel and J. G. Hawkes. Cambridge: Cambridge University Press.

Gabrielson, Ira N. 1942. *Wildlife Conservation*. New York: Macmillan.

Hanson, Herbert C. 1939. "Check-Areas as Controls in Land Use." *Scientific Monthly* 48(2): 130–146.

Hutchinson, G. E. 1959. "Homage to Santa Rosalia; or, Why Are There So Many Kinds of Animals?" *American Naturalist* 93 (May–June): 145–159.

Jenkins, Robert B. 1972. "A Natural Areas Inventory." *Nature Conservancy News* 22(3): 16–18.

Kloppenburg, Jack Ralph. 1988. *First the Seed: The Political Economy of Plant Biotechnology, 1492–2000*. New York: Cambridge University Press.

Lehmann, V. W. 1938. "Some Values of Natural Areas." *Bird Lore* 40 (September): 310–314.

Leopold, Aldo. 1933. *Game Management*. New York: Charles Scribner's Sons.

MacArthur, Robert H. 1955. "Fluctuations of Animal Populations and a Measure of Community Stability." *Ecology* 36(3): 533–536.

MacArthur, Robert H., and Edward O. Wilson. 1967. *The Theory of Island Biogeography*. Princeton, NJ: Princeton University Press.

Margelef, Ramon. 1963. "On Certain Unifying Principles in Ecology." *American Naturalist* 97(897): 357–374.

May, Robert. 1971. "Stability in Multi-species Community Models." *Mathematics Biosciences* 12: 59–79.

May, Robert. 1972. "Will a Large Complex System Be Stable?" *Nature* 238: 413–414.

May, Robert. 1973a. "Qualitative Stability in Model Ecosystems." *Ecology* 54(3): 638–641.

May, Robert. 1973b. *Stability and Complexity in Model Ecosystems*. Princeton, NJ: Princeton University Press.

Metcalf, Lee, and Keith Sebalius. 1978. "A Program for Preserving America's Natural Diversity." *Nature Conservancy News* 28(1): 6–12.

Milton, John P. 1966. "Retrospect." In *Future Environments of North America*, ed. F. Fraser Darling and John P. Milton. Garden City, NY: Natural History Press.

National Research Council. 1978. *Conservation of Germplasm Resources: An Imperative*. Washington, DC: National Academy of Sciences.

Norse, Elliott A. 1996. "A River That Flows to the Sea: The Marine Biological Diversity Movement." *Oceanography* 9(1): 5–9.

Norse, Elliott A. 1999. Interview with author by telephone. March 31.

Norse, Elliott A., and Roger E. McManus. 1980. "Ecology and Living Resources: Biological Diversity." In *Environmental Quality 1980: The Eleventh Annual Report of the Council on Environmental Quality*. Washington, DC: Council on Environmental Quality.

Odum, Eugene P. 1953. *Fundamentals of Ecology*. Philadelphia, PA: W. B. Saunders.

Odum, Eugene P. 1963. *Ecology*. New York: Holt, Rinehart, and Winston.

Office of Technology Assessment. 1987. *Technologies to Maintain Biological Diversity*, OTA-F-330. Washington, DC: Government Printing Office.

Pistorius, Robin, and Jereon van Wijk. 1999. *The Exploitation of Plant Genetic Information: Political Strategies in Crop Development*. New York: CABI.

Plucknett, Donald L., Nigel J. H. Smith, J. T. Williams, and N. Murthi Anishetty. 1987. *Gene Banks and the World's Food*. Princeton, NJ: Princeton University Press.

Rush, William. 1937. "What are Wildlife Values?" *Nature Magazine* 30(1): 40–43.

Scovell, E. L. 1938. "Overlook No Living Thing." *Recreation* 32 (August): 295–296.

Shelford, Victor E., ed. 1926. *Naturalist's Guide to the Americas*. Baltimore, MD: Williams and Wilkins.

Skutch, Alexander. 1948. "Earth and Man." *Audubon Magazine* 50 (November): 356–359.

Stone, Peter. 1973. *Did We Save the Earth at Stockholm?* London: Earth Island.

Tansley, Arthur G. 1935. "The Use and Abuse of Vegetational Concepts and Terms." *Ecology* 16(3): 284–307.

The Nature Conservancy. 1975. *The Preservation of Natural Diversity*. Washington, DC: The Nature Conservancy.

UNESCO. 1970. *Use and Conservation of the Biosphere*. Paris: UNESCO.

UNESCO. 1974. *Programme on Man and the Biosphere: Task Force on Criteria and Guidelines for Choice and Establishment of Biosphere Reserves*. Paris: UNESCO.

Wilkes, Garrison. 1983. "Current Status of Crop Plant Germplasm." *CRC Critical Reviews in Plant Sciences* 1(2): 133–181.

Wilson, E. O., ed. 1988. *Biodiversity*. Washington, DC: National Academy Press.

Witt, Steven C. 1985. *Biotechnology and Genetic Diversity*. San Francisco: California Agricultural Lands Project.

Worthington, E. B., ed. 1975. *The Evolution of IBP*. Cambridge: Cambridge University Press.

Van Name, Willard G. 1919. "Zoological Aims and Opportunities." *Science* 50(1282): 81–84.

Van Name, Willard G. 1941. "Need for the Preservation of Natural Areas Exemplifying Vegetation Types." *Science* 93(2418): 423.

2
EXTINCTION AND BIODIVERSITY
A historical perspective

David Sepkoski

Why care about biodiversity? This is a question that could be answered from many perspectives – economic, philosophical, pragmatic, aesthetic, ethical, etc. – and many of the essays in this volume explore those arguments that can be and have been mustered in favor of preserving the diversity of life on Earth. But it is also a question that demands a historical answer: not so much why *should* we care about biodiversity so much as why *do* we now, why has biodiversity preservation only emerged as a fairly recent topic of global political and scientific concern, and what has changed over the past 150 years or so in cultural awareness and biological understanding that has brought this about? It would be tempting to think that concern for the diversity of life is a self-evident value, requiring little justification. It may well be that arguments can be made that biodiversity has just such intrinsic value from a philosophical standpoint, but the simple historical fact is that Western society has not always recognized this. That is to say, the past 200 years or so of biological thought reveals a shifting landscape of opinion concerning the value and even the very existence of the category we would now call "biodiversity." It is only in the past fifty years or so that scientists have recognized that the study of biological diversity merits special attention and methods – whether in ecological or geological context – and even more recently that its current preservation has become a political issue.

It turns out that one of the key components of the biological understanding of diversity – and of changes in its valuation – has been how biologists and paleontologists have conceptualized and understood extinction. One of the most effective rhetorical tools in the biodiversity preservation movement has been the association of the current biodiversity crisis with a "sixth mass extinction" (Leakey and Lewin 1995; Kolbert 2014). This idea developed out of paleontologists' identification of five major mass extinctions – the so-called "big five" – in the history of life, events during which anywhere from 70 to 98 percent of standing diversity was extinguished in a geological instant (Raup and Sepkoski 1982). But the concept of mass extinction itself is a relatively new feature of biology: before the 1970s it was an idea more associated with the lunatic fringe than with serious science, and it was not until the 1980s (prompted by the discovery of physical evidence of the bolide impact that killed the dinosaurs some 65 million years ago) that paleontologists have been able to present concrete proof that catastrophic extinctions have occurred, or to model their ecological and evolutionary effects (Sepkoski 2012). For most of the history of modern biology, it has been assumed that extinction is a slow, gradual process that operates over very long periods of time, and which is in equilibrium with replacement via the evolution of new species. We have become so accustomed, in recent decades, to think of the foothold of life on Earth (including our own species') as tenuous that it is easy to forget that we have not always perceived it so.

The simple reason that extinction is so closely related to beliefs about the value of biodiversity is that, as a culture, we tend to value most what we fear losing. Concern for individual endangered species or regret over their passing – the California condor, the passenger pigeon, the great auk – dates well back into the nineteenth century (Barrow 2009). But the preservation of particular species (which, not coincidentally, are usually associated with romantic or aesthetic attachments) is not the same thing as valuing *all* organisms, or regarding diverse environments as being essential for our own survival. Anxiety about the loss of such individual species has long motivated conservation efforts, but it is fear over losing *many* species in a short period of time – and not only the pretty or valuable ones – that motivates the recent biodiversity movement (Takacs 1996; Farnham 2007). This concern has, in recent decades, been formalized in agreements such as the 1992 UN Convention on Biological Diversity, where more than 150 nations affirmed "the intrinsic value of biological diversity" and pledged to foster its protection (UN 1992: 1). That strong association of biological diversity with "intrinsic value" also animates discussions of other kinds of diversity – linguistic, ethnic, cultural – that have gained momentum since the early 1990s as well (Sepkoski 2015). A decade after the UN Convention on Biological Diversity, UNESCO produced a "Universal Declaration on Cultural Diversity" which framed cultural diversity in exactly the same language as the earlier document had presented biological diversity: as "a living, and thus renewable treasure that must not be perceived as unchanging but as a process guaranteeing the survival of humanity …. [C]ultural diversity is as necessary for humankind as biodiversity is for nature" (UNESCO 2002).

If, as I have argued, biodiversity is only a fairly recently recognized value, then how did this transformation come about, and what earlier value system did it replace? These are the questions I will address in this chapter. I will begin with the nineteenth-century acceptance of extinction as a natural – and fairly common – occurrence, placing Victorian-era biological views of extinction in their cultural context. I argue that, for most of the nineteenth century, biological diversity was not recognized as a phenomenon worth studying or preserving in its own right because biologists understood the natural world to be in constant equilibrium where, as Carl Linnaeus had put it in 1762, "the death and destruction of any one thing should always be subservient to the restitution of another" (Linnaeus 1762: 40). Perhaps not surprisingly, this view was often taken as implicit justification for activities by Europeans and their colonists that led to the eradication of native peoples, flora, and fauna during an era of imperialist expansion (Brantlinger 2003).

It was only during the second half of the twentieth century that the idea of *mass* extinctions became scientifically respectable, and consequently that the notion of a biodiversity "crisis" could emerge. This change in scientific consensus was preceded and accompanied by a series of dramatic cultural and political shifts that brought about a new climate of "threat" or anxiety in the West. Two world wars, Hiroshima and Auschwitz, the Doomsday Clock, *Silent Spring* and Three Mile Island, Vietnam and the student protests of the 1960s, and a host of other factors helped produce what historian Eric Hobsbawm has called "the Age of Catastrophe" (Hobsbawm 1994). In the mid-1980s, this heightened sense of cultural anxiety was a receptive context for the introduction of two important scientific developments. The first was the discovery in 1980 by Luis and Walter Alvarez that the dinosaurs had become extinct suddenly as the result of a catastrophic collision with an asteroid or comet. This received considerable media interest and led to further study of mass extinctions in the geological past, which elevated mass extinction to dramatic public attention (Sepkoski 2012: ch. 9). Parallels between the Alvarez impact hypothesis and the projected aftermath of a nuclear exchange were missed neither by the public nor scientists: in fact, the much-discussed "nuclear winter" scenario was developed on the basis of climate models first proposed for impact events (Badash 2009).

The second event was the official introduction of "biodiversity" into scientific and political vocabulary. In 1986, the botanist Walter G. Rosen and the entomologist and ecologist E. O. Wilson organized a "National Forum on BioDiversity" in Washington, DC. Sponsored by the National Academy of Sciences and the Smithsonian Institution, this conference brought together experts in biology, ecology, paleontology, economics, and public policy for a summit to consider broad threats to biological diversity caused by human action (or inaction). While this was not the first call to protect the environment or endangered species from the consequences of industrialization, pollution, overpopulation, and the like, it dramatically changed both the tone of the conversation and the political stakes involved. Not only did this event introduce the term "biodiversity" to common currency, it also made the link between extinction and biodiversity conservation clear from the very start. Writing in the introduction to the published conference proceedings, Wilson commented that "The current reduction of diversity seems destined to approach that of the great natural catastrophes at the end of the Paleozoic and Mesozoic eras – in other words, the most extreme in the past 65 million years" (Wilson 1988a: 11–12). From this point forward, our current biodiversity crisis became inextricably linked to our understanding of the major mass extinctions of the geological past – the "sixth extinction" trope was born (Leakey and Lewin 1995).

Victorian extinction

The phenomenon of extinction first received serious scientific attention through the work of the French naturalist Georges Cuvier (1769–1832) in the first decades of the nineteenth century. Prior to this point, European and American naturalists and savants had debated whether god would allow any species to be destroyed, as such an event was seen to potentially violate the divinely ordained balance of nature. Famously, when confronted with recently discovered fossils of the strange beast dubbed "mammoth," Thomas Jefferson expressed confidence that living specimens would be found in the unexplored American continent, since, as he put it in his *Notes on the State of Virginia*, "such is the balance of nature, that no instance can be produced, of her having permitted any one race of her animals to become extinct; of her having formed any link in her great work so weak as to be broken" (Jefferson 1801: 77).

Mounting fossil evidence, however, quickly convinced naturalists that the former world had indeed been populated by creatures that no longer existed, but the question of extinction was not immediately settled. Jean-Baptiste Lamarck (1744–1829), for example, rejected the idea of extinction, favoring instead an evolutionary process that had molded organisms into their current forms. Lamarck's position was directly challenged by Cuvier, who interpreted huge fossil deposits in the quarries outside of Paris as evidence of great "revolutions" or catastrophic mass extinctions that he believed punctuated episodes in Earth's history (Rudwick 1997). One irony for our story, then, is that while the first major theory of extinction recognized mass extinctions as a regular feature of the history of life, Cuvier's "catastrophism" was decisively rejected by most of the scientific community for the next 150 years or more (Rudwick 2005). The reasons for this are complex, but one major consideration was likely the fact that Cuvier's theory of episodic catastrophes upset longstanding beliefs about the balance or economy of nature found in Judeo-Christian and even Classical teachings.[1] It is notable that, even when Darwin introduced his supposedly revolutionary notion of evolution via natural selection, he nonetheless preserved a fairly conservative view of the equilibrium of nature (as I will discuss below) (Cuddington and Ruse 2004).

In Britain, especially, it was taken for granted throughout most of the nineteenth century that when extinction did take place (and by the 1820s and 1830s there was mounting evidence for this fact) the finely tuned balance of nature was never upset (Parkinson 1804). This commitment to a stable natural equilibrium prevented most naturalists from granting the possibility of mass extinctions. Charles Lyell (1797–1875), the great Scottish pioneer of modern geology and close friend of Darwin's, helped install the view that extinction was a natural and common occurrence, but also staunchly maintained that extinctions of individual species were always equally balanced by the appearance of new ones. In his famous *Principles of Geology* (1830–1833) he argued, on the one hand, that since "species are subject to incessant vicissitudes … it will follow that the successive destruction of species must now be part of the regular and constant order of nature," while on the other that "the addition of any new species, or the *permanent* numerical increase of one previously established, must always be attended either by the local extermination or the numerical decrease of some other species" (Lyell 1830–1833, vol. 1: 141–142). In short, while Lyell envisioned an Earth that was subject to constant fluctuation and variation, he nonetheless held firm in the belief that "the successive extinction of terrestrial and aquatic species … [is] part of the economy of our system" (Lyell 1830–1833, vol. 2: 168).

It was this basic conception that informed Darwin's own views on the role of extinction in natural selection and evolution. Throughout the *Origin of Species*, Darwin treated the relationship between extinction and speciation as a dynamic equilibrium, where extinction was understood simply as the failure of a species to adapt to its environment. Darwin believed that the total number of species had remained stable throughout Earth's history, since the constant competition between organisms tended to average out between the winners and losers in the struggle for life. As he put it, "it inevitably follows, that as new species in the course of time are formed through natural selection, others will become rarer and rarer, and finally extinct" (Darwin 1859: 110), and that since a species is "maintained by having some advantage over those with which it comes into competition … the consequent extinction of less-favoured forms almost inevitably follows" (Darwin 1859: 320). In the *Origin*, Darwin envisioned the struggle for existence as a Malthusian zero-sum game that operated directly only on individuals, meaning that extinction would also tend to be a slow and gradual process, and that "the old notion of all the inhabitants of the earth having been swept away at successive periods by catastrophes, is very generally given up" (Darwin 1859: 317).

The Lyell/Darwin view of extinction as a gradual process that supported a stable equilibrium was, in general terms, the dominant view in Europe and America during the nineteenth century. It also both drew from and supported contemporary cultural and political beliefs about progress and European natural superiority that lent very little support to concerns about biological or cultural diversity. In Darwin's scheme, extinction was the inevitable result of failure to compete in a "fair game"; extinct species were simply nature's losers, and it was easy to transpose this perspective onto European political concerns. Even before Darwin, it was fairly common to invoke biological justifications for the consequences of European imperial expansion. According to Patrick Brantlinger, such expansion was often justified by the belief that the subjugation and extermination of so-called "savage" races – and their native floras and faunas – was the natural outcome of contact between superior and inferior civilizations, the latter of whom were "doomed" to inevitable extinction (Brantlinger 2003). It is hardly surprising that if extinction was seen to be an inevitable and natural occurrence, little concern was expressed for preserving cultural or biological diversity.

Lyell, for example, wrote in the *Principles* that "We must at once be convinced, that the annihilation of species has already been effected, and will continue to go on hereafter, in certain regions, in a still more rapid ratio, as the colonies of highly-civilized nations spread themselves over unoccupied lands" (Lyell 1830–1833, vol. 2: 156). This occasioned little regret, he argued, since "if we wield the sword of extermination as we advance, we have no reason to repine at the havoc committed, nor to fancy, with the Scottish poet, that 'we violate the social union of nature,'" but rather merely should "reflect, that in thus obtaining possession of the earth by conquest ... we exercise no exclusive prerogative. Every species which has spread itself from a small point over a wide area, must, in like manner, have marked its progress, by the diminution, or the entire extirpation, of some other" (Lyell 1830–1833, vol. 2: 156). Lyell emphasized that this explanation applied equally to "the extirpation of savage tribes of men by the advancing colony of some civilized nation." While he did pause to express regret for this fact, he offered no apology, since, as he was quick to note, this was the natural and *inevitable* course of nature: "few future events are more certain than the speedy extermination of the Indians of North America and the savages of New Holland in the course of a few centuries, when these tribes will be remembered only in poetry and tradition" (Lyell 1830–1833, vol. 2: 175).

Such expressions were overwhelmingly common during the Victorian era. Darwin himself had remarked in his account of his voyage on the HMS *Beagle* that "wherever the European has trod, death seems to pursue the aboriginal," and concluded that "the varieties of man seem to act on each other in the same way as different species of animals – the stronger always extirpating the weaker" (Darwin 1909: 459). Likewise, the physician and ethnographer James Cowles Prichard (1786–1848) wrote, in an 1840 essay in the *Edinburgh Philosophical Journal* titled "On the Extinction of the Human Races," that "Wherever Europeans have settled, their arrival has been the harbinger of extermination to the native tribes ... and it may happen that, in the course of another century, the aboriginal nations of most parts of the world will have ceased entirely to exist" (Prichard 1840: 168–170). And Darwin's good friend and co-discoverer of natural selection, Alfred Russell Wallace (1823–1913), quite openly justified racial extinction as the result of natural selection applied to human beings. In an 1864 essay titled "The Origin of Human Races and the Antiquity of Man Deduced from the Theory of Natural Selection," Wallace wrote that:

> It is the same great law of *"the preservation of favored races in the struggle for life,"* which leads to the inevitable extinction of all those low and mentally undeveloped populations with which Europeans come in contact ... [J]ust as the more favorable increase at the expense of the less favorable varieties in the animal and vegetable kingdoms, just as the weeds of Europe overrun North America and Australia, extinguishing native production by the inherent vigour of their organization, and by their greater capacity for existence and multiplication.
>
> (Wallace 1864: clxiv–clxv)

Darwin's theory of evolution, then, merely reinforced an already well-established discourse of extinction that regarded the phenomenon to be part of the orderly balance of nature. It also showed remarkably little awareness of biological diversity as a phenomenon worthy of study or conservation – and why should it, if Darwin imagined that diversity was a property kept in intrinsic natural equilibrium? As he put it in the *Origin*:

> Battle within battle must ever be recurring with varying success; and yet in the long-run the forces are so nicely balanced, that the face of nature remains uniform for long periods of time, though assuredly the merest trifle would often give victory to one organic being over another. Nevertheless so profound is our ignorance, and so high our presumption, that we marvel when we hear of the extinction of an organic being;

and as we do not see the cause, we invoke cataclysms to desolate the world, or invent laws on the duration of the forms of life!

(Darwin 1859: 73)

While he understood the diversity of life to be in constant fluctuation, Darwin did not regard it to be affected or threatened by extinction: "Everyone has heard that when an American forest is cut down," he remarked, "a very different vegetation springs up; but it has been observed that the trees now growing on the ancient Indian mounds, in the Southern United States, display the same beautiful diversity and proportion of kinds as in the surrounding virgin forests" (Darwin 1859: 112). In other words, nature's inherent fecundity ensures that there will always be new forms standing by to replace the old ones, and that those new species will survive if they maintain a competitive advantage with their environments. He concluded ultimately that "Thus the appearance of new forms and the disappearance of old forms, both those naturally and those artificially produced, are bound together … we know that species have not gone on indefinitely increasing, at least during the later geological epochs, so that, looking to later times, we may believe that the production of new forms has caused the extinction of about the same number of old forms" (Darwin 1872: 296).

Ultimately, then, to the extent that Darwin recognized something like biological diversity in nature, he regarded it as an endlessly renewing resource. This attitude reflects both the older notion of "plentitude" in nature associated with Linnaeus and other theologically inspired naturalists, as well as Lyell's interpretation of geological history as a dynamic equilibrium. What Darwin added was the regular cycle of extinction and speciation, which made Darwin's view of nature considerably more transient than earlier conceptions of a static balance or economy. But beneath this constant change is a fundamental, underlying stability, provided thanks to nature's capacity for endless self-generation of more diversity. The issue, then, isn't whether Darwin recognized or thought natural variety was important – he certainly did – but rather whether he thought that diversity itself could be diminished by extinction, and whether the stability of nature could be threatened by a loss of diversity, which he did not. Competition and replacement were, for Darwin, the engine that drives the progressive improvement of the natural system, and which maintains the economy of nature. Far from seeing diversity as something to be conserved, he viewed it as essentially the fuel for that engine, the source of continued competition, selection, and extinction. The idea that nature exists in a harmonious, unchanging balance may have been upset, at the end of the eighteenth century, by authors such as Malthus and Cuvier, who suggested that competition and the specter of extinction are an inherent part of the natural order. But Darwin's message was, essentially, that struggle and even extinction are positive forces – in the long view – thus soothing Victorian anxieties about their own impact on the world. The world may be subject to constant change, but faith in the ultimate constancy of nature was not shaken.

Extinction and biodiversity in the Age of Catastrophe

Viewed from an early twenty-first-century perspective, an extraordinary shift has taken place in the way we understand extinction and value biological diversity. It is now generally held that biodiversity is a precious "resource" for the health of our planet; mass extinctions are widely regarded to be events that have shaped evolutionary and ecological development of life at a number of points in the Earth's history; and human diversity of all kinds – linguistic, ethnic, cultural – is considered a cherished asset. The transition from the nineteenth-century attitude to the one generally held today is an extremely complex story that can only be sketched in brief in this essay. Generally, however, the major factors responsible for this change encompass the narrowly scientific to the broadly cultural. In the first place, beginning in the 1950s, a new science of mass extinction developed – primarily in paleontology – which both conclusively established the reality of major mass extinctions in life's past, as well as explicitly reoriented paleontology towards the quantitative analysis of patterns of biological diversity over time. Second, during roughly the same period, the discipline of ecology underwent a conceptual transformation that overturned earlier ideas about the intrinsic balance of nature. Ecosystems came to be seen as potentially much more vulnerable than had been previously suspected, and ecological diversity reconceptualized as a resource that could hedge against major environmental disruption (much as, slightly earlier, population geneticists advanced the notion that genetic diversity contributes to stability in populations). Third, awareness of the vulnerability of the environment was heightened by studies and popular accounts (such as Rachel Carson's 1962 *Silent Spring*) that linked modernization with increased risk of extinction of plants, animals, and even humanity itself. And finally, the political and cultural climate changed quite dramatically, from Victorian optimism about limitless progress to a much more pessimistic tenor colored by experiences with world wars, political instability, nuclear proliferation, and rapid social change. This last feature contributed to a broadly foreboding atmosphere in much of the West, with threats of catastrophes of all kinds capturing the public (and scientific) imagination. These factors reached an apotheosis in the late 1970s and early 1980s, in what became the perfect storm of scientific and popular anxiety that produced the modern biodiversity consciousness.

Our current understanding of historical mass extinctions has roots in both immediate postwar paleontology and ecology. While Darwin's explanation that most cases of extinction could be attributed to gradual competitive replacement appeared satisfactory to most subsequent observers, some paleontologists continued to be troubled by the more spectacular and apparently abrupt departures from the fossil record – the trilobites, for instance, or, especially, the dinosaurs (Packard 1886; Schuchert 1924). Nonetheless, for most of the scientific community, major events such as the extinction of the dinosaurs remained a mystery, and the subject of catastrophic mass extinctions was generally treated as acceptable for idle speculation, but not serious scientific study. This began to change only in the 1950s, and largely as the result of the efforts of a single paleontologist, the American invertebrate specialist Norman D. Newell. In a series of papers from the mid-1950s to the late 1960s, Newell helped to establish the legitimacy of mass extinction by taking a new approach: due to the rapid expansion of the marine invertebrate fossil record, thanks both to increased collecting efforts and new methods of fossil preparation, Newell realized he had an enormous quantity of data that could be mathematically analyzed to reveal broad evolutionary patterns (Sepkoski 2012: ch. 2). The results of these investigations were graphs that clearly showed major mass events as distinctive spikes that stood out against the background of "normal" extinction. Despite skepticism from many colleagues, Newell pushed forward with his work, and in 1967 produced a definitive study of mass extinction that concluded that "modern paleontology must incorporate certain aspects of both catastrophism and uniformitarianism while rejecting others" (Newell 1967: 64).

While Newell declined to characterize these mass extinctions as true "catastrophes" (in the sense of a sudden event – he rather favored gradual changes in sea level as the culprit), his work set the stage for further study by legitimating the topic. Importantly, his research also helped draw a closer connection between the study of mass extinctions and the study of diversification. In Newell's quantitative approach (which was refined over the next several decades by students and admirers), extinctions were detected as major drops in standing biological diversity. Absent other kinds of physical evidence about their causes, paleontologists began to see extinctions as, literally, spikes (or troughs) in diversity graphs. As biology increasingly turned to computers and a "data–driven" approach, such statistical estimates of diversity and extinction became valuable tools both for historical studies and for modern conservation efforts. Ultimately, mass extinction came to be defined as a statistical aberration in diversity data.

The study of biological diversification and extinction at this time was also significantly influenced by new ecological approaches, many of which originated in the circle of the Yale ecologist G. Evelyn Hutchinson. Some of this impact was methodological: Hutchinson and his students (especially Robert A. MacArthur) developed heuristic mathematical models for studying phenomena such as species abundance, migration, and population limits as fluctuations in ecological diversity (Kingsland 1985, Dritschilo 2008). Some of these techniques, such as MacArthur and E. O. Wilson's "species–area effect" and their theory of island biogeography had a major impact on paleontology by providing models that could be applied to diversification and extinction over geological time (Sepkoski 2012: ch. 4). But Hutchinson himself also helped spark a new conceptual orientation towards ecological diversity by closely identifying niche diversity with ecological stability. As he put it in a landmark 1959 paper "Homage to Santa Rosalia; or Why Are There So Many Kinds of Animals?," "Modern ecological theory appears to answer our initial question at least partially by saying that there is a great diversity of organisms because communities of many diversified organisms are better able to persist than are communities of fewer less diversified organisms" (Hutchinson 1959: 150). In other words, diverse communities are more resistant to unexpected environmental change because they have a greater number of "options" to cope with that change.

This idea essentially reflects the wisdom of "not putting all of one's eggs in one basket," and follows closely the logic of the population geneticist Theodosius Dobzhansky, who in 1937 had argued that high genetic diversity in a population constitutes "a store of concealed, potential, variability," and warned that "A species perfectly adapted to its environment may be completely destroyed by a change in the latter [its environment] if no hereditary variability is available in the hour of need" (Dobzhansky 1937: 127). This ecological/genetic argument for the importance of diversity as a "storehouse" against unforeseen change is, I argue, a vital component of modern biodiversity discourse. It emerged largely in the post-World War II context, and influenced both emerging environmental movements as well as debates about race and cultural diversity. This moment also reveals quite clearly how scientific debates became entwined with cultural and political problems around diversity. For Dobzhansky and his students (especially the geneticist Richard Lewontin), studies of genetic variability and diversity became the basis for political stances against racism and in favor of protecting cultural diversity (Dobzhansky 1962, Lewontin *et al.* 1984). For Hutchinson, it was a call to arms that would reverberate across the decades to the modern biodiversity conservation movement. At the end of "Homage to Santa Rosalia" Hutchinson described the "indiscriminate" reduction of diversity caused by human actions, and opined that "we may hope for a limited reversal of this process when man becomes aware of the value of diversity no less in an economic than in an esthetic and scientific sense" (Hutchinson 1959: 156).

The political and cultural sources of change in the postwar period are harder to quantify, though seemingly ubiquitous. The ways in which anxieties about nuclear annihilation seeped into popular consciousness in the West in the 1950s and beyond have been well documented, producing, in the words of historian Spencer R. Weart, "a world of fear, suspicion, and almost inevitable catastrophe" (Weart 1988: 115). Above all, threat of nuclear Armageddon signaled a shift in thinking about human progress and permanence towards a much darker and more pessimistic direction. The images of the Trinity test site or the mushroom clouds over Hiroshima and Nagasaki also focused attention on the spectacular, or catastrophic. This was reflected in popular culture (movies such as *Planet of the Apes* and *On the Beach*) that dramatized the onset or aftermath of nuclear conflagration; in politics, where events such as the Cuban Missile Crisis paralyzed the American public; and in emerging environmental awareness, where works such as Rachel Carson's *Silent Spring* warned that "Along with the possibility of the extinction of mankind by nuclear war, the central problem of our age has therefore become the contamination of man's total environment" (Carson 1962: 8). As Weart broadly summarizes,

> Nuclear weapons gave the twentieth century's nihilism a dismal solution. Immediately upon hearing the news from Hiroshima, sensitive thinkers had realized that doomsday – an idea that until then had seemed like a religious or science-fiction myth, something outside worldly time – would become as real a part of the possible future as tomorrow's breakfast.
>
> (Weart 1988: 392)

This general climate of anxiety also contributed to public and scientific interest in mass extinctions. While it was pilloried by virtually all experts, Immanuel Velikovsky's account of celestial catastrophe *Worlds in Collision* (1950) was a worldwide bestseller and lodged itself in the popular imagination for decades. Although it may have done more harm than good to the scientific acceptance of catastrophic mass extinctions (by further associating "catastrophism" with pseudo-science), it has recently been described by a historian as "one of postwar America's most culturally significant works about the natural world" (Gordin 2012: 135). Even as he dismissed it as "a pathetic, ominous, and superstitious piece of work," *New Yorker* reviewer Alfred Kazin nonetheless acknowledged that *Worlds in Collision* "fits only too well into the intellectual melodrama of this period" (Kazin 1950: 103). That "melodrama" had, by the early 1970s, joined a nearly omnipresent threat of political breakdown and nuclear war with a host of additional existential threats and morbid fascinations: catastrophic environmental pollution, dire warnings of an immanent population explosion, political radicalism and terrorism in Europe and the US, and daily scenes of body bags being loaded onto planes in Southeast Asia. For many observers, the end of the world did indeed appear nigh.

When, in 1980, the father-and-son team of Luis and Walter Alvarez published a paper claiming to have discovered evidence of a massive bolide impact some 65 million years ago that triggered a worldwide catastrophe, they did more than just provide a potential answer to a longstanding scientific mystery (Alvarez *et al.* 1980). The particular scenario they envisioned – a cataclysmic impact, followed by worldwide firestorms, deadly acid rains, months of near-total darkness, a massive greenhouse effect that lasted for thousands of years, and extinction of the dinosaurs – spoke directly to the public's consciousness of our own species' tenuous grip on the planet. Indeed, the very scenario adopted for the Alvarez impact hypothesis was borrowed by the astronomer Carl Sagan and others to provide the basis for models of the so-called nuclear winter that would follow a global thermonuclear exchange (Badash 2009). The impact of the theory on popular consciousness is probably best exemplified in a contemporary newspaper column written in 1984, whose author, Ellen Goodman, asked "I wonder whether every era gets the dinosaur story it deserves," noting that "the scientists of the 19th century – a time full of belief in progress – saw evolution as part of the planet's plan of self-improvement … Those who lived in a competitive economy valued the 'natural' competition of species. The best man won." But, she continued, "surely we are now more sensitive to cosmic catastrophe, to accident," adding "in that sense, the latest dinosaur theory fits us uncomfortably well. 'Our' dinosaurs died together in some meteoric winter, the victims of a global catastrophe. As humans, we fear a similar shared fate." However, as Goodman observed in closing her column, "the difference is that their world was hit by a giant asteroid while we – the large-brained, adaptable creatures who inherited the earth – may produce our own extinction" (Goodman 1984).

Along with the spectacular publicity the Alvarez theory attracted, the extinction research initiated by Newell reached maturity during the late 1970s and early 1980s. Building on the equilibrial models of ecological theorists like MacArthur and Wilson, a small group of paleontologists – including David M. Raup, J. John Sepkoski, Jr., and David Jablonski – began producing more refined quantitative studies of the history of diversification and extinction over the past 500 million years. What this research showed was that while there has been (contrary to Darwin's expectation) a general trend towards increased diversity over the history of life, the pattern of diversification has been perturbed a number of times in what paleontologists determined were major mass extinctions (Sepkoski 1984). Those extinctions – Raup and Sepkoski had detected five such major events – had a profound impact on subsequent evolution by producing massive evolutionary "bottlenecks" (steep reductions in remaining genetic diversity) and also by opening up new ecological space – and creating new ecological conditions – for natural selection to work in (Raup and Sepkoski 1982). One implication of this pattern of extinction and diversification was that while the taxonomic groups that survived mass extinctions eventually did recover and diversify, the total amount of what Stephen Jay Gould has called "disparity" (i.e. major morphological or genetic differentiation between higher taxa) was permanently reduced. In other words, what are left after mass extinctions are a greater number of species and genera (and fewer families and orders) that are more similar to one another. Another implication explored by Raup and Jablonski is that in the aftermath of catastrophic mass extinctions "normal" evolutionary rules do not apply: while normal "background" extinction is probably as gradual and selective as Darwin had envisioned, mass extinctions may target taxonomic groups that had been perfectly well-adapted to previous conditions and who become extinct through no "fault" of their own (Jablonski 1986). As Raup put it in a popular book some years later, many cases of extinction may simply be the result of "bad luck" (Raup 1991).

This was the setting out of which the biodiversity conservation movement emerged. While conservation biologists and ecologists had been warning for years that human activity was putting many species at risk (Myers 1979), paleontological study of extinction and diversification provided a quantitative metric for assessing the present crisis and the future consequences. As Wilson put it in his 1992 *The Diversity of Life*, "the laws of biological diversity are written in the equations of speciation and extinction" (Wilson 1992: 220). In fact, Wilson and other biodiversity activists immediately began using Raup's calculation of the normal or "background" rate of extinction for the past 500 million years – two to three species per year, in Raup's estimate – as the figure against which to compare current rates of species loss. The direct source, then, for considering the modern biodiversity crisis to be a potential "mass extinction" are paleontological studies of mass extinctions and diversification from the 1970s and 1980s. It is in those terms that the crisis is sometimes called a "sixth extinction" (number five being the one that eliminated the dinosaurs), and predictions about the evolutionary, ecological, and environmental consequences of a current mass extinction are explicitly informed by paleontology (Leakey and Lewin 1995).

While biodiversity activists like Wilson certainly actively sought out the participation of paleontologists – Raup was a featured presenter at the first BioDiversity conference in 1986 – paleontologists have also actively sought to contribute their expertise to the discussion. Raup warned at the initial meeting that "without consideration of the time perspective available from the geological record, a full evaluation of the contemporary extinction problem may prove as difficult as would be the case … if an epidemiologist were to treat an infectious disease without medical records" (Raup 1988: 57). This statement echoed Wilson's own contribution, which compared the current crisis to "the great natural catastrophes at the end of the Paleozoic and Mesozoic eras," cautioning that "the modern episode exceeds anything in the geologic past" (Wilson 1988b: 11–12). Many of the paleontologists most directly responsible for uncovering the ecological dynamics of mass extinctions would subsequently comment on the contemporary crisis as well. For example, Sepkoski argued in 1997 that "it may indeed be possible that we are on the brink of the greatest of all mass extinctions," and encouraged activism among his colleagues by noting that "we are the only scientists who have ever seen biodiversity crises to their end … and have some idea of what happens in their aftermath" (Sepkoski 1997: 536). Likewise, Michael Benton has concluded that "extinction events in the fossil record … can give indications of what might, or might not, happen in the future," and pointed out that "comparison of present crises with documented ancient examples at least allows scientists and policy makers to work with real facts and figures" (Benton 2003: 8 and 16). As Jablonski has put it, "in order to understand the dynamics of biodiversity … we need to understand extinctions and their complex aftermath" (Jablonski 2004: 174), reminding us that "whatever the exact magnitude of present-day diversity losses, rebounds in the fossil record suggest that they will not be recouped in the next thousand years" (Jablonski 1991: 755).

Conclusion

I have tried to emphasize two related points in this chapter. First, valuations of biodiversity are closely tied to biological understanding of extinction. Much of the urgency surrounding discussions of the current state of biodiversity are dependent on a particular conception of mass extinction that has emerged only in the past few decades. Second, these values have shifted over time, partly due to changing biological views of extinction, and in part because of broader cultural transformations. It is undoubtedly the case that science has an important role in shaping public perceptions of the natural world, but scientists are also participants in their cultures and to some degree are influenced by broader social and political values. Victorian naturalists like Lyell and Darwin – though politically liberal for their day – shared the same easy optimism about limitless future progress that their contemporary statesmen and captains of industry held. They would have been as unlikely to credit the possibility that humanity might extinguish itself in a few short generations as they would to acknowledge that the spread of European culture might have a harmful effect. The evidence of mass extinction had already been presented – thanks to Cuvier – but it was rejected in favor of another, equally plausible explanation (for its time) that fit more comfortably with middle-class European values.

Likewise, mid-twentieth-century biologists and paleontologists were quite explicitly conditioned by the same Cold War climate of fear and anxiety that affected their non-scientist contemporaries. Paleontologists like Raup and Sepkoski did, to be sure, have access to data and tools (such as computers) that their predecessors lacked. But even so, they faced an uphill battle to convince their colleagues that major mass extinctions had taken place, and the ultimate acceptance of their ideas was attributable, at least in part, to living in a society where the threat of catastrophe had become culturally ingrained. It is certainly the case that the modern biodiversity movement has capitalized on the convergence of science and culture around extinction, as evidenced by the fact that the notion that we are living through a potential "sixth extinction" has earned widespread cultural currency. This is not to say that without the input of paleontologists there would be no biodiversity movement, but rather that the particular discourse that exists around biodiversity conservation has been shaped very directly by recent reconceptualizations of extinction.

Ultimately, this suggests that even some of our culture's deepest values and assumptions – the idea of the "balance of nature," for example – are subject to revision. Little more than a hundred years ago most biologists would have described the natural world as an endlessly renewing resource incapable of being destabilized by mere human actions. Today, we think quite otherwise. In response to "some apologists for development [who] have argued that extinction at any scale … poses no biological worry but, on the contrary, must be viewed as part of an inevitable natural order," Stephen Jay Gould has commented that "capacity for recovery at geological scales has no bearing whatever upon the meaning of extinction today" (Gould 1990). Nor does Wilson believe that current data on biodiversity loss should comfort "anyone who believes that what *Homo sapiens* destroys, Nature will redeem," at least "within any length of time that has meaning for contemporary humanity" (Wilson 1992: 31). Diversity – whether biological or cultural – is now seen as a resource that must be actively preserved, since once lost it is not easily recovered. On geological scales, the Earth may indeed be self-sustaining, but that may come as little comfort to us human beings. After all, as Gould reminds us, "our planet will take good care of itself and let time clear the impact of any human malfeasance" (Gould 1990).

Note

1 A longstanding historical chestnut holds that Cuvier's theory of catastrophic upheaval was aligned with a biblical chronology of the Earth. As Martin Rudwick has shown, this interpretation was promoted mostly by nineteenth-century British geologists who mixed Cuvier's decidedly naturalistic theory with their own scriptural geology (Rudwick 2005).

References

Alvarez, Luis W., Walter Alvarez, Frank Asaro, and Helen V. Michel. 1980. "Extraterrestrial Cause for the Cretaceous-Tertiary Extinction." *Science* 208(4448): 1095–1108.

Badash, Lawrence. 2009. *A Nuclear Winter's Tale: Science and Politics in the 1980s.* Cambridge, MA: MIT Press.

Barrow, Mark V. 2009. *Nature's Ghosts: Confronting Extinction from the Age of Jefferson to the Age of Ecology.* Chicago, IL: University of Chicago Press.

Benton, Michael J. 2003. *When Life Nearly Died: The Greatest Mass Extinction of All Time.* London and New York: Thames and Hudson.

Brantlinger, Patrick. 2003. *Dark Vanishings: Discourse on the Extinction of Primitive Races, 1800–1930.* Ithaca, NY: Cornell University Press.

Carson, Rachel. 1962. *Silent Spring.* Cambridge, MA: Houghton Mifflin.

Cuddington, Kim, and Michael Ruse. 2004. "Biodiversity, Darwin, and the Fossil Record." In *Philosophy and Biodiversity*, ed. Marku Oksanen and Juhani Pietarinen, 101–118. Cambridge: Cambridge University Press.

Darwin, Charles. 1859. *On the Origin of Species.* London: J. Murray.

Darwin, Charles. 1872. *On the Origin of Species.* London: J. Murray.

Darwin, Charles. 1909. *The Voyage of the Beagle.* Vol. 29, Harvard Classics. New York: P. F. Collier & Son.

Dobzhansky, Theodosius. 1937. *Genetics and the Origin of Species.* Columbia Biological Series. New York: Columbia University Press.

Dobzhansky, Theodosius. 1962. *Mankind Evolving: The Evolution of the Human Species.* New Haven, CT and London: Yale University Press.

Dritschilo, William. 2008. "Bringing Statistical Methods to Community and Evolutionary Ecology: Daniel S. Simberloff." In *Rebels, Mavericks, and Heretics in Biology*, ed. Oren Harman and Michael Dietrich, 356–371. New Haven, CT: Yale University Press.

Farnham, Timothy J. 2007. *Saving Nature's Legacy: Origins of the Idea of Biological Diversity.* New Haven, CT: Yale University Press.

Goodman, Ellen. 1984. "Musings of a Dinosaur Groupie." *The Washington Post*, January 3, A17.

Gordin, Michael D. 2012. *The Pseudoscience Wars: Immanuel Velikovsky and the Birth of the Modern Fringe.* Chicago, IL: University of Chicago Press.

Gould, Stephen Jay. 1990. "The Golden Rule: A Proper Scale for Our Environmental Crisis." *Natural History* 99(9): 24–28.

Hobsbawm, E. J. 1994. *The Age of Extremes: A History of the World, 1914–1991.* New York: Pantheon Books.

Hutchinson, G. Evelyn. 1959. "Homage to Santa Rosalia, or Why Are There So Many Kinds of Animals?" *American Naturalist* 93: 145–159.

Jablonski, David. 1986. "Background and Mass Extinctions; The Alternation of Macroevolutionary Regimes." *Science* 231(4734): 129–133.

Jablonski, David. 1991. "Extinctions; A Paleontological Perspective." *Science* 253(5021): 754–757.

Jablonski, David. 2004. "The Evolutionary Role of Mass Extinctions: Disaster, Recovery and Something Inbetween." In *Extinctions in the History of Life*, ed. P. D. Taylor, 151–177. New York: Cambridge University Press.

Jefferson, Thomas. 1801. *Notes on the State of Virginia*, 8th American edn. Boston, MA.

Kazin, Alfred. 1950. "On the Brink." *New Yorker*, April 29.

Kingsland, Sharon E. 1985. *Modeling Nature: Episodes in the History of Population Ecology*. Chicago, IL: University of Chicago Press.

Kolbert, Elizabeth. 2014. *The Sixth Extinction: An Unnatural History*. London: Bloomsbury Press.

Leakey, Richard, and Roger Lewin. 1995. *The Sixth Extinction: Patterns of Life and the Future of Humankind*. New York: Doubleday.

Lewontin, Richard C., Steven Rose, and Leon J. Kamin. 1984. *Not in Our Genes: Biology, Ideology, and Human Nature*. New York: Pantheon Books.

Linnaeus, Carolus. 1762. "The Oeconomy of Nature." In *Miscellaneous Tracts Relating to Husbandry and Physick*, 39–129. London: J. Dodsley.

Lyell, Charles. 1830–1833. *Principles of Geology; Being an Attempt to Explain the Former Changes of the Earth's Surface, by Reference to Causes Now in Operation*. London: J. Murray.

Myers, Norman. 1979. *The Sinking Ark: A New Look at the Problem of Disappearing Species*. New York: Pergamon Press.

Newell, Norman D. 1967. "Revolutions in the History of Life." In *Uniformity and Simplicity*, 63–91. Boulder, CO: Geological Society of America (GSA).

Packard, A. S. 1886. "Geological Extinction and Some of Its Apparent Causes." *The American Naturalist* 20(1): 29–40.

Prichard, James Cowles. 1840. "On the Extinction of Human Races." *The Edinburgh New Philosophical Journal* 28: 166–170.

Parkinson, James. 1804. *An Examination of the Mineralized Remains of the Vegetables and Animals of the Antediluvian World*, Vol. 1. London: C. Whittingham.

Raup, David M. 1988. "Diversity Crises in the Geological Past." In *Biodiversity*, ed. E. O. Wilson, 51–57. Washington, DC: National Academy Press.

Raup, David M. 1991. *Extinction; Bad Genes or Bad Luck?* New York: W. W. Norton & Co.

Raup, David M., and J. John Sepkoski, Jr. 1982. "Mass Extinctions in the Marine Fossil Record." *Science* 215(4539): 1501–1503.

Rudwick, Martin J. S. 1997. *Georges Cuvier, Fossil Bones, and Geological Catastrophes; New Translations & Interpretations of the Primary Texts*. Chicago, IL: University of Chicago Press.

Rudwick, Martin J. S. 2005. *Bursting the Limits of Time: The Reconstruction of Geohistory in the Age of Revolution*. Chicago, IL: University of Chicago Press.

Schuchert, Charles. 1924. *Outlines of Historical Geology*. New York: J. Wiley & Sons.

Sepkoski, David. 2012. *Rereading the Fossil Record: The Growth of Paleobiology as an Evolutionary Discipline*. Chicago, IL: University of Chicago Press.

Sepkoski, David. 2015. "Extinction, Diversity, and Endangerment." In *Endangerment, Biodiversity, and Culture*, ed. Fernando Vidal and Nélia Dias, 62–86. New York: Routledge.

Sepkoski, J. John, Jr. 1984. "A Kinetic Model of Phanerozoic Taxonomic Diversity. III. Post-Paleozoic Families and Mass Extinctions." *Paleobiology* 10(2): 246–267.

Sepkoski, J. John, Jr. 1997. "Biodiversity; Past, Present, and Future." *Journal of Paleontology* 71(4): 533–539.

Takacs, David. 1996. *The Idea of Biodiversity: Philosophies of Paradise*. Baltimore, MD: Johns Hopkins University Press.

UN. 1992. *Convention on Biological Diversity*. New York: UN.

UNESCO. 2002. *UNESCO Universal Declaration on Cultural Diversity*. Paris: UNESCO.

Wallace, Alfred Russell. 1864. "The Origin of Human Races and the Antiquity of Man Deduced from the Theory of Natural Selection." *Journal of the Anthropological Society of London* 2: clviii–clxxxvii.

Weart, Spencer R. 1988. *Nuclear Fear: A History of Images*. Cambridge, MA: Harvard University Press.

Wilson, E. O. 1988a. *Biodiversity*. Washington, DC: National Academy Press.

Wilson, E. O. 1988b. "The Current State of Biological Diversity." In *Biodiversity*, ed. E. O. Wilson, 3–18. Washington, DC: National Academy Press.

Wilson, E. O. 1992. *The Diversity of Life*. Cambridge, MA: Belknap Press.

PART II

What is biodiversity?

3
APPROACHES TO BIODIVERSITY

Sahotra Sarkar

Introduction

The aim of this chapter is to propose a tentative classification of different approaches to defining "biodiversity." While it will elaborate and implicitly defend one particular approach – what will be called "weak" or "local normativism" later – it will leave critique of other approaches (or sustained argument for this approach over the others) for some other occasion. It is thus largely an expository chapter, more akin to critical reviews in the sciences than to usual philosophical treatments which present sustained arguments for or against a position. There is one important caveat: unlike a full-fledged review, there is no attempt here to include all contributions to the problem of explicating biodiversity. Rather, citations are to representative instances in the literature on the topic and what appear to be the most innovative or influential sources.

The context of "biodiversity"

Scientific terms and concepts do not emerge in a vacuum with their meanings definitively established through explicit definition. Rather, they emerge because they have roles to play in active research programs, most importantly, cognitive roles in the relevant theoretical structures and the design of experiments. These truisms are often forgotten by many of those who have expounded on how "biodiversity" should be defined. Most importantly, the term "biodiversity" and the associated concept(s) were introduced in the context of the institutional establishment of conservation biology as an academic discipline in the 1990s (Takacs 1996, Sarkar 1998b). Subsequently, the term and the concept were embraced by other disciplines, particularly by taxonomists. A cynical view of this move would be that the introduction of "biodiversity" and concern for its conservation served as a conduit for funding that taxonomists wanted to exploit. It will be seen below [e.g. in the differences between Maclaurin and Sterelny (2008) and Sarkar (2005)] that whether the concept of biodiversity is restricted to conservation biology makes a strong difference in how the problem of definition is approached.

As is now well-known, the term "biodiversity" was only coined by Walter G. Rosen at some point during the organization of a 1986 National Forum on BioDiversity held under the auspices of the United States National Academy of Sciences and the Smithsonian

Institution (Takacs 1996, Sarkar 2002). The new term was intended as nothing more than a shorthand for "biological diversity"; by the time the proceedings of the forum were published (Wilson 1988), it had become the title of the book. The BioDiversity forum was held only shortly after the founding of the US Society for Conservation Biology in 1985 (Sarkar 2002). Soulé's (1985) manifesto for the new discipline of conservation biology and Janzen's (1986) exhortation to tropical ecologists to undertake the political activism necessary for conservation had appeared in the previous two years. A sociologically synergistic interaction between the use of biodiversity and the growth of conservation biology as a discipline occurred and it led to a reconfiguration of environmental studies with the conservation of biodiversity as a central concern.

What has often been forgotten in discussions of "biodiversity" is that there was a vigorous and extended debate about diversity in ecology long before the introduction of that term, a history that has been reconstructed by Sarkar (2007). This debate included the questions of how ecological diversity should be defined, measured, and operationalized besides that of the relationship between diversity and stability. Insights included the realization that richness alone is not a good measure of diversity (though this is the only measure that taxonomists seem to have in mind), that equitability matters, and that there may well be other features that an adequate definition of ecological diversity should capture (Sarkar 2010). One standard measure of biodiversity, as shall be discussed in more detail below, is complementarity, what new features some unit brings to those already in a set of selected units. In 2003 Magurran finally pointed out that this measure was closely related to what ecologists have called β-diversity since the early 1960s (Magurran 2003). The point is that terminological choices in fields other than conservation biology, for instance in taxonomy and those other fields considered by some commentators (e.g. Maclaurin and Sterelny 2008), may be of little relevance to understanding the specific role of "biodiversity."

Normativity

Leaving aside the general question whether all scientific concepts are at least implicitly value-laden, an important point to note is that concepts of ecological diversity are not intended to be or interpreted as being normative (like most concepts in the sciences). In contrast, many philosophers have argued that biodiversity is also a normative concept (Callicott *et al.* 1999, Norton 2008, Sarkar 2008) though many scientists do not endorse such a claim (Gaston 1996c, Takacs 1996). The rationale behind claims of normativity is that the conservation of biodiversity is the explicit goal of conservation biology – recall why the term was introduced – and this goal is justified because of ethical reasons.

The question whether the concept of biodiversity is normative provides a good framework to classify the different approaches that have been taken in the literature. At one extreme are approaches that ignore normative concerns altogether – these will be treated under "Scientism" below. At the other extreme are approaches that embrace the normativity of biodiversity but are largely agnostic about scientific disputes on how it should be defined; however, even these normative approaches typically do not entirely deny the importance of some of the scientific issues that have been raised when various proposed definitions of biodiversity have been debated. Somewhere in the middle are deflationary accounts that are relatively pluralistic but, in particular, do not accept that there are particularly salient scientific or normative issues. Off the scale, but in different ways, are suggestions that the term be abandoned altogether – this is referred to as eliminativism below.

What is biodiversity?

It will be assumed throughout this paper that, if there is a defensible concept of biodiversity (that is, eliminativism as discussed below is unwarranted), then this concept must play some role in conservation biology, perhaps most saliently in systematic conservation planning with its focus on the prioritization of potential conservation areas on the basis of their biodiversity content. Thus, a perhaps conceptually appealing move such as defining biodiversity as the diversity of life at all levels of structural, taxonomic, and functional organization will not suffice since it cannot be operationalized in practice (Sarkar 2005).[1]

The views treated in the "Scientism" subsection all assume that there is a well-defined and relevant empirical question to be settled when a decision is made regarding the biodiversity content of a unit. Those treated in the following subsections do not make this assumption, though not all of them deny any role for empirical data.

Scientism

All definitions in this group are primarily non-normative; some entirely so. Differences in the extent of normativity will be noted below as the discussion progresses. The ones that are entirely non-normative uniformly had counterparts in ecology prior to the establishment of conservation biology, though this fact has seldom been recognized in the literature within the latter discipline.

Richness

This is the most commonly used measure of biodiversity, though it is widely recognized as fundamentally problematic – Gaston (1996b) as well as Maclaurin and Sterelny (2008) are somewhat unique in defending its utility. The richness of a unit is the number of entities in it. The unit in question is typically an area potentially targeted for conservation [a "conservation area" *sensu* Sarkar (2003)]. The relevant entities are often species (all species or those belonging to some specified group) but could also be habitat types or entities of some other kind deemed relevant to biodiversity. The common use of richness reflects the situation that the required data (e.g. in the form of species lists for an area) are very often available for an area and are often the only biodiversity data available (Williams *et al.* 2002, Margules and Sarkar 2007).

Issues connected with richness have been systematically studied in ecology since at least the 1950s when the possible diversity–stability relationship began to be explored; in this work "diversity" was almost always interpreted as species richness.[2] However, also since the 1950s it has been recognized that richness does not fully capture what is meant by diversity: equitability (also called evenness) matters. Consider two units Φ and Ψ. Let Φ consist of 90 percent type α entities and 10 percent type β entities. In contrast, let Ψ consist of 50 percent type α entities and 50 percent type β entities. Though both Φ and Ψ have a richness of 2, there is an obvious sense in which Ψ is more diverse than Φ. A variety of diversity measures (including Shannon's and Simpson's indices) have been devised within ecology to integrate richness and equitability (Magurran 2003, Sarkar 2006). These measures capture what has been called α-diversity within ecology since 1960.[3] Many other adequacy criteria beyond richness and equitability have been proposed within ecology for a satisfactory definition of diversity – these include average abundance rarity, abundance transfer, and distinctiveness.[4]

This entire body of work was ignored within conservation biology when the concept of biodiversity was introduced and defined, a curious fact that begs explanation. While not enough historical attention has yet been focused on the origins of conservation biology, its apparent disdain for this earlier work on ecological diversity must partly be a result of the discipline's self-perception of breaking new ground, of the practitioners creating what has cyncially been described as a "brave new science" (Sarkar 1998b). However, part of the explanation probably also lies in an implicit realization by conservation biologists that what they were after is a concept of diversity that is also normative.

Difference

For a concept of biodiversity that can be used in practice, for instance, in the selection of conservation areas, richness was shown to be inadequate in the 1980s. Consider three units, Φ, Ψ, and Λ. Let Φ have entities of 100 types, Ψ have entities of 90 types, and Λ have entities of 20 types. Suppose that only two of these units can be slated for conservation (e.g. due to cost considerations). If richness is used as the criterion for selection, Φ and Ψ would be selected. However, suppose that Φ and Ψ share 80 of the types of entities in them whereas Φ and Ψ share none. Selecting Φ and Ψ would include 110 types whereas selecting Φ and Ψ would include 120 types. The conclusion to be drawn is that biodiversity value should be measured by the difference a unit makes, how many new types of entities a unit brings into play. This is the "principle of complementarity" – its role in conservation biology, especially systematic conservation planning, is hard to overstate.[5]

In particular, complementarity was used to design algorithms for the selection of conservation areas since the early 1980s (Kirkpatrick *et al.* 1980, Kirkpatrick and Harwood 1983) though the term was only introduced in the 1990s (Vane-Wright *et al.* 1991). Decision support software implementing this approach to biodiversity go back to a seminal paper by Margules and collaborators in 1988 (Margules *et al.* 1988).[6]

As pointed out earlier, it was only as late as 2003 that Magurran noted how closely complementarity is related to what ecologists had been studying as β-diversity since the 1960s. While β-diversity quantifies the differences between the composition of two units, and is thus symmetric, complementarity is an asymmetric use of this difference by treating one of the units as already given (or "selected") and measuring the difference potentially introduced by the second.

Entities: The entities used to determine complementarity have often been subsets of species as in the case of richness. However, habitat types have also very often been used (e.g. Sarkar *et al.* 2007); organismic traits have also been proposed (Vane-Wright *et al.* 1991).[7] These entities are usually referred to as biodiversity "surrogates" (Sarkar 2002, Sarkar and Margules 2002). In many cases heterogeneous surrogate sets, such as both habitat types and species (or at-risk species), have been proposed (Faith *et al.* 2001, Sarkar *et al.* 2009).

Disparity: Even if the units chosen are to be taxonomic groups, there is no reason other than familiarity and convenience that the relevant taxon must be species.[8] That higher-level taxa may be more salient than species has been suggested and often seems *prima facie* reasonable – differences between these higher taxa are sometimes called disparity and it is open to question whether disparity should be used as a measure of biodiversity, whether it is different from biodiversity, or whether it should be a conservation goal along with and beyond biodiversity (Sarkar 2002, Sarkar and Margules 2002, Maclaurin and Sterelny 2008).

Richness and difference

Irrespective of whether complementarity is taken to be the natural measure of difference, the use of difference along with richness has been a standard approach to defining and measuring biodiversity, perhaps even the most common one (Gaston 1996a, 1996c, Maclaurin and Sterelny 2008).

In fact, some ways of using complementarity can be interpreted as using both richness and difference for two reasons: (1) when no units have yet been selected, the logic of complementarity reduces to the selection of the first unit by richness because the unit with the highest richness adds most things to the null set (Sarkar *et al.* 2004); (2) the argument given earlier as to why complementarity is better than richness in the iterative extension of the set of units in a conservation area network uses the criterion that the *total* richness within the set of units would be maximized better when the cardinality of the set is constrained.[9] (Note that this total richness is a measure of ecologists' γ-diversity, the total diversity of a region.)

Uniqueness

The uniqueness (or, equivalently, the distinctiveness) of the (typically biotic) entities in a unit is often taken to be indicative of its biodiversity content.

Irreplaceability: In particular, Pressey (e.g. 1999) has argued for prioritization of areas for the "irreplaceability" of their contents rather than complementarity; for him, conservation area selection should use two criteria, irreplaceability and vulnerability. Here, irreplaceability refers to the level of difficulty of conserving biodiversity features (the entities) using other potential units of conservation. Vulnerability refers to the probability that a unit would be irreversibly changed if conservation action is delayed to some other time (if any) in the future. While this idea has been attractive, it has proved to be difficult to operationalize, that is, use in practice in decision support tools (Margules and Sarkar 2007).

Rarity: Uniqueness has also been interpreted as rarity and invoked in concepts of biodiversity. But rarity has also been defined in a variety of ways using the size of the geographical range (which is supposed to be inversely correlated with rarity) or habitat specificity (correlated with rarity) or (local) population size (inversely correlated with rarity) (Rabinowitz *et al.* 1986). There does not appear to be a case in which rarity alone has been used to define biodiversity; typically it has been used along with difference (for an example, see below). However, the use of rarity alone to prioritize areas on the basis of their biodiversity content has been explored (Csuti *et al.* 1997, Sarkar *et al.* 2004).

Endemism: Geographical rarity becomes endemism when the range of an entity (almost always a species or somewhat higher taxon) is limited to a specified geographical region and that alone. (The contrast here is when some entity has a total restricted range but is widely distributed in small pockets over a large biogeographic zone.) When the geographic zone is highly restrictive, the entities are said to be microendemics. (Neither "endemic" nor "microendemic" have precise accepted definitions.) Though endemism is implicitly widely accepted as a criterion to determine the biodiversity content of units, like rarity, there appears to be no endorsement of using it by itself to define biodiversity.

Richness and uniqueness

However, once both richness and endemism are used together, one of the most influential implicit definitions of biodiversity ever used emerges – this is the "biodiversity hotspot" approach (Myers 1988, Myers *et al.* 2000) that has not only had very high visibility but has been the focus of interest of wealthy big non-governmental organizations ("BINGOs" *sensu* Dowie 2009) that have raised and spent (often very controversially) large sums of money for biodiversity conservation. Moreover, the designation of certain countries of the world as "megadiverse" also relies implicitly on using richness and uniqueness together (Sarkar *et al.* 2009).

This approach to defining biodiversity is also obviously attractive. But the problems of using richness rather than difference that were discussed earlier (subsection "Difference") remain unresolved. The adoption of the hotspot approach has not been accompanied by serious intellectual discussion of its merits, which may well suggest some cynicism about whether BINGOs are concerned with scientific credibility or merely with organizing successful political campaigns for conservation. In the latter case, this approach belongs better with normative approaches to biodiversity but BINGOs, especially Conservation International (which was perhaps the one that was criticized most strongly by Dowie 2009), have not presented their normative reasoning to academic (or any public) scrutiny.

Difference and uniqueness

One solution is to use difference and rarity. Indeed, interpreting difference as complementarity and rarity as geographical rarity, this approach has long been advocated by Sarkar and collaborators on both conceptual and algorithmic grounds (Sarkar *et al.* 2002). The conceptual argument consists of two parts: (1) because of the flaws in the iterative use of richness to construct a set of potentially conserved units, as shown by example earlier ("Difference"), complementarity should be used instead; (2) rarity also is important, typically more important if what is to be achieved using complementarity only leads to the addition of common entities (such as species) that are not at risk of extinction. The conclusion is that both complementarity and rarity should have a role in any definition of biodiversity.

The algorithmic reason goes back to the design of software decision support tools for the selection of conservation area networks initiated by Margules *et al.* (1988). In the context of the performance of one of these tools, the ResNet software package, Sarkar *et al.* (2002) compared an algorithm which used rarity followed by complementarity to one which used richness followed by complementarity. The former performed better than the latter in the sense that fewer units were required to include the same amount of biodiversity (measured by the total number of species represented in the set of units).

The important point to note is that this approach is broader than the potential use of endemism with complementarity but subsumes the latter (including the case of microendemics, which necessarily receive prioritized treatment).

Eliminativism

The wide variety of approaches to defining biodiversity – and there will be even more below – makes it plausible to believe that there is no compelling reason to accept that "biodiversity" is a legitimate scientific term (or concept). The concept of biodiversity plays no role in any theoretical *representational* construct that has emerged from conservation biology and this has not only long been acknowledged in the historical, philosophical, and sociological literature (Takacs 1996, Sarkar 1998b, 2002), but has rarely, if ever, been claimed in the scientific literature within conservation biology. It has been used in the theoretical work for the selection of conservation area networks and in the design of decision support algorithms but only after it has been operationally replaced by complementarity, rarity, or (with little success) richness.

Three types of reaction to this situation have appeared in the literature: eliminativism, deflationism, and normativism. The first is the focus of this section, the other two of the next two sections. Eliminativism demands that the concept (and term) be abandoned altogether mainly because of there being no prospect that any one of the approaches discussed in the last section will be convincingly shown to be better than the others. Most notably, this argument is part of the objections to "biodiversity" by Santana (2014) and Morar *et al.* (2015). This is a strong argument, but how compelling it is depends on whether these various approaches are completely incompatible with each other. The use of richness alone is incompatible with complementarity, but richness is the degenerate case of complementarity when the biodiversity of the first unit of interest is estimated. The other criteria used in those definitions are largely compatible with each other. Thus the conflicts that worry the eliminativists may well not be that significant.[10]

Morar *et al.* (2015) also acknowledge that many proponents and most analysts of the concept of biodiversity are explicit about the normative role of the concept; they object that the normative use of the term does not capture all the ways in which natural features should be valued, perhaps not even the most important ones.[11] This is a rather serious misrepresentation of those who accept some role for the concept of biodiversity in attempts to conserve natural values. Contrary to Morar *et al.*, biodiversity conservationists do not argue that biodiversity is the *only* natural value. Perhaps those who are infatuated with the catholic use of sustainability, and view biodiversity as reducible to an aspect of sustainability [as the Rio declaration of 1992 did on a global scale (Sarkar 2012b)] may have this kind of single-criterion vision. However, those comfortable with "biodiversity" sometimes explicitly endorse a pluralism of natural values (Sarkar 2012b, Norton 2015); some even endorse the explicit use of multi-criteria analysis to navigate trade-offs between biodiversity and other values, both socio-political ones and other natural values such as wild nature (Faith 1995, Sarkar 2012b).

Morar *et al.* (2015) also claim that the use of "biodiversity" is politically inappropriately misleading because it is perceived to be a factual/scientific term rather than a normative one. This may be true, and would be normatively problematic if it were true, but they present no evidence (e.g. survey-based empirical data) to defend this claim. It remains an open question whether societal perceptions of biodiversity regard it as a scientific concept rather than one designed to organize socio-political campaigns to conserve some part of natural value beyond more traditional ones such as wild nature and at-risk or charismatic species.

Deflationism

Eliminativism may seem too strong. However, the problems noted by eliminativists about the "scientistic" approaches to biodiversity remain unresolved. Deflationary strategies accept the use of "biodiversity" but show considerable latitude about how it should be defined.

Strong deflationism

Sarkar (2002) argued that what should be regarded as biodiversity is not a question to be determined through empirical observations, that is, it is not a scientific question. Rather, it must be settled by convention; moreover, the easiest way to determine what is being conceptualized as biodiversity is to see what is implicitly protected during the construction of conservation area networks. This view was dubbed a "deflationary" account of biodiversity by Santana (2014).[12]

Sarkar and Margules (2002) embedded this approach into the protocol of systematic conservation planning (Margules and Pressey 2000). They distinguished between constituents of biodiversity and surrogates for them.[13] What the constituents are is not an empirical question (as noted earlier). However, the adequacy of the surrogates for them *is* an empirical question. Techniques of surrogacy analysis were devised to answer this question. The crucial test was whether conservation plans devised using surrogates were concordant with those devised using consitutents (Sarkar *et al.* 2005). Surrogates in the terminology of systematic conservation planning correspond to "proxies" in some other areas of conservation biology (Caro 2010). In a positive development for conservation planning, Sarkar *et al.* (2005) showed that small sets of (abiotic) environmental variables were adequate surrogates for several reasonable constituent sets. This was a desirable result because such environmental data sets are available for almost the entire world. It meant that systematic conservation planning would not stall because of a paucity of available data [as The Nature Conservancy (TNC) used to argue (Redford *et al.* 1997)].

Weak deflationism

Nevertheless, viewing the adequacy of biodiversity constituent sets purely as a matter of convention was deemed unsatisfactory by the same individuals who had championed the move. A weaker deflationary position emerged in their later work. Sarkar (2008) argued that the choice of the desired constituents of biodiversity should be based on normative considerations. This move finally established consistency between how the concept of biodiversity originated in a normative context and how it is used in the practice of conservation biology. But it has a debatable implication: what biodiversity should be is relativized to the context, typically to the norms of the relevant culture – this issue will be further treated in the "Normativism" section below. The weakening of deflationism, including the introduction of adequacy conditions discussed next, does not necessitate a commitment to a normative definition of biodiversity constituents.

Sarkar (2012b) went on to make the concept of biodiversity even less deflationary by adding three necessary adequacy conditions for any set of biodiversity constituents: (1) that the entities be biotic; (2) that variability in these entities should be captured in the chosen set; and (3) that taxonomic spread be important. These adequacy conditions were supposed to capture much of the use of "biodiversity" in the practice of conservation. The second constraint is the one that does critical conceptual work: it shows why biodiversity is *diversity* at all. (A fourth adequacy condition, that not all biodiversity constituents be immediately useful resources, was presented as desirable but not necessary – it was motivated by a desire to accommodate the fact that many conservationists were uncomfortable with even a weak form of anthropocentrism in environmental ethics.) This weak deflationism was intended to answer eliminativists who argued that what is meant by biodiversity within conservation biology (or in a broader socio-political context of protecting natural values) is entirely unconstrained by either biology or diversity.

Normativism

When conservation biology was originally institutionally established, its principles included explicitly normative assumptions (Soulé 1985). It was implicitly presumed in the early literature that the concept of biodiversity had a normative component. The philosophical literature – some of which predated the introduction of the term "biodiversity" – explicitly noted that the normative goal of conservation was the variety of life, which was later referred to as "biodiversity" (Norton 1987).

After the term "biodiversity" was embraced by other parts of biology, especially those that are not goal-directed in the same way as conservation biology, for instance, taxonomy, the normative origins of the concept began to be ignored in the scientific though not in the philosophical literature (Callicott *et al.* 1999, Sarkar 2005, Norton 2008). Earlier (at the beginning of this chapter) some cynicism was expressed about the embrace of "biodiversity" by taxonomy and related disciplines. Whether or not that cynicism is warranted, the expansion of the use of the concept helped fuel many of the debates about scientistic definitions (though these have not been emphasized in this chapter because of its explicit focus on conservation biology and the definitions most relevant to it). The move to a weak deflationism accompanied by normativism about the choice of constituents returns to an acknowledgment of the normativity of "biodiversity." The critical question is: whose norms?

Global heritage

Conservation biology, as institutionalized in the United States, presumed that biodiversity was a global heritage and, therefore, conservationists from the global North should have a stake in what happens in the global South including, in particular, the tropics. Given the latitudinal diversity gradient, with diversity interpreted as richness (Sarkar 2005), conservation biologists from the North felt confident in the ethics of their desire to intervene in the South, sometimes making such intervention a part of the research priorities of conservation biology (Soulé and Kohm 1989).

This is not the place to review the issues raised by this dubious move in any detail. Suffice to note that: (1) here, conservation biologists were offering, consciously or not, prescriptions that continued European colonial policies in the South for a century directed towards wildlife protection for colonial hunters and their nature-enthusiast descendants (Fitter and Scott 1978); (2) this attitude was severely criticized even by environmentalists from the South working in Northern contexts (Guha 1989, 1997, Sarkar 1998a) from what they called a "social ecology" framework (Guha 1994, Sarkar and Montoya 2010); and (3) its legitimacy was routinely denied by academics in the North working (loosely) within the political ecology tradition (Brechin *et al.* 2002, Neumann 2004, West 2006).

These critiques make it clear that normativity about biodiversity based on universal norms applicable across cultures requires a more convincing defense than what conservation biology has offered so far. For such a defense, it will not be enough only to provide ethical arguments for the conservation of biodiversity without addressing what constitutes biodiversity. Such arguments abound in the discipline of environmental ethics, but more is necessary. Any credible defense must also address which biodiversity *constituents* have the ethical status that makes their conservation a duty for all humans across cultural differences. (In these statements, it is being assumed that ethical claims have transcultural relevance which is itself open to dispute.)

Local culture

The prominent Indian biologist Madhav Gadgil (personal communication in 2003) may have been the first to point out that what matters for selecting biodiversity constituents must be fundamentally based on the local commitments of people living in their habitats. That there was a conflict between local values and global values had been recognized for at least a decade and had just been reviewed by Vermeulen and Koziel (2002). That there was no uncontroversial scientific answer must have been obvious to Gadgil (as it was to most of the scientific community). What is crucial to this position is that skepticism about global values did not necessarily lead to eliminativism with respect to biodiversity.

Since then, Sarkar (2002, 2012a) has continued to develop this (possibly weak) normativism about the concept of biodiversity in conjunction with the weak deflationism discussed earlier. Whether this position is tenable, as well as whether it satisfies the role "biodiversity" plays in the community of conservationists, remains to be seen. It has three implications: (1) biodiversity conservation must embrace more than what normally is practiced as science – it must embrace the kind of normative analysis that is typically found in the humanities. Thus the recent inter-disciplinary move from conservation biology to conservation science (Kareiva and Marvier 2012), though positive, is insufficient; (2) philosophy, with its disciplinary inclusion of norma-tive ethics, obviously has a role in this expansion of the intellectual and practical circle of how conservation is envisioned; (3) but so has cultural anthropology – local norms must be under-stood in their own context with attention to detail. Most importantly, there is a compelling need to ensure that power relations do not result in the global North continuing to impose its cultural proclivities on the South. Discussion of the extent to which biodiversity conservation as a political movement has been exclusionary in the global North is important – and will be left for another occasion.

Final remarks

These will be brief since this chapter is intended primarily as a survey and not as a defense of any particular position:

1. There is no consensus about how biodiversity should be defined or operationalized among the positions lumped under "Scientism" earlier even if the appropriate units and entities within them can be uncontroversially designated. Does richness matter? If so, to what extent? Does difference matter? To what extent? Does uniqueness matter? To what extent? Should several of these criteria be used simultaneously? If so, which ones? How should trade-offs between them be navigated?
2. The units and relevant entities matter – this was emphasized in discussions of appropriate choice of biodiversity constituents (Sarkar 2002, Sarkar and Margules 2002, Sarkar *et al.* 2006, Margules and Sarkar 2007). The problem is that of the appropriate units of conservation in two senses: (i) should the focus be on potential conservation areas (or sites) as is almost always presumed in systematic conservation planning (Margules and Pressey 2000, Margules and Sarkar 2007)? Or should the focus really be on taxonomic groups? Or trait groups?[14] (ii) Which levels of organization (structural, taxonomic, or functional) are the most pertinent? This old question has no scientific answer.
3. The arguments for eliminativism are largely based on critiques of positions falling under the rubric of scientism. Once the discussion is broadened, in particular, to fundamentally normative conceptions of biodiversity, especially if accompanied by some deflationism, it is unclear – to say the least – that these arguments can be plausibly recast.
4. The political – and, in this context, *ipso facto* ethical – divide between advocates of biodi-versity as a global value and those embracing local self-determination is serious, of immense importance to the livelihoods and futures of people in the global South, and will not be resolved unless institutions in the global North address the disparities of power when valua-tions of natural variety are attempted.[15]

Acknowledgments

For discussions, in many cases over several decades, thanks are due to J. Baird Callicott, Michael Ciarleglio, Trevon Fuller, Madhav Gadgil, Justin Garson, Chris Margules, Bryan Norton, Chris Pappas, Samraat Pawar, Anya Plutynski, and Victor Sánchez-Cordero. Garson and Plutynski also commented on an earlier version of this chapter.

Notes

1 For the same type of reason it deems biodiversity as consisting of diversity of genes, species, and eco-systems as an irrelevant definition of the concept. This particular proposal has the additional demerit of *ex cathedra* canonization of just structural differences and those, too, only at three levels.

2 See the discussion in Sarkar (2005) and the references therein for this history.

3 The distinctions between α-, β-, and γ-diversity, which will be relevant in many remarks below in the text, all go back to Whittaker (1960).

4 For an extended philosophical discussion of these criteria and the problems associated with their use, see Sarkar (2007).

5 The issues connected with this are discussed in great depth by Margules and Pressey (2000), Margules and Sarkar (2007), and Sarkar and Illoldi-Rangel (2010). Justus and Sarkar (2002) reconstruct the history of the principle of complementarity; Sarkar (2012a) that of the algorithms that were developed in the 1980–2000 period mentioned later in the text.

6 Subsequent conceptual advances along the lines of this research program are summarized by Sarkar and collaborators (Sarkar *et al.* 2006).

7 See also Faith's contribution to this volume.

8 Of course, in some contexts the socio-political context strongly suggests the use of species, for instance, in the United States when the legal background is that of the Endangered Species Act of 1973. But, even here, note that the law includes subspecies besides species.

9 From the point of view of the theory of algorithms, the use of richness is what is called a "greedy algorithm." This does not guarantee a completely successful maximization – for a technical discussion of this problem, see Sarkar *et al.* (2006).

10 See, however, Santana's contribution to this volume.

11 Since they philosophize without giving any concrete examples, it is quite hard to decipher what types of situation they have in mind. However, even Callicott *et al.* (1999) have expressed similar worries in their discussion of the superiority of "biocomplexity" over "biodiversity."

12 See also his contribution to this volume.

13 In the early work (Sarkar 2002, Sarkar and Margules 2002, Margules and Sarkar 2007) the constituents were referred to as "true" surrogates (presumably for biodiversity) and the surrogates were referred to as "estimator" surrogates. However, calling the constituents "surrogates" suggests that there are referents of biodiversity beyond them. Sarkar (2008 and subsequently) changed the terminology to the one being used here.

14 See Faith's contribution to this volume.

15 See also the contribution by Plutynski and Fujita-Lagerqvist to this volume.

References

Brechin, S. R., P. R. Wilshusen, C. L. Fortwangler, and P. C. West. 2002. "Beyond the Square Wheel: Toward a More Comprehensive Understanding of Biodiversity Conservation as Social and Political Process." *Society and Natural Resources* 15: 41–64.

Callicott, J. B., L. B. Crowder, and K. Mumford. 1999. "Current Normative Concepts in Conservation." *Conservation Biology* 13: 22–35.

Caro, T. 2010. *Conservation by Proxy: Indicator, Umbrella, Keystone, Flagship, and Other Surrogate Species.* Washington, DC: Island Press.

Csuti, B., S. Polasky, P. H. Williams, R. L. Pressey, J. D. Camm, M. Kershaw, A. R. Kiester, B. Downs, R. Hamilton, M. Huso, and K. Sahr. 1997. "A Comparison of Reserve Selection Algorithms Using Data on Terrestrial Vertebrates of Oregon." *Biological Conservation* 80: 83–97.

Dowie, M. 2009. *Conservation Refugees: The Hundred-Year Conflict between Global Conservation and Native Peoples.* Cambridge, MA: MIT Press.

Faith, D. P. 1995. *Biodiversity and Regional Sustainability Analysis.* Technical report, Commonwealth Scientific and Industrial Research Organisation, Lyneham.

Faith, D. P., C. R. Margules, and P. Walker. 2001. "A Biodiversity Conservation Plan for Papua New Guinea Based on Biodiversity Trade-offs Analysis." *Pacific Conservation Biology* 6: 304–324.

Fitter, R., and P. Scott. 1978. *The Penitent Butchers: The Fauna Preservation Society, 1903–1978.* London: Fauna Preservation Society.

Gaston, K. J., ed. 1996a. *Biodiversity: A Biology of Numbers and Difference*. Oxford: Blackwell.

Gaston, K. J. 1996b. "Species Richness: Measure and Measurement." In *Biodiversity: A Biology of Numbers and Difference*, ed. K. J. Gaston, 77–113. Oxford: Blackwell.

Gaston, K. J. 1996c. "What is Biodiversity?" In *Biodiversity: A Biology of Numbers and Difference*, ed. K. J. Gaston, 1–9. Oxford: Blackwell.

Guha, R. 1989. "Radical American Environmentalism and Wilderness Preservation: A Third World Critique." *Environmental Ethics* 11: 71–83.

Guha, R., ed. 1994. *Social Ecology*. New Delhi: Oxford University Press.

Guha, R. 1997. "The Authoritarian Biologist and the Arrogance of Anti-humanism: Wildlife Conservation in the Third World." *Ecologist* 27: 14–20.

Janzen, D. H. 1986. "The Future of Tropical Ecology." *Annual Review of Ecology and Systematics* 17: 305–324.

Justus, J., and S. Sarkar. 2002. "The Principle of Complementarity in the Design of Reserve Networks to Conserve Biodiversity: A Preliminary History." *Journal of Biosciences* 27(S2): 421–435.

Kareiva, P., and M. Marvier. 2012. "What is Conservation Science?" *BioScience* 62: 962–969.

Kirkpatrick, J. B., and C. E. Harwood. 1983. "Conservation of Tasmanian Macrophytic Wetland Vegetation." *Papers and Proceedings of the Royal Society of Tasmania* 117: 5–20.

Kirkpatrick, J. B., M. J. Brown, and A. Moscal. 1980. *Threatened Plants of the Tasmanian Central East Coast*. Hobart: Tasmanian Conservation Trust.

Maclaurin, J., and K. Sterelny. 2008. *What is Biodiversity?* Chicago, IL: University of Chicago Press.

Magurran, A. E. 2003. *Measuring Biological Diversity*. Oxford: Blackwell.

Margules, C. R., and R. L. Pressey. 2000. "Systematic Conservation Planning." *Nature* 405: 245–253.

Margules, C. R., and S. Sarkar. 2007. *Systematic Conservation Planning*. Cambridge: Cambridge University Press.

Margules, C. R., Nicholls, A. O., and R. L. Pressey. 1988. "Selecting Networks of Reserves to Maximize Biological Diversity." *Biological Conservation* 43: 63–76.

Morar, N., T. Toadvine, and B. J. M. Bohannan. 2015. "Biodiversity at Twenty-five Years: Revolution or Red Herring?" *Ethics, Policy & Environment* 18(1): 16–29.

Myers, N. 1988. "Threatened Biotas: 'Hot Spots' in Tropical Forests." *Environmentalist* 8: 187–208.

Myers, N., R. A. Mittermeier, C. G. Mittermeier, G. A. B. Da Fonseca, and J. Kent. 2000. "Biodiversity Hotspots for Conservation Priorities." *Nature* 403: 853–858.

Neumann, R. P. 2004. "Moral and Discursive Geographies in the War for Biodiversity in Africa." *Political Geography* 23: 813–837.

Norton, B. G. 1987. *Why Preserve Natural Variety?* Princeton, NJ: Princeton University Press.

Norton, B. G. 2008. "Toward a Policy-relevant Definition of Biodiversity." In *Saving Biological Diversity*, ed. G. D. Dreyer, G. R. Visgilio, and D. Whitelaw, 11–20. Dordecht: Springer.

Norton, B. G. 2015. *Sustainable Values, Sustainable Change: A Guide to Environmental Decision Making*. Chicago, IL: University of Chicago Press.

Pressey, R. L. 1999. "Applications of Irreplaceability Analysis to Planning and Management Problems." *Parks* 9: 42–51.

Rabinowitz, D., S. Cairns, and T. Dillon. 1986. "Seven Forms of Rarity and Their Frequency in the Flora of the British Isles." In *Conservation Biology: The Science of Scarcity and Diversity*, ed. M. Soulé, 182–204. Sunderland, MA: Sinauer.

Redford, K., M. Andrews, D. Braun, S. Buttrick, S. Chaplin, M. Coon, R. Cox, L. Ellis, D. Grossman, C. Groves, D. Livermore, S. Pearsall, J. Shopland, P. Tabas, K. Wall, D. Williamson, and N. Rousmaniere. 1997. *Designing a Geography of Hope: Guidelines for Ecoregion Conservation in The Nature Conservancy*. Arlington, VA: The Nature Conservancy.

Santana, C. 2014. "Save the Planet: Eliminate Biodiversity." *Biology and Philosophy* 29(6): 761–780.

Sarkar, S. 1998a. *Genetics and Reductionism*. Cambridge: Cambridge University Press.

Sarkar, S. 1998b. "Restoring Wilderness or Reclaiming Forests?" *Terra Nova* 3: 35–52.

Sarkar, S. 2002. "Defining 'Biodiversity'; Assessing Biodiversity." *Monist* 85: 131–155.

Sarkar, S. 2003. "Conservation Area Networks." *Conservation and Society* 1: v–vii.

Sarkar, S. 2005. *Biodiversity and Environmental Philosophy: An Introduction*. Cambridge: Cambridge University Press.

Sarkar, S. 2006. "Ecological Diversity and Biodiversity as Concepts for Conservation Planning." *Acta Biotheoretica* 54: 133–140.

Sarkar, S. 2007. "From Ecological Diversity to Biodiversity." In *The Cambridge Companion to the Philosophy of Biology*, ed. D. L. Hull and M. Ruse, 388–409. Cambridge: Cambridge University Press.

Sarkar, S. 2008. "Norms and the Conservation of Biodiversity." *Resonance* 13: 627–637.

Sarkar, S. 2010. "Diversity: A Philosophical Perspective." *Diversity* 2: 127–141.

Sarkar, S. 2012a. "Complementarity and the Selection of Nature Reserves: Algorithms and the Origins of Conservation Planning, 1980–1995." *Archive for History of Exact Sciences* 66: 397–426.

Sarkar, S. 2012b. *Environmental Philosophy: From Theory to Practice.* Oxford: Wiley-Blackwell.

Sarkar S., and P. Illoldi-Rangel. 2010. "Systematic Conservation Planning: An Updated Protocol." *Natureza & Conservação* 9: 19–26.

Sarkar S., and C. R. Margules. 2002. "Operationalizing Biodiversity for Conservation Planning." *Journal of Biosciences* 27(S2): 299–308.

Sarkar S., and M. Montoya. 2010. "Beyond Parks and Reserves: The Ethics and Politics of Conservation with a Case Study from Perú." *Biological Conservation* 144(3): 945–1174.

Sarkar, S., A. Aggarwal, J. Garson, C. R. Margules, and J. Zeidler. 2002. "Place Prioritization for Biodiversity Content." *Journal of Biosciences* 27(S2): 339–346.

Sarkar, S., C. Pappas, J. Garson, A. Aggarwal, and S. Cameron. 2004. "Place Prioritization for Biodiversity Conservation Using Probabilistic Surrogate Distribution Data." *Diversity and Distributions* 10: 125–133.

Sarkar, S., J. Justus, T. Fuller, C. Kelley, J. Garson, and M. Mayfield. 2005. "Effectiveness of Environmental Surrogates for the Selection of Conservation Area Networks." *Conservation Biology* 19: 815–825.

Sarkar, S., R. L. Pressey, D. P. Faith, C. R. Margules, T. Fuller, D. M. Stoms, A. Moffett, K. Wilson, K. J. Williams, P. H. Williams, and S. Andelman. 2006. "Biodiversity Conservation Planning Tools: Present Status and Challenges for the Future." *Annual Review of Environment and Resources* 31: 123–159.

Sarkar, S., M. Mayfield, S. Cameron, T. Fuller, and J. Garson. 2007. "Conservation Area Networks for the Indian Region: Systematic Methods and Future Prospects." *Himalayan Journal of Sciences* 4(6): 27–40.

Sarkar, S., V. Sánchez-Cordero, M. C. Londoño, and T. Fuller. 2009. "Systematic Conservation Assessment for the Mesoamerica, Chocó, and Tropical Andes Biodiversity Hotspots: A Preliminary Analysis." *Biodiversity and Conservation* 18: 1793–1828.

Soulé, M. E. 1985. "What is Conservation Biology?" *BioScience* 35: 727–734.

Soulé, M. E., and K. A. Kohm. 1989. *Research Priorities for Conservation Biology.* Washington, DC: Island Press.

Takacs, D. 1996. *The Idea of Biodiversity: Philosophies of Paradise.* Baltimore, MD: Johns Hopkins University Press.

Vane-Wright, R. I., C. J. Humphries, and P. H. Williams. 1991. "What to Protect? Systematics and the Agony of Choice." *Biological Conservation* 55: 235–254.

Vermeulen, S., and I. Koziell. 2002. *Integrating Global and Local Values: A Review of Biodiversity Assessment.* Technical report, International Institute for Environment and Development, London.

West, P. 2006. *Conservation is Our Government Now: The Politics of Ecology in Papua New Guinea.* Durham, NC: Duke University Press.

Whittaker, R. H. 1960. "Vegetation of the Siskiyou Mountains, Oregon and California." *Ecological Monographs* 30: 279–338.

Williams, P. H., C. R. Margules, and D. W. Hilbert. 2002. "Data Requirements and Data Sources for Biodiversity Priority Area Selection." *Journal of Biosciences* 27(4): 327–338.

Wilson, E. O., ed. 1988. *BioDiversity.* Washington, DC: US National Academy Press.

4

IS BIODIVERSITY A NATURAL QUALITY?

James Maclaurin

Why do we need to know what biodiversity is?

Is biodiversity an objective feature of the natural world akin to a natural kind? Does it need to be in order to fulfill its dual role in science and public policy? This chapter evaluates competing philosophical theories about the nature of biodiversity and finds each corresponds to a recent theory about the nature of biological natural kinds. I argue that this tells us something important both about the nature of biodiversity and about the way we understand natural kinds in the life sciences. First, it demonstrates an inherent vagueness in a recent characterization of biological natural kinds by P. D. Magnus (2012). More importantly, it demonstrates a fundamental incoherence in the assertion in the UN Convention on Biological Diversity that biodiversity be both "intrinsically valuable" and that its scope should extend to all diversity in all living systems. I begin by explaining why we should care whether biodiversity is a natural kind.

Biodiversity is an extraordinary scientific concept. There are at least five distinct reasons why it is difficult to define.

The explosion in biodiversity measurement strategies

Given the large and open-ended number of ways in which organisms and ecosystems vary, there is a large and seemingly open-ended number of types of measure of biological diversity. The nineteen substantive chapters of Magurran and McGill's recent *Biological Diversity: Frontiers in Measurement and Estimation* (2011) summarize whole domains of biodiversity measurement, each replete with competing theories about measurement, estimation, the identification of suitable surrogates. As far back as 1978, Southwood noted an "explosive speciation" in diversity indices (p. 421) whose authors habitually condemn their predecessors. Faith and Baker (2006: 127) describe this proliferation as the "curse of biodiversity informatics." It is a curse which shows no sign of abating (for more detail, see Lean and Maclaurin 2016).

High level of usage across a wide variety of contexts

The term "biodiversity" has an extraordinary level of penetration, appearing in many and various contexts in the corpus of written English. Google's Ngram tool shows how often a word appears in the comprehensive Google Books database as well as how its frequency changes from year to year. At the time of writing, "biodiversity" was more common in written English than "nuclear power," "cholera," "insomnia," "recession" and just slightly less common than "malaria," "climate change," "famine," and "heart disease." Of these terms, only "biodiversity" and "climate change" are less than forty years old. Academically, as of August 2015, Google Scholar lists over 1.4 million articles that make reference to biodiversity, over 100,000 of which have "biodiversity" in their titles.

Breadth of scientific applicability

The idea of biodiversity has a great breadth of applicability within biology. While it first came to prominence in the mid-1980s in the context of biological conservation, it has since become clear that diversity of biological form, function, and interaction is important in many life science disciplines including ecology, genetics, taxonomy and systematics, evolutionary theory, developmental biology, morphometrics, etc. (for a survey of the relevant science, see chapters 1–6 of Maclaurin and Sterelny 2008).

Metaphysical intractability and practical constraints

Characterizing all the respects in which organisms and ecosystems vary is philosophically problematic. It is difficult to determine what is to count as a property in such a broad context and it is similarly difficult to determine how so many potentially incommensurable properties should be aggregated (Maclaurin and Sterelny 2008, section 1.3). Even if we could agree on a principled characterization of biodiversity understood as diversity with respect to all properties of biological systems, there is widespread acceptance that we could not measure all the components of biodiversity of any area. In practice, therefore, we are always either measuring some particular component (such as genetic diversity) or employing a biodiversity surrogate, that is, a partial measure of biodiversity which is thought to correlate well with a wide variety of biodiversity components (the idea that surrogacy is fundamental to our understanding of biodiversity is well set out in section 4 of Sarkar and Margules 2002).

Legal status

Biodiversity is the focus of much political debate and legal regulation. There are presently seven major international treaties that aim primarily at its conservation. These are the Convention on Biological Diversity (196 countries), the Convention on International Trade in Endangered Species of Wild Fauna and Flora (181 countries), the Convention on the Conservation of Migratory Species of Wild Animals (120 countries), the International Treaty on Plant Genetic Resources for Food and Agriculture (130 countries), the Convention on Wetlands (169 countries), the World Heritage Convention (191 countries), and the International Plant Protection Convention (178 countries). These treaties, particularly the Convention on Biological Diversity, make substantive claims about the nature of biodiversity and are very specific about the scientific and social roles that the concept is to play. The preamble to the Convention begins "Conscious of the intrinsic value of biological diversity and of the ecological, genetic, social, economic, scientific, educational, cultural, recreational and aesthetic values of biological diversity and its components." So biodiversity is characterized both as a characteristic of the biosphere which is good in and of itself and as a set of resources belonging to nation-states. "States have, in accordance with the Charter of the United Nations and the principles of international law, the sovereign right to exploit their own resources pursuant to their own environmental policies" (Article 3) and signatories should "Encourage cooperation between its governmental authorities and its private sector in developing methods for sustainable use of biological resources" (10 (e)). Such uses would include recreation and customary harvesting by indigenous peoples, but they would also extend to "results and benefits arising from biotechnologies based upon genetic resources" (19 (e)). While the conservation legislation of individual states is often not couched in terms of biodiversity, the idea of biodiversity nonetheless plays a significant political role in the conservation endeavours of many countries. For example, while New Zealand's main conservation legislation (the Environment Act, the Conservation Act, and the Resource Management Act) makes no reference to biodiversity, a search on the website of New Zealand's Department of Conservation finds over 700 web pages and policy documents containing the term.

The combination of all these factors has led to considerable difficulty in providing a philosophical characterization of biodiversity which both addresses its importance in conservation ethics, and its centrality in the public understanding of conservation aims and which provides advice to scientists about how they might best measure biodiversity across different contexts and use biodiversity in the setting of both large- and small-scale conservation goals.

In *Philosophies of Paradise: The Idea of Biodiversity*, David Takacs (1996) asks a group of well-known biologists for their preferred definitions of "biodiversity." Even taking into account the fact that these definitions are brief and given in an interview situation, there is nonetheless a distinct lack of uniformity (pp. 46–50). Some stress that biodiversity is hierarchical or multi-level (Peter Brussard, Donald Falk, Daniel Janzen, Thomas Lovejoy, Jane Lubchenco, Gordon Orians, Michael Soulé, E. O. Wilson). Others see it as fundamentally phylogenetic (Thomas Eisner, Gordon Orians, G. Carelton Ray). Only a few talk about interactions as well as types of entity (e.g. Terry Erwin, Jerry Franklin, Daniel Janzen, Peter Raven). Vickie Funk ties it to conservation and particularly to triage. A few tie it to species (David Pimentel, David Woodruff) or to species plus the genes they contain (S. J. Mcnaughton, Peter Raven) and finally David Ehrenfeld and Walter Rosen cast doubt on the idea that definitions of "biodiversity" are helpful at all.

The variety of actual definitions, measurement strategies, and theoretical contexts in which biodiversity is discussed might lead one to think that biodiversity science would never have got off the ground. However, in reality the result has been threefold. There has (as noted above) been a growth industry in the development of new measures of diversity. There has also been "... a large number of papers [that] open with the recognition that species richness is only one measure of biodiversity but proceed to treat it as if it were *the* measure of biodiversity ... typified by the focus on species extinction in discussion of biodiversity loss" (Gaston 1996: 4). Finally, and most importantly, many biologists have simply got on with the job of conservation, developing metrics and algorithms which aggregate various biodiversity surrogates for the identification of appropriate conservation areas (for a good overview and review of computational biodiversity conservational planning tools, see Sarkar *et al.* 2006).

Because of disagreement among biologists and philosophers about the nature and use of biodiversity as a scientific concept, there remains a suspicion that biodiversity might really be not much more than a rallying cry (as suggested by some of Tacaks's interviewees) or even that biodiversity conservation might be a sort of "bait and switch" enterprise in which funders are attracted by the images of the charismatic megafauna which conservation biologists see as mere umbrella species (Roberge and Angelstam 2004) motivating the conservation of a vastly wider array of entities and processes. Such concerns have surfaced most recently in Donald Maier's *What's So Good About Biodiversity? A Call For Better Reasoning About Nature's Value* (2012) and Carlos Santana's "Save the Planet: Eliminate Biodiversity" (2014). Maier worries that "Maybe, just maybe, biodiversity is the wrong hook on which to hang the value of nature" (2012: 3), while Santana notes that "In both public and scientific discourse, biodiversity has been reified and is treated as a real property of the natural world" (2014: 768). This, he thinks, is a mistake for:

> Biodiversity is generally the assumed target of conservation biology, but the biological world is composed of a number of distinct types of diversity, which only loosely correlate with each other and with biological value … we should therefore consider eliminating biodiversity from its privileged position in conservation theory and practice.
>
> (p. 778)

So there is scepticism about the effects of a focus on biodiversity on the methodological integrity of conservation biology. This, we are told, will have flow-on effects on the fundability of particular conservation projects and on the way in which we prosecute these projects (e.g. on the design of reserves). In short, biodiversity should not be our central or fundamental principle for saving the planet. Whether or not one shares the scepticism of Maier and Santana, it seems incumbent upon philosophers to establish the ontological facts here. Are philosophers, biologists, and conservationists right to treat it as an objective and quantifiable feature of the world?

Theories about the nature of biodiversity

Article 2 of the UN Convention on Biological Diversity famously defines biodiversity as "the variability among living organisms from all sources including, inter alia, terrestrial, marine and other aquatic ecosystems and the ecological complexes of which they are part; this includes diversity within species, between species and of ecosystems." Despite having a certain amount of currency with some biologists, for reasons set out in the previous section, the holism suggested by the UN Convention is neither operantionalizable, nor is it metaphysically coherent.

A weaker but more tractable claim would be a form of pluralism regarding the wide variety of current measurement strategies. But the pluralist must be able to (1) determine which measures are appropriate in which circumstances; (2) provide principled weighting procedures for cases in which we use multiple measures; and (3) demonstrate that the measures we will want to use are commensurable and hence that such weighted aggregation of data from multiple measures is at least possible in principle. There are presently two philosophical theories that provide solutions to the problem of determining which measures are appropriate in which circumstances. One is a surrogate-based, value-focused approach and the other is based on regularities in nature.

The surrogate-based approach is most clearly articulated in the systematic conservation planning approach promoted by Sahotra Sarkar, Chris Margules, and others (Sarkar *et al.* 2006, Margules and Sarkar 2007, Sarkar 2012). Central to this approach is the enumeration and analysis of the various quantitative procedures that measure types of biological diversity relevant to stakeholders in a given conservation context. Crucially, this procedure rests on the ontological premise that biodiversity surrogates are fundamental to conservation biology and that they must necessarily rest on a consensus that is negotiated between scientists and stakeholders about what is to count as biodiversity in any particular context:

> Surrogates that are supposed to represent total or general biodiversity are sometimes called "true" surrogates … Usually, species or other taxa are used as true surrogates. However, because general biodiversity is too diffuse a term to be precisely defined, the choice of a true surrogate set appeals at least implicitly to some convention or

consensus about what constitutes the relevant features of biodiversity in a given context. Thus, choosing a true surrogate set amounts to accepting an operational definition of biodiversity.

(Sarkar *et al.* 2006: 130)

But what are these constituents, the "real" components of biodiversity? ... Once we must select among biotic entities, there is no obvious answer and, more importantly, no technical answer to the question. Rather, any answer will involve a social choice based on cultural values.

(Sarkar 2012: 115)

Jay Odenbaugh notes that this approach makes conservation biology "an explicitly socio-ecological, value-oriented discipline" (Odenbaugh, 2014: 93). Taking a surrogate-based, value-focused approach has some obvious benefits. It leverages the tools of social science to address the central ontological problem of determining which aspects of biological diversity matter and which of those aspects matter most. It also promises to address the concerns of the sceptics about the relationship between biodiversity and conservation value. By tying what is to count as biodiversity in any particular context to stakeholder judgements about value, the socio-ecological approach also makes good on the assertion in the preamble of the Convention on Biological Diversity that biodiversity is intrinsically valuable (at least in the sense that "biodiversity" here is valuable *by definition*).

On the other hand, surrogate-based, value-focused biodiversity will, in some circumstances, differ significantly from biodiversity as measured by common scientific metrics such as species diversity and phylogenetic diversity. In particular, the biodiversity goals and targets produced via systematic conservation planning will often be much more nuanced than, for example, the maximization of species diversity that biologists so often think of as a reasonable proxy for the biodiversity of an ecosystem. This approach to biodiversity is also inherently local and contextual and hence might not be well suited to large-scale scientific projects such as the measurement of the effects on biodiversity of global climate change. Finally, in tying biodiversity more closely to value, this account risks falling foul of the second objection of sceptics such as Maier and Santana, namely that biodiversity is not really a scientifically established feature of the world. Odenbaugh thinks of surrogate-based, value-focused biodiversity as a pragmatist or instrumentalist characterization of conservation biology, but for the sceptics, Sarkar "is close to acknowledging that 'biodiversity' means nothing at all" (Santana 2014: 965). What one makes of these objections depends of course on the question central to this chapter, namely, what should the ontological aspirations of good biological science be? I return to this question in the next section.

In section 1.4 of *What is Biodiversity?* (2008) Maclaurin and Sterelny argue that all scientific analysis of biodiversity essentially rests on one of two types of regularity. Characterization of biodiversity is either forward-looking, focusing on the typical *effects* of the presence or absence of certain types of biological diversity, or backward-looking, focusing on the typical *causes* of the presence or absence of certain types of biological diversity. While the central theme of that book is the idea that the scientific importance of biodiversity extends far beyond the bounds of conservation biology, the authors note that both biodiversity as cause and biodiversity as effect have been centrally important to conservation projects (pp. 21–24). A typical forward-looking conservation principle is that diversity adds redundancy and hence robustness at various biological scales. A typical backward-looking conservation principle is that phylogenetic diversity is a crucial source of evolutionary potential (Forest *et al.* 2007, Mooers 2007) which in turn is a significant source of value to human populations (Faith *et al.* 2010, Lean and Maclaurin 2016).

So for Maclaurin and Sterelny, aspects of biodiversity will be important both as tools of conservation and as justifications for conservation goals (p. 2), but in either role, real aspects of biodiversity are all and only those diversity-based regularities that appear in well-supported biological theories. In contrast to surrogate-based, value-focused biodiversity, the regularity approach is deliberately intended to flow from our current scientific ontology. To be sure, many conservation projects will only focus on a small number of aspects of biodiversity and hence this regularity-based approach could be harnessed to a socio-ecological consensus-based method for determining which regularities to count in which contexts. However, both Maclaurin and Sterelny (pp. 154–157) and more recently Lean and Maclaurin (2016) recommend that choices about which regularities to count in particular contexts should be consequentialist, maintaining the ability of well-functioning ecosystems to provide ecosystem services (broadly construed). Being motivated more by scientific ontology and less by social and cultural considerations, the regularity-based approach is of course prone to sceptical concerns regarding the relationship between biodiversity and conservation value.

A tempting analogy is to think of the surrogate-based, value-focused approach as seeing biodiversity as something like art (fundamentally defined in terms of the perception and desires of human beings), while the regularity-based account sees it as something more like climate (crucial to human existence but ultimately not defined in terms of human interests). The aim of this chapter then is to determine whether surrogate-based or regularity-based accounts of biodiversity describe a scientific concept that picks out a type of thing in the world as opposed to merely reifying an idea that is ultimately political. I will argue that both the accounts described in this section genuinely ground biodiversity as a scientific concept, but that they have different consequences for the way biodiversity connects up to conservation value and ultimately for the way in which conservation biology works.

Natural kinds and natural qualities

Questions about the ontological status of features of the natural world have traditionally been dealt with in the complex philosophical literature on natural kinds. However, in thinking about whether the scientific concept *biodiversity* picks out a real feature of the natural world I am not suggesting that it is a kind of thing like a black hole or a cell. Rather, I am considering whether biodiversity is a (perhaps complex) property that picks out a real and scientifically significant feature of natural systems just as "selection pressure" and "memory" pick out real and scientifically significant features of evolving systems and cognitive systems, respectively. I shall call such features "natural qualities."[1] Despite biodiversity patently not being a natural kind, I will argue in the next section that two current theories about natural kinds serve as a useful basis for interpreting biodiversity as a natural quality. However, given that the concept of a natural kind has a long history and has performed many roles in diverse philosophical contexts (Hacking 1991), I will begin by discussing two ideas, namely similarity and essence, that have been important in the broad natural kinds literature. I will argue that neither are well suited to understanding the ontology of biological systems and hence they will not be further considered in the investigation of whether biodiversity is a natural quality.

In *A System of Logic*, John Stuart Mill (1874) argued that whether or not a group of objects is to count as a kind depends fundamentally on the number of ways in which those objects are similar to one another, particularly with regard to their causal powers. Things composed of sulphur have in common many important similarities. White things do not. So in thinking about whether or not biodiversity is a real feature of the natural world, it is tempting to think that if similarity in general tells us what biological natural kinds are, its converse, diversity in general, also tells us a fundamental and important fact about biological systems, namely their biodiversity. Two important problems stand in the way of harnessing Mill's insight about kinds and similarity in the context of biology. First, Mill's characterization of kinds does not work well in the life sciences. Mill proposes that science will detect a relatively small number of discrete and highly explanatory kinds of entity. This makes sense when considering chemistry's periodic table, but it seems an implausible goal for biological ontology considering the vast array of evolved biological species and adaptations. Natural selection produces variation that is continuous rather than discrete and hence there is no biological equivalent of the chemical elements and the differences that characterize biological groups are often explanatory only at a local scale. Second, the idea of similarity is less helpful than one might think as a means for assessing the ontological merit of putatively scientific concepts (Goodman 1972: 437). There are simply too many ways in which we can characterize similarity and so, not surprisingly, the development of all sciences sees *similarity in general* becoming less and less important as a means of detecting the sort of kinds that matter (Quine 1969: 138). Indeed, the idea that similarity in general is a useful determinant of biological ontology was tried and found wanting in mid-twentieth-century "numerical" taxonomy (Sober 1988: 37–67). Mature life sciences focus on differences that have particular explanatory and predictive significance, not on difference in general. So given what we know about biological ontology, it seems plausible that if biodiversity is a natural quality it will be one based on important differences, not biological difference *in general*.

The idea that biological natural kinds (particularly biological species) must be united by essential, intrinsic (and perhaps micro-structural) properties has been a seductive one for philosophers. Some philosophers have simply assumed that science will eventually discover essential properties that justify our intuitive belief in, for example, biological species (Kripke 1971, 1972, Putnam 1975). If, indeed, some fundamental feature of biological systems explained their macroscopic characteristics in the way that atomic/molecular structure explains the characteristics of elements and compounds, then this fundamental feature might happily be harnessed in an ontological characterization of the nature of biodiversity. However, in practice, ontological essentialism has proved much more complex than is suggested by the chemical example and the search for a fundamental biological essence has not met with success. The idea that ontology should rest on essences is both older (coming to us via Aristotle and particularly Locke) and more disparate in its characterization than the idea that successful science detects natural kinds. The variety of historical usages is helpfully set out by P. D. Magnus:

> In the Lockean tradition, a kind is said to have a "real essence" just if there are some facts about the world that correspond to the unity of the kind … Another usage identifies essences with sets of necessary and sufficient conditions for kind membership … Yet another usage identifies essences with properties which are held together by laws of nature … Another assumption associated with essentialism is that the defining character of a kind must be *intrinsic* rather than relational … Other usages describe essences ontologically, as abstract objects which exist over and above the entities which constitute the kind.
>
> (Magnus 2012: 18–19)

I agree with Magnus that the concept of essence in debates about natural kinds is often used as a way of determining whether natural kinds have some special sort of metaphysical lustre (2012: 19), rather than as a tool for determining which things in the world are natural kinds and which are not. Moreover, even where essence is being used diagnostically, it is of very little use in the life sciences, given (as noted above) that natural selection has produced a biosphere in which there are few sharp boundaries between different types of biological entities. This has led to claims that particular basic biological kinds such as species (Mischler and Donoghue 1982) and biological individuals (Hull 1978) should not be characterized as sets of entities united by some essential property. John Dupré (1993) has championed the more general claim that biology is unlike the so-called hard sciences in that it does not have essence-based natural kinds at all. So, as with similarity, while the idea of essence has played an important role in the tradition of natural kinds, it is of limited use in characterizing successful biological science and hence is unhelpful in determining whether biodiversity is a natural quality.

Where does this leave the current investigation? In the next two sections I set out two current theories about natural kinds which both seek to accommodate the sort of biological kinds that proved problematic for earlier theories. One is the successor to Mill's claim about similarity which rests biological ontology on predictive and explanatory similarity (as opposed to similarity in general). The second claims that biological ontology should maximize the explanatory and predictive success of science and that, particularly in the life sciences, this necessitates abandoning the assumption that biological kinds must rest on similarity.

Homeostatic property clusters

Mill's view about similarity, also found in the work of William Whewell (1860), rested on the idea that scientific ontology ought to serve as an effective basis for induction. Twentieth-century debate about the logical status of induction in science notwithstanding, there is still support for the idea that science should identify kinds that allow inductive inference to facilitate prediction and explanation. The modern expression of this view is found most clearly in the idea of a Homeostatic Property Cluster (Boyd 1991). Richard Samuels and Michael Ferreira write that "philosophers of science have, in recent years, reached a consensus – or as close to a consensus as philosophers ever get – according to which natural kinds are Homeostatic Property Clusters" (2010: 222).

> The natural definition of one of these homeostatic property cluster kinds is determined by the members of a cluster of often co-occurring properties and by the ("homeostatic") mechanisms that bring about their co-occurrence. It is an a posteriori theoretical question which of these properties and which of the homeostatic mechanisms count, and to what extent they count, in determining membership in the kind.
>
> (Boyd 1991: 140)

So of biological species, which he takes to be a paradigm case of a homeostatic property cluster, Boyd notes:

> The appropriateness of any particular biological species for induction and explanation in biology depends upon the imperfectly shared and homeostatically related morphological, physiological and behavioral features which characterize its members. The definitional role of mechanisms of homeostasis is reflected in the role of interbreeding in the modern species concept; for sexually reproducing species, the exchange of genetic material between populations is thought to be essential to the homeostatic unity of the other properties characteristic of the species and it is thus reflected in species definitions.
>
> (Boyd 1991: 141)

So Boyd proposes an ontology for science based on causal processes which act to maintain groups of similar entities. Because of the "imperfect" nature of such homeostatic mechanisms, the similarity of the members of such groups cannot be characterized in terms of sufficient conditions and hence we should not interpret it as resting on the possession of essential properties. So this gives us a picture of biological natural kinds as groups whose members need not be identical to one another with respect to the possession of any particular property and whose similarity matters only to the extent that it is the product of, and maintained by, homeostatic mechanisms.

This characterization of biological natural kinds accommodates much of the ontology of modern life sciences but, as P. D. Magnus has recently argued, one of its central features appears to be inessential, perhaps even arbitrary. A process-based ontology such as this need not limit itself to detecting groups of *similar* entities. Magnus describes this element of Boyd's theory as "similarity fetishism" which results "primarily from the narrow inductivist focus on induction as projective inference. For similarity fetishists like Mill, Goodman, and Quine, a kind could only be a collection of similar things" (Magnus 2012: 55). His concern is with cases in which homeostatic property clusters contain members that are radically dissimilar to one another such as species that undergo metamorphosis or that include radical sexual dimorphism (his favoured example being male and female angler fish *Linophryne arborifera*). Leaving aside the plausibility of describing angler fish as clusters of similar individuals, it is clear that the sort of mechanisms that Boyd sees as grounding biological ontology will in some cases produce not similarity but diversity. Moreover, diversity that is reliably and regularly produced or consumed by biological processes will be just as important in prediction and explanation as the clusters of similarity at the heart of Boyd's theory. It is this idea that sits behind Maclaurin and Sterelny's theory about the nature of biodiversity. So it seems reasonable that if Boyd's homeostatic property clusters plausibly pick out biological natural kinds, then Maclaurin and Sterelny's account of biodiversity should be seen as describing a natural quality (i.e. a real and scientifically significant feature of natural systems). However, Magnus does not stop short at attacking the so-called similarity fetishism of the homeostatic property cluster idea. He has in mind a more radical revision of the the idea of a biological natural kind.

Magnus on natural kinds

In light of his rejection of a scientific ontology founded on similarity-fuelled induction, in *Scientific Enquiry and Natural Kinds: From Planets to Mallards*, Magnus (2012) proposes a much broader characterization of the idea of a natural kind. He proposes, inter alia, that we should see natural kinds as those that make science successful, leaving open exactly what it is about such kinds that helps them to fuel the success of science. For Magnus:

> A category k is a natural kind for domain d if (1) k is part of a taxonomy that allows the scientific enquiry into d to achieve inductive and explanatory success, and (2) any alternative taxonomy that excluded k would not do so.
>
> (p. 48)

He calls (1) the success clause and (2) the restriction clause. Unlike Boyd's account, according to which you might think there is some unique (albeit very complicated) best description of biology in terms of kinds and the processes that produce them, for Magnus:

the important idea behind the account is not that there be some unique taxonomy that can support successful enquiry. Rather, the idea is that the world condemns a great many taxonomies to failure. Constraint from the world is what makes identifying natural kinds the discovery of structure in the world, rather than merely the imposition of a set of labels onto things that are undifferentiated in nature.

(p. 50)

Crucially, according to Magnus, natural kinds are domain-relative and, given his views about the homeostatic property cluster idea, they need not rest on clumps of similarity. They depend on the actual explanatory and predictive purposes of particular scientific endeavours. Although the search for natural kinds is ultimately about identifying structure in the world, the success clause allows natural kinds to be motivated by features of practical utility to people. Whether or not we treat some entity as a natural kind depends as much upon our purposes in doing science with the relevant domain as it does on the domain's causal structure. Just as the homeostatic property cluster account seemed convivial to Maclaurin and Sterelny's regularity-based account of biodiversity, Magnus's account of natural kinds appears convivial to the surrogate-based, value-focused account of biodiversity, according to which what we count as biodiversity is fundamentally contextual, depending as much on our purposes in undertaking a conservation biology as it does on the structure and function of biological systems.

In summary, regularity-based biodiversity is a natural quality in the sense that it results from the production and consumption of diversity by natural processes. Surrogate-based, value-focused biodiversity is a natural quality in the sense that its focus on value is inferred from an interpretation of the basic goals of conservation biology. Can both theories be right?

So is biodiversity a natural quality?

In the first section I argued that biodiversity is inherently difficult to characterize because as a concept it does an extraordinary amount of work in a very wide variety of scientific, legal, and social contexts. Moreover, despite being metaphysically intractable, it is seemingly open to estimation using a large and perhaps open-ended number of scientific procedures. These facts about the role and history of biodiversity give rise to a number of concerns about its status in both science and public policy. While there is much good science demonstrating the importance of various types of diversity in many different contexts within the life sciences, one might wonder whether we gain anything by talking as if all this work is really the study of one feature of the world, namely its biodiversity. Even if it makes sense to aggregate all this science, what does the resulting aggregation have to do with the original context of biodiversity as a goal for conservation? Sceptics like Santana and Maier worry that couching conservation in terms of biodiversity might really be a very expensive sort of scientific window-dressing that might occasionally lead to biodiversity-driven outcomes that do not square with the political and ethical desiderata that ought to motivate conservation.

The aim of this chapter has been to address these concerns from an ontological perspective. I have asked whether current scientific and philosophical theories of biodiversity justify us in thinking of it is a natural quality, that is, a real and scientifically significant feature of the world. While acknowledging that biodiversity is putatively a type of property rather than a type of entity, I have harnessed the natural kinds literature for its longstanding focus on scientific ontology. I argued in the last section that two recent accounts of natural kinds justify the idea that two leading philosophical theories of the nature of biodiversity each pick out a distinct type of natural quality.

If we think of biological natural qualities as akin to homeostatic property clusters, Maclaurin and Sterelny's regularity-based account of biodiversity picks out all and only those types of diversity that are inputs or outputs of homeostatic mechanisms. So Maclaurin and Sterelny's regularity-based account of biodiversity is guaranteed scientific significance because of its intimate connection with biological processes and it is at least as real as the homeostatic property clusters so often identified as the natural kinds of the life sciences. However, this ontological sturdiness is purchased at a price. There is, at best, a complicated connection between regularity-based biodiversity and conservation. Perhaps it makes sense to think of this type of biodiversity as intrinsically valuable according to biocentrist accounts of environmental ethics such as Leopold's Land Ethic (1949), but no similar inference demonstrates all biodiversity to be valuable by the lights of non-anthropocentric consequentialist environmental ethics. For consequentialists, there will be occasions in which it is rational to conserve regularity-based biodiversity. This is particularly true where we are acting under uncertainty (Lean and Maclaurin 2016) or where we can demonstrate a link between particular aspects of biodiversity and crucial ecosystem services (Faith *et al.* 2010).

If we think of biological natural qualities as those qualities required to achieve the goals of particular sciences, then we might infer that the value-infused aims of conservation biology require that we ought to interpret biodiversity as a value-laden concept. There is certainly no doubt that the systematic conservation planning promoted by Sarkar and Margules *et al.* succeeds as a decision procedure in the complex and sometimes contentious world of conservation planning. It is, in short, a more practical toolkit than the consequentialist conservation of regularity-based biodiversity recommended by Maclaurin and Sterelny. So should we conclude that both interpretations of biodiversity are natural qualities and that both therefore are good responses to scepticism about the ontological credentials of biodiversity?

To apply Magnus's success clause, we must apparently evaluate the truth of counterfactuals involving alternative ways of doing science and then carry out potentially fine-grained comparisons of their relative success. In other words, we must ask whether a given science would have been just as successful (or less) had it not included a certain kind k (Slater 2014). Moreover, one might worry about my interpretation of Magnus's "success clause" as it applies specifically to this case. Are the types of diversity valued by stakeholders really part of the ontology of conservation biology, or should we think of them more like the institutions that support the scientists and the funding provided by granting bodies, that is as necessary for the success of the science but not part of its ontology? If we interpret Magnus's success clause narrowly as licensing ontology that allows induction within the relevant domain, then at least in the life sciences Magnus's view is hard to distinguish from the homeostatic property cluster account of biological ontology. If we interpret it more broadly, it subsumes within the relevant science all sorts of practical, mathematical, and philosophical entities and properties that we would not normally think of as part of the ontology of science.

Ultimately, the existence of these two distinct interpretations of biodiversity do nothing to diminish its importance, but they do say something important, about the persistent debate about how we should characterize and hence conserve biodiversity. The preamble to the UN Convention on Biodiversity states that it is "intrinsically valuable," while Article 2 states that biodiversity is "the variability among living organisms from all sources including, inter alia, terrestrial, marine and other aquatic ecosystems and the ecological complexes of which they are part; this includes diversity within species, between species and of ecosystems." What we now know is that the Convention is fundamentally inconsistent. The best available definition based on the "all variability" clause (the regularity-based account) produces a characterization of biodiversity that is important but not "intrinsically valuable." Conversely, the best available definition based on the "intrinsically valuable" clause picks and chooses between the diversity of regions and biota and hence does not respect the "all variability" clause.

Note

1 I have used "quality" rather than "property" to avoid confusion with David Lewis's use of "natural property" (1986: 60).

References

Boyd, R. 1991. "Realism, Anti-Foundationalism and the Enthusiasm for Natural Kinds." *Philosophical Studies* 61: 127–148.

Dupré, J. 1993. *The Disorder of Things: Metaphysical Foundations of the Disunity of Science.* Cambridge, MA: Harvard University Press.

Faith, D. P., and A. M. Baker. 2006. "Phylogenetic Diversity (PD) and Biodiversity Conservation: Some Bioinformatics Challenges." *Evolutionary Bioinformatics Online* 2: 121–128.

Faith, D. P., S. Magallon, A. P. Hendry, E. Conti, T. Yahara, and M. J. Donoghue. 2010. "Evosystem Services: An Evolutionary Perspective on the Links between Biodiversity and Human Well-being." *Current Opinion on Environmental Sustainability* 2: 66–74.

Forest, F., R. Grenyer, M. Rouget, T. J. Davies, R. M., Cowling, D. P. Faith, A. Balmford, J. C. Manning, S. Proche, M. van der Bank, G. Reeves, A. J. Hedderson, and V. Savolainen. 2007. "Preserving the Evolutionary Potential of Floras in Biodiversity Hotspots." *Nature* 445: 757–760.

Gaston, K. 1996. *Biodiversity: A Biology of Numbers and Difference.* Oxford: Blackwell Science.

Goodman, N. 1972. "Seven Strictures on Similarity." In *Problems and Projects*, ed. N. Goodman, 437–447. Indianapolis, IN: Bobbs Merrill.

Hacking, I. 1991. "A Tradition of Natural Kinds." *Philosophical Studies* 61: 109–126.

Hull, D. 1978. "A Matter of Individuality." *Philosophy of Science* 45: 335–360.

Kripke, S. 1971. "Identity and Necessity." In *Identity and Individuation*, ed. M. K. Munitz, 135–164. New York: New York University Press.

Kripke, S. 1972. "Naming and Necessity." In *Semantics of Natural Language*, ed. G. Harman and D. Davidson, 253–355. Dordrecht: Reidel.

Lean, C., and J. Maclaurin. 2016. "The Value of Phylogenetic Diversity." In *Biodiversity Conservation and Phylogenetic Systematics: Preserving Our Evolutionary Heritage in an Extinction Crisis*, ed. R. Pellens and P. Grandcolas, 19–38. New York: Springer.

Leopold, A. 1949. *A Sand County Almanac.* New York: Oxford University Press.

Lewis, D. 1986. *On the Plurality of Worlds.* Oxford: Blackwell.

Maclaurin, J., and K. Sterelny. 2008. *What is Biodiversity?* Chicago, IL: University of Chicago Press.

Magnus, P. D. 2012. *Scientific Enquiry and Natural Kinds: From Planets to Mallards.* Basingstoke and New York: Palgrave Macmillan.

Magurran, A., and B. McGill. 2011. "Challenges and Opportunities in the Measurement and Assessment of Biological Diversity." In *Biological Diversity: Frontiers in Measurement and Assessment*, ed. A. Magurran and B. McGill, 1–10. Oxford: Oxford University Press.

Maier, D. S. 2012. *What's So Good About Biodiversity? A Call For Better Reasoning About Nature's Value.* Dordrecht: Springer.

Margules, C., and S. Sarkar. 2007. *Systematic Conservation Planning.* Cambridge: Cambridge University Press.

Mill, J. S. 1874. *A System of Logic.* New York: Harper Brothers.

Mishler, B. D., and M. J. Donoghue. 1982. "Species Concepts: A Case for Pluralism." *Systematic Zoology* 31: 491–503.

Mooers, A. 2007. "The Diversity of Biodiversity." *Nature* 445: 717–718.

Odenbaugh, J. 2014. "Environmental Philosophy 2.0: Ethics and Conservation Biology for the 21st Century." *Studies in History and Philosophy of Biological and Biomedical Sciences* 45: 92–96.

Putnam, H. 1975. "The Meaning of 'Meaning'." *Minnesota Studies in the Philosophy of Science* 7: 215–271.

Quine, W. v. O. 1969. "Natural Kinds." In *Ontological Relativity and Other Essays.* New York: Columbia University Press.

Roberge, Jean-Michael, and Per Angelstam. 2004. "Usefulness of the Umbrella Species Concept as a Conservation Tool." *Conservation Biology* 18: 76–85.

Samuels, R., and M. Ferreira. 2010. "Why Don't Concepts Constitute a Natural Kind?" *Behavioral and Brain Sciences*, 33: 222–223.

Santana, C. 2014. "Save the Planet: Eliminate Biodiversity." *Biology and Philosophy* 29(6): 761–780.

Sarkar, S. 2012. *Environmental Philosophy: From Theory to Practice*. Oxford: Wiley-Blackwell.

Sarkar, S., and C. R. Margules. 2002. "Operationalizing Biodiversity for Conservation Planning." *Journal of Biosciences* 27 (S2): 299–308.

Sarkar, S. *et al.* 2006. "Biodiversity Conservation Planning Tools: Present Status and Challenges for the Future." *Annual Review of Environmental Resources* 31: 123–159.

Slater, M. H. 2014. "Review of Scientific Enquiry and Natural Kinds: From Planets to Mallards." *Notre Dame Philosophical Reviews*. Available online at https://ndpr.nd.edu/news/40779-scientific-enquiry-and-natural-kinds-from-planets-to-mallards/

Sober, E. 1988. *Reconstructing the Past: Parsimony, Evolution and Inference*. Cambridge, MA: MIT Press.

Southwood, T. R. E. 1978. *Ecological Methods*. London: Chapman and Hall.

Takacs, D. 1996. *The Idea of Biodiversity: Philosophies of Paradise*. Baltimore, MD: Johns Hopkins University Press.

Whewell, W. 1860. *On the Philosophy of Discovery*. London: Chapters Historical and Critical.

5

A GENERAL MODEL FOR BIODIVERSITY AND ITS VALUE

Daniel P. Faith

Introduction

This chapter addresses philosophical issues related to the definitions, and values, of biodiversity. The term "biodiversity" – a contraction of "biological diversity" – logically will reflect some notion of diversity, or living variation. Thus, the Convention on Biological Diversity (CBD, article 2; www.cbd.int/sp/) provides this definition: "the variability among living organisms from all sources … This includes diversity within species, between species and of ecosystems." The Conceptual Framework of the Intergovernmental Platform on Biodiversity and Ecosystem Services (IPBES; Díaz *et al*. 2015: 12) also defines biodiversity as variation. However, it also includes "changes in abundance and distribution over time and space within and among species, biological communities and ecosystems." Including *change* may capture "variability" more than variety: oddly, the extinction of a species would imply an *increase* in biodiversity (because abundance changes in going to zero). These issues are avoided when "variability" is interpreted to mean "living variation" (e.g. Mace *et al*. 2012).

While living variation or variety is the core of these definitions, in practice, the term "biodiversity" continues to be used in many different ways. Current usage ranges from being very specific (e.g. equating it with a single species), very general (e.g. equating it with the "fabric of life"), or somewhat tangential (e.g. equating it with any ecological factor relevant to ecosystems). For example, Díaz *et al*. (2009: 55) described "biodiversity" as "the number, abundance, composition, spatial distribution, and interactions of genotypes, populations, species, functional types and traits, and landscape units in a given system." Partly, this usage reflects interest in "ecosystem services" – the benefits that humans obtain from natural ecosystems (e.g. fresh water, timber; see Daily 1997). There now are at least ten different definitions of "ecosystem services" (Polasky *et al*. 2015). Ecosystem services may be delivered from transformed land, from the whole planet as an "ecosystem", and from non-biotic components. In this context, "biodiversity" typically has ecological definitions (similar to that of Diaz *et al*. quoted above), in order to characterize it as something whose importance arises by underpinning ecosystem services (Gasparatos and Willis 2015).

Reflecting the various definitions of "biodiversity," discussions of the value of biodiversity sometimes have focused on the value of all of nature, or of the value of specific elements, such as individual species or traits. Individual elements often will have a clear value – but it is less clear how we describe the value of biodiversity, particularly given the range of different definitions. Recent reviews of the history of the term "biodiversity" consequently have expressed some disenchantment with the idea of assigning value to biodiversity. For example, Morar *et al.* (2015: 24) concluded that biodiversity "is arguably of relatively minor importance in comparison with the actual species, processes, habitats, and so on."

In this chapter, I will address some of these issues by constraining the "biodiversity" problem. "Biodiversity" will be defined as living variation, at multiple levels, and the corresponding values of *biodiversity* then will be those that relate specifically to the values of maintaining such *variety*. I note that conventional use of the term, for example, in "the biodiversity of Madagascar," also can be a reference to the *set* of elements providing that total variation. As will be seen below, this may help clarify some attributions of value to "biodiversity."

My "simplified" focus does raise complications. How do we operationalize the idea of multiple levels of living variation? Is there a general model or framework that makes sense of it all? Can a general framework also make sense of the numerous existing calculations and indices that are referred to as "diversity" measures? Here, I will discuss a general biodiversity framework (Faith 1994) that uses many possible kinds of units of variation – species, features, genes, etc. – to describe different levels of living variation. I will show how this seemingly complex interpretation of biodiversity, with many kinds of units, actually simplifies things. "Biodiversity" is simply a measure of variety, expressed by counting up the number different units in a given set of "objects" (e.g. "how many different species are in this set of areas?"). Thus, for any choice of the *kind* of unit (e.g. species), there are many different units (e.g. many different species) and we want to count these up. The good news is that we need only saddle the term "biodiversity" with this one important idea – the counting-up of the different units. Many other calculations based on those same units are possible (dissimilarity, abundance-weighted ecological "diversity" measures, complementarity, etc.), but we do not have to try to accommodate any of these as measures of "biodiversity."

In this chapter, I also will argue that if we "tie down" biodiversity as the counting-up of some nominated units, then we gain another advantage. Just as "biodiversity" always is simply a "counting-up," with different kinds of units substituted in, other calculations (e.g. of dissimilarity) that traditionally have focused on one kind of unit (e.g. species) immediately expand to cover the corresponding calculations for all other kinds of units (e.g. dissimilarity based on genetic variants). Thus, that spirit of substituting one kind of unit for another not only provides a definition of biodiversity spanning all levels of variation (kinds of units), but also provides a useful companion "calculus" that is shared across all the different kinds of units.

My simplification also will mean that the "value of biodiversity" can focus on the (often neglected) values of maintaining living variation. Familiar debates about species conservation have focused on intrinsic values (inherent values, independent of human perceptions) plus various anthropocentric use values. My unconventional conclusion will be that the value of *variety* may not include intrinsic value; such intrinsic value typically is assigned to *individual* objects or elements of biodiversity, but appears not to be a value attributable to "variety" per se. I'll argue that the core value of biodiversity-as-variety is anthropocentric, and traces back to the earliest discussions of the value of biodiversity. It is a form of "option value," capturing the value that variation has in potentially providing unanticipated benefits for humans in the future.

With these issues in mind, I will first introduce the idea of countable units in a general framework for biodiversity (Faith 1994, 2013, Faith and Walker 1996, Faith *et al.* 2003). I will present some examples of the framework including the well-known "phylogenetic diversity" ("PD"; Faith 1992) and "environmental diversity" ("ED"; Faith and Walker 1996) biodiversity measures. These examples have been the focus of some philosophical discussions and I will use these to highlight properties of the general framework. The examples also will be the basis for discussion of the idea of a general calculus based on the choice of any kind of unit of variation. I will then turn to the issue of the possible values of biodiversity-as-variation. This will allow some reconciliation of biodiversity values with the values of ecosystem services. I will finish this chapter by discussing how values of biodiversity can be integrated with other values of society through multi-criteria analyses and related methods – and discuss how this is another important opportunity to usefully extend existing analyses to cover many possible kinds of units and objects.

A general framework for biodiversity as variation at multiple levels

Consider a common assessment of biodiversity-as-variety. An evaluation of a set of protected areas asks "how many different species are represented by that set of areas?" We can convert this to a more general biodiversity question: "how many different units are represented by that set of objects?" Thus, "species" corresponds to just one *kind* of "unit" of variation (with different species as different "units"). Also, the areas more generally are any "objects." Biodiversity assessment considers a wide range of these possible objects for decision-making – not just areas, but also species, populations, and other entities (Faith 1994). Biodiversity therefore can be quantified in general as a count of the number of different units represented by a given set of objects. Examples of other objects/units combinations include species/traits (or features) and species' populations/genetic variants.

Typically, many of the units that we would like to count are unknown (many species are still unknown to science; many features of species are undescribed). We may directly observe some objects (say, species) and want to quantify the relative number of un-observed units (say, features) that are represented by those objects. The relative number of units for any object or set of objects therefore has to be estimated through the use of an inferential model or a surrogate of some kind (Faith 1994; see also Sarkar and Margules 2002). Faith and Walker (1996: 405) proposed that: "A model that successfully reflects the underlying processes that determine the distribution of units among objects (a pattern–process model) may tell us enough about the relationships among the objects to enable inference of relative numbers of units represented by any set of those objects." Faith (1994: 45) described this as "a general framework for using pattern to quantify diversity at a level below that of the original objects."

Thus, at any nominated level of variation, a model describes relationships among objects. These relationships among different objects (typically pictured as some kind of pattern) help us to get at what is of real interest: biodiversity expressed as the number of units represented by those objects. The model allows, by inference, a "counting-up" of the lower-level units. This means that we can compare different objects – or sets of objects – with respect to the count of the number of different units represented.

The phylogenetic diversity measure (PD; Faith 1992) illustrates this framework. Here, the objects are species (or other taxa) and the units are features (or characters). The use of phylogeny (the "tree of life") to make inferences about the relative feature diversity of different sets of species is an attempt to overcome our lack of knowledge about all the features of species. The PD, and the estimated relative feature diversity, of a set of species is calculated as the minimum total length of all the phylogenetic branches required to connect all the species in that set on the phylogenetic tree (Faith 1992). This definition follows from an evolutionary model in which branch lengths reflect evolutionary changes (new features), and shared ancestry accounts for shared features among species (Fig. 5.1a). The phylogenetic model implies that PD in effect counts up the features represented by a set of species (or other taxa). Further, the amount of branch length

71

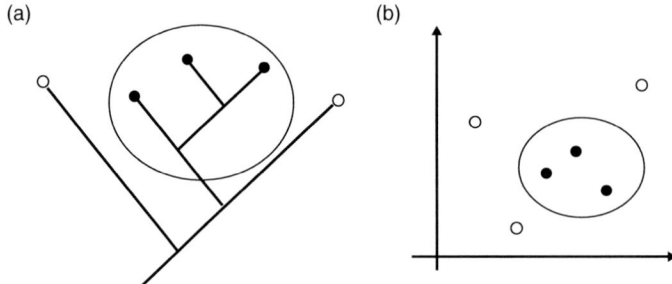

(a)

(b)

Figure 5.1 Two examples of pattern–process models that allow inferences about shared units among objects. Circles are different objects. Black circles, in contrast to white circles, are those objects sharing the same ("black") unit.

(a) for a "tree" pattern, shared ancestry among the objects (e.g. species) predicts shared units (e.g. features).

(b) for a spatial pattern, shared environment among the objects (populations, species, or localities) predicts shared units (genetic variants, traits or species, respectively).

added when a species joins an existing set of species indicates the relative number of additional features gained. PD therefore provides a basis for biodiversity conservation at the level of feature diversity – we can compare the PD of different sets of species and talk about gains and losses.

PD now is much applied (for review, see Faith 2013), and often is considered to be a measure of variation that is more informative than species-level biodiversity. The same number of species can provide large or small PD (depending on branch lengths and degree to which ancestral branches are shared among the species). The extra information about feature diversity may be regarded as more informative about insurance and option values of biodiversity (Faith 1992; see below).

PD is a useful exemplar for the general framework because it has been critiqued in some studies on the philosophy of biodiversity; addressing these issues highlights aspects of the general framework. For example, Maier, in his book *What's So Good About Biodiversity?* (2012: 96), concluded that PD involves "demoting species to a secondary role," and that it rejects "a multi-dimensional conception of biodiversity" (p. 99). This incorrect interpretation highlights the need to emphasize that biodiversity includes many possible choices of units of variation (Faith 1994). Other critiques have focused on the nature of the PD model and the use of "features" as units. Maclaurin and Sterelny (2008: 148) concluded that "the Achilles heel of the phylogenetic approach is the assumption that all character changes are equally important." Maier argued that PD is unsatisfactory in not allowing us to list all of the features for extant species. These concerns are answered by noting that the biodiversity framework seeks to make inferences about unknown units (here, features), and these naturally will have equal "importance" – we do not yet know which ones will provide benefits in the future. Maier (2012) also argued that, because of convergent evolution (features arising independently in two or more species), the PD assumption of shared ancestry explanations for shared features (Fig. 5.1a) is inappropriate. Faith (1992) considered the issue of convergent evolution and argued that the phylogenetic model attempts *general* inferences over a large number of known and unknown features, and may well "fail" for any individual feature. Below, I describe how convergent evolution has been considered in an alternative shared habitat/shared features model proposed by Faith (1992).

Several other examples of this general framework are listed in Table 5.1. Functional trait diversity among species (including convergently derived traits) may be inferred based on the

Table 5.1 Examples of biodiversity units and objects under the Faith (1994) framework

Name	Units	Objects	Model	Reference
PD	features	species, populations	shared ancestry	Faith (1992)
ED	species	sites, clusters of sites	unimodal response	Faith and Walker (1996)
ED (microbial)	lineages	samples	unimodal response	Faith *et al.* (2009)
ED (populations)	genetic variants, traits	populations	unimodal response	Faith *et al.* (2009); Mimura *et al.* (in prep)
EDf	traits	species	unimodal response	Faith (1996, 2015)
Split diversity	genetic variants	populations	turnover on networks	Minh *et al.* (2007)

assumption that shared *habitat* explains shared features or traits (providing a measure of functional "resilience"; Faith and Walker 1996). This model nicely parallels PD's assumption that shared *ancestry* explains shared features/traits (Faith 1992), and illustrates how the same objects and units may be linked by alternative pattern–process models (Fig. 5.1a, b).

Alternative models also have been useful when different populations of a species are the objects, with traits or genetic variants as the units. While Faith (1992) applied the PD framework in this case (with terminal taxa as haplotypes), Minh *et al.* (2007) used a pattern model based on networks, rather than phylogenetic trees, in order to make inferences at the level of units equal to genetic variants among populations.

Sometimes the pattern may not change, but we may adopt a different model for inferring the counts of units. For example, Horn *et al.* (1996) retained species, features, and phylogeny, but proposed an alternative model for the inference of feature diversity.

I referred above to functional trait diversity among species or populations based on models in which objects that share environmental space are expected to share units (Fig. 5.1b). The oldest and best-known example of a biodiversity measure based on this model is the ED ("environmental diversity") method (Faith and Walker 1996; for review, see Beier and Albuquerque 2015). Here, ecosystems or localities are the objects and species typically are the different units. Again, different sets of objects will vary in how well they capture the total number of different units. The assumption that shared environment implies shared species (as units), is plausible as a model reflecting the adaptation of species to environmental gradients. The pattern of interest is a configuration of localities along multiple environmental gradients (Fig. 5.1b). The relative biodiversity of a set of localities (objects), at the level of species (units) is inferred by applying a p-median criterion to this pattern (see Faith and Walker 1996). This calculation in effect counts up the relative number of units represented by a nominated set of objects.

How the choice of units of variation also provides many other useful calculations

The ability to count up units has attractive properties beyond the quantification of biodiversity (Faith 1994). If we can count up units, we can carry out many other useful calculations for decision-making. Further, we can plug-in any nominated kind of units, not just for our measure of biodiversity, but also for other calculations, such as dissimilarity indices among objects (see Faith *et al.* 2009).

An important companion calculation to biodiversity is what is called "complementarity" (Kirkpatrick 1983) – the gains and losses in biodiversity as objects are gained or lost (discussed above for PD and ED). Faith (1994: 48) referred to the important idea that inferred counts of units provide other useful calculations, including complementarity:

> The problem of prioritising areas illustrates how pattern (specifically environmental pattern) can be used as a surrogate for biodiversity, in predicting the same components of biodiversity that would be used at the species level directly, notably complementarity [see below]. This approach to using environmental pattern to quantify and estimate biodiversity can be generalised to cover other patterns as predictors of lower-level biodiversity.

Faith (1994) noted the capacity to consider many calculations, but focused on complementarity (Kirkpatrick 1983). While biodiversity is quantified by an inferred count of number of different units, the PD and ED examples above illustrated how we "work" with biodiversity by making various calculations based on those units. The notable example is the calculation of marginal changes – the number of units lost, or the increase in the number represented by an added protected area. A core biodiversity conservation challenge arises from the fact that individual localities typically are the focus of decision-making, but assessment of total biodiversity depends on sets of localities. Every place has biodiversity, but its contribution to regional or global biodiversity is indicated by its complementarity value (Kirkpatrick 1983; for review, see Justus and Sarkar 2002), not its total diversity (see also Sarkar 2010). In early work, complementarity was something directly calculated for the surrogate or proxy information (for example, asking how many additional species were added by an additional protected area; Justus and Sarkar 2002). Faith (1992, 1994) extended the idea to consider what he called "pattern complementarity." Here, the count of the number of additional units is not made directly but instead is inferred by the associated pattern and model. Under the ED model, the gain in species from the addition of a locality to a set is given by the pattern complementarity, equivalent to the degree of increased representation of the pattern provided by that locality (see Faith and Walker 1996).

The calculation of complementarity nicely illustrates how the counting-up of units also enables various calculations that are useful for decision-makers. Many of the other common species-level biodiversity-related calculations in ecology can be transformed into corresponding measures based on other kinds of units (for examples, see Faith 2013). For example, the various species-level indices surveyed in Whittaker *et al.*'s (2001) general theory of species diversity can be converted to corresponding indices for other units. Similarly, weighted endemism (for an object, we count up its units, each inverse-weighted by the total number of objects that have that unit) is well known for species and for PD, but extends to genetic units (for objects defined as the different populations within a given species).

A wide family of PD calculations has been derived through the interpretation of PD as counting-up features (Faith *et al.* 2009, Faith 2013). PD-complementarity can measure biodiversity gain not only for species but also for added localities. Other useful PD calculations for geographic localities include PD-endemism and PD-dissimilarities between localities (Faith 2013). "Expected PD" calculations are based on either probabilities of extinction or presence–absence. For example, estimated extinction probabilities indicate amounts of "expected PD loss" (e.g. Faith 2013).

All of the PD indices mentioned above equally could apply to any other kinds of units in the general framework. For example, Faith (2015) described calculations of the expected loss in functional trait diversity. These applications highlight how counting-up the relative number of units remains the core measure of "biodiversity," but many other calculations based on the inferred units may be useful for decision-making (see also Chao *et al.* 2014).

Other general frameworks for biodiversity

The framework based on counting-up units contrasts with other proposals for general frameworks for biodiversity, including those proposals that have attempted to include a variety of calculations within the definition of biodiversity. Sarkar's (2014a: 3) consideration of "units" other than species appears compatible with the general framework of Faith (1994). However, his proposal differs in integrating other additional calculations into the quantification of biodiversity. For example, Sarkar proposed that biodiversity must include a number of aspects beyond richness. At the species level, Sarkar (2014a: 3) argued that a measure of biodiversity should reflect complementarity, rarity, endemism, and also "equitability" (reflecting relative abundances). Another aspect to be included was "disparity," reflecting taxonomic distance between species.

This proposal recalls how "biodiversity" is sometimes expanded to include a variety of ecological indices (often referred to as "biodiversity" measures), adding to the complexity of possible meanings of the term "biodiversity" (for related discussion, see Sarkar 2008). For example, Noss (1990) regarded biodiversity as including composition, structure, and function, reflecting the range of "diversity" measures in ecology.

The unlimited possibilities raised by these proposals highlight the advantages of a simpler framework where "biodiversity" focuses on the *number* of units, while recognizing that the same units can be part of numerous other calculations that include standard ecological indices. Thus, complementarity, endemism, and dissimilarities and many traditional ecological "diversity" measures all can be calculated, but are not measures of "biodiversity" (Faith 1994).

Sarkar's consideration of a taxonomic measure of *difference* between species as part of "biodiversity" echoes a popular general strategy. Weitzman (1992) presented a general framework for biodiversity based on the idea of objects, and measures of *difference* between objects. The biodiversity of a given set of objects then is reflected in the amount of difference represented by the set. Weikard (2002) similarly argued that any operational concept of diversity requires some measure of dissimilarity between appropriately defined objects (see also Maclaurin and Sterelny 2008, Morgan 2010).

This strategy assumes that we can define meaningful differences among the initial objects, but there are many ways to define "differences." A review of phylogenetic diversity and conservation (Winter *et al.* 2012) interpreted "phylogenetic diversity" as derived from any between-species differences or relationships based on phylogeny. Winter *et al.* did not recognize PD as a model-based measure of feature diversity, and so did not consider the possible advantages of an alternative pattern-model units approach. Maclaurin and Sterelny (2008: 20) also incorrectly interpreted PD as an application of objects and differences, with species as objects and differences given by "genealogical depth."

The general framework of Faith (1994) avoids weaknesses of the objects–differences strategy. Morgan (2010) pointed out the core weakness: even if one has some agreed natural measure of differences, a remaining problem is how to trade off more objects for fewer differences (or vice versa) to assess biodiversity. The general framework of Faith (1994) sidesteps this dilemma, because it uses the inferred relative number of units, without any need for an "objects and differences" trade-off. Using PD again as an example, a set of four species logically may have lower PD (lower feature diversity) than a set of three species.

The above discussion sets the stage for a better understanding of the values of biodiversity. By recognizing other calculations as useful, but outside the realm of "biodiversity," and by sidestepping the classic objects–differences framework, we can focus on counting-up different units, and focus on the value of having many different units (variation).

Option values of biodiversity

Frustrations in applying an objects–differences approach have contributed to recent arguments that "biodiversity" does not seem to capture what we value. In their review of the history of the use of the term "biodiversity," Morar *et al.* (2015: 24) argued that "Insofar as biodiversity emphasizes the differences between kinds, in distinction from the concrete individuals and places that instantiate these kinds, it risks occluding what we value most: not life's variety but rather life itself in its concrete richness."

In contrast, when biodiversity is interpreted as variety (counting the number of units), then the value of biodiversity better resonates with the idea that many individual elements (units) have known current values. In fact, the best argument for what we call the option value of biodiversity is that we see many currently beneficial units, and maintaining a large number of units (biodiversity) for the future will help maintain a steady flow of such beneficial units (see also the "storehouse" analogy in Faith *et al.* 2010). In accord with this idea, the Millennium Ecosystem Assessment (MEA; 2005a: 32) described option value as: "the value individuals place on keeping biodiversity for future generations." Option value refers not only to the unknown future benefits from known units of biodiversity, but also to the unknown benefits from unknown units. Biodiversity option value therefore links "variation" and "value": providing a fundamental relational value of biodiversity reflecting our degree of concern about benefits for future generations. Each choice of a kind of units (Table 5.1) defines a measure of option value of biodiversity (Faith 1994).

Faith (1992) explicitly characterized PD and feature diversity as capturing biodiversity option value. Forest *et al.* (2007) later explored PD and option value in their examination of the phylogenetic distribution of angiosperm plants having known human uses (classified as medicinal, food, and all other uses). They studied an estimated phylogenetic tree for genera found in the Cape hotspot of South Africa. Forest *et al.* (2007) concluded that, if we did not know about those medicinal, food, and other uses, then preserving sets of species with high PD would be a good way to preserve these unknown benefits. This argument supports the link between PD and biodiversity option value at the level of features (see also Bernstein 2014).

I noted above that a good argument for maintaining biodiversity is based on the listing of the many units that currently provide known benefits. However, this does not mean that the best way to conserve biodiversity is to focus conservation only on these known beneficial units. Faith and Pollock (2014) showed how conservation focused on currently useful species was not necessarily a good way to conserve overall PD and its option values (we may spend our conservation budget just on one part of the tree). This can be called the "biodiversity conservation paradox." Currently useful units provide a good way to promote the importance of maintaining biodiversity, but we must conserve more than those units, because other units may provide unanticipated benefits in the future.

While PD now is a popular expression of option value, this link between biodiversity and human well-being actually traces back to the "pre-history" of "biodiversity" (roughly, the history of the term before it was invented). Haskins (1974: 646) summarized an important discussion meeting where participants called for "an Ethic of Biotic Diversity in which such diversity is viewed as a value in itself and is tied in with the survival and fitness of the human race." Haskins (1974: 646) warned, "Plants and animals that may now be regarded as dispensable may one day emerge as valuable resources" and that extinction "threatens to narrow down future choices for mankind." Roush (1977: 9) similarly argued that "diversity increases the possibility of future benefits" (for review, see Farnham 1997). IUCN's (1980: section 3) arguments for the conservation of diversity (referring to "the range of genetic material found in the world's organisms") echoed Haskins: "we may learn that many species that seem dispensable are capable of providing important products, such as pharmaceuticals, or are vital parts of life-support systems on which we depend."

Later philosophical discussions supported these perspectives. Norton (1986) argued that *diversity itself* has utilitarian value. Randall (1986: 103) similarly considered unit species and proposed that all species not already distinguished in having recognized human-use values "would be treated as having a positive but *unknown expected value*." These ideas flowed on to discussions around the new term "biodiversity." McNeely (1988) and Reid and Miller (1989) referred to "option values" of biodiversity. E. O. Wilson (1988) highlighted values for biodiversity reflecting our lack of knowledge about the components of life's variation and their importance to humankind.

The MEA (2005a: 32) concluded that "the value individuals place on keeping biodiversity for future generations – the option value – can be significant." Gascon *et al.* (2015) reviewed the many, sometimes surprising, benefits of species to argue for the importance of option value (and pointed to PD as a candidate measure of option value). The Encyclical Letter "On Care for Our Common Home" (Francis 2015: para. 32) addressed the loss of biodiversity, arguing for the importance of not only intrinsic values of species, but also the option values of biodiversity:

> The loss of forests and woodlands entails the loss of species which may constitute extremely important resources in the future, not only for food but also for curing disease and other uses. Different species contain genes which could be key resources in years ahead for meeting human needs and regulating environmental problems.

Maclaurin and Sterelny concluded: "The crucial point about option value is that it makes diversity valuable. As we do not know in advance which species will prove to be important, we should try to conserve as rich and representative a sample as possible" (2008: 154). This argument extends beyond the species level; Maclaurin and Sterelny's statement that option value "links variation and value" quoted Faith (2003), which reviewed the general pattern framework and considered option value across all levels of biodiversity.

It is important to highlight how this use of option value applies to biodiversity-as-variety, not simply to specific elements. Maier (2012) criticized Maclaurin and Sterelny's (2008) arguments for biodiversity option value and interpreted "option value" as associated with a given resource or ecosystem service. Quantification of value then requires a range of calculations – including estimates of reliability of stock, of risk aversion, and of willingness to pay. Maier complained that these basics were missing in Maclaurin and Sterelny's arguments. However, contrary to Maier, reference to "option value of biodiversity" does not have to be interpreted to mean that the actual value of the options is determined. Instead, over a range of possible conservation outcomes, there are more or fewer options represented, as indicated by counts of units.

Other possible values of biodiversity

Option value of biodiversity clearly is anthropocentric, and discussion of this value has a long history. However, a sampling of the current biodiversity literature would give the impression that the earliest core value of biodiversity was intrinsic (inherent values, independent of human perceptions), with the ecosystem services movement forging new links from biodiversity to human well-being (for discussion, see Faith 2012). For example, Hardy (2008: 3) concluded that "the idea of ecosystem services allows for acknowledging more than the 'intrinsic' value of biodiversity by expanding the breadth of the conservation argument to include the 'utilitarian' values of nature." Frishkoff *et al.* (2014: 1343) similarly ignored the existing links from PD to option value, characterizing PD as having intrinsic value, with additional value supposedly based only on its contribution to ecosystem services. IPBES (IPBES/3/INF/7 2014) has misrepresented PD in the same way.

Some even promote the idea that there is now a "deep divide" in conservation corresponding to a conservation focus on intrinsic value versus an ecosystem services focus on human well-being (Polasky *et al.* 2012: 140). Faith (2012) reviewed the numerous other studies similarly setting up a false intrinsic value versus ecosystem services dichotomy.

I showed above how many early studies, preceding the ecosystem services movement, already had attributed anthropocentric option value to biodiversity as living variation. Below, I will argue that intrinsic value in contrast has a weak link to biodiversity because the attribution of intrinsic value is largely to individual elements, not *variation*.

Intrinsic value was part of the early discussions related to biodiversity – but largely this has referred to individual elements (species), not variety. Early debates about the values of individual species influenced the development of ideas about the value of biodiversity. Sober (1986) argued that species should be preserved for reasons other than their known values as resources for human use. Callicott (1986) discussed philosophical arguments regarding non-utilitarian value and concluded that species have some intrinsic value "in and for itself" (Callicott 1986: 140; see also Soulé 1985, McShane 2007). Callicott (1989) argued that elements at all levels of biological organization have intrinsic value (see also McShane 2007, Vucetich *et al.* 2015). Despite common reference to the intrinsic value of "biodiversity," it is nowhere clear that this is about intrinsic value of *variation*. Later philosophical discussions of biodiversity and intrinsic value also clearly focus on individual elements, not variation (e.g. Justus *et al.* 2008, Maguire and Justus 2008; see also Sarkar 2014a).

Soulé's (1985) presentation of the basis for an emerging discipline of conservation biology emphasized that "biotic diversity has intrinsic value," and is cited as a foundational link from biodiversity to intrinsic value. However, Soulé's discussion referred to species and other elements of biodiversity as having intrinsic value, not variety itself. Confusion can be avoided by noting that the term "biodiversity" sometimes will be used to refer to the collection of individual elements ("biospecifics"; Faith 1997). Thus, "the biodiversity of Madagascar" (as a set of objects and units representing some amount of biodiversity) can be linked to intrinsic value, but that is reference to the individual elements, not variation itself.

These ideas can help to reconcile two perspectives on the value of biodiversity and ecosystem services. We have, on the one hand, the rich history of the term "biodiversity" as all about human well-being. On the other hand, we have the more recent history presenting ecosystem services as introducing consideration of human well-being. Perhaps "ecosystem services" are to refer, by definition, to *any* benefits for humans, and so the term has been retrospectively fitted to biodiversity option value. However, some authors appear to exclude this, seeing links to human well-being as countering any counting-up of units that is required for assessment of option value. Mace (2014: 1559) argued that "metrics that link nature to human well-being … are very different from those of species and protected areas." This echoes Norton's (2001: 88) argument that a focus on maintaining functions of healthy ecosystems would totally replace the "increasingly obsolete" inventory/items perspective of biodiversity. The framework reviewed here for counting-up units, with links to option value and human well-being, effectively counters both of those surprising perspectives.

Another difficulty in retrofitting biodiversity option value as ecosystem services is that, historically, ecosystem services have been distinguished from biodiversity option values. For example, Ehrlich and Ehrlich (1992: 219) justified biodiversity conservation by presenting an option value argument: "humanity has derived many direct economic values from biodiversity … The potential of nature's genetic library for providing more of these benefits is enormous," and then made a separate ecosystem services argument about elements of biodiversity. The MEA (2005a) made this same distinction.

The IPBES conceptual framework (Díaz *et al*. 2015: 14) referred to both option value associated with particular benefits ("nature's benefits") and also "the 'option values of biodiversity', that is, the value of maintaining living variation in order to provide possible future uses and benefits." Put simply, there are nature's benefits (ecosystem services), but, through biodiversity, there also are benefits from "living variation in order to provide possible future uses and benefits." We could use the term "ecosystem services" for any of the unanticipated future uses and benefits that eventuate through biodiversity option value (Gascon *et al*. 2015 used this terminology).

An additional consideration helps to reconcile values of biodiversity and values of ecosystem services. Biodiversity option value may be relevant both within an ecosystem and globally (for example, PD is applied at both scales; Faith 2015). Option value provided by local biodiversity may focus on local processes and services. Option value of biodiversity is all about unknown, unanticipated, future benefits from the units. However, there are other closely related, even overlapping, values of biodiversity-as-variation. One might be called "proxy value." Here, the focus is less on unanticipated future benefits and more on the many current existing benefits. Given that we may not be able to list all of these, biodiversity stands in for these as a proxy, and so is a valuable target for conservation. Such proxy value may be an important consideration within ecosystems, where the multitude of functions and services seem to draw on the full variety in the system (Lefcheck *et al*. 2015). Thus, PD may have a "proxy role" – management to maintain PD might be a proxy strategy, helping to maintain all the *current* benefits that objects/units are providing in the ecosystem.

This value within ecosystems also links to what is sometimes called the "insurance value" of biodiversity. Here, the variety of units is seen as providing functions and other benefits over changing environmental conditions. Turner (1999) used the term "insurance value" to refer to a kind of insurance that diversity provides against the failure of ecosystems to provide goods and services (see also Awasthi *et al*. 2015). Different units may be known to provide benefits under different conditions. Of course, future conditions may mean that some units have unanticipated benefits – thus, this variation again has option value in providing unknown, unanticipated, benefits from one or more units. It is apparent that there is no strict boundary between proxy, option, and insurance values of biodiversity. Perhaps option value will be most compelling at the global scale, with insurance value most appreciated at the more local scale.

The contrast between local and global contexts reminds us of the importance of complementarity. West (2015; see also Mace *et al*. 2012) argued that preserving elements of biodiversity within an ecosystem can be a service, supporting what amounts to biodiversity option values. The ecosystem service is the protection of particular elements, not overall biodiversity – but these elements do contribute (through complementarity) to the preservation of regional/global variation and option values.

I conclude that biodiversity is not an ecosystem service; instead, both of these terms separately represent benefits that may be valued by society. There are two ways to reconcile biodiversity option values with the popular claim that "ecosystem services" covers all of conservation for human well-being: (1) option value of biodiversity represents possible unanticipated future benefits – and each one of those future benefits might be called an "ecosystem service";

(2) ecosystems may preserve, as an ecosystem service, individual *elements* of biodiversity, and so make a contribution to overall global preservation of option values of biodiversity.

Systematic conservation planning (SCP) and multi-criteria analyses

There are three major ways in which the ecosystem services movement has not served the conservation of biodiversity-as-variety. One problem has been the neglect of the important arguments for option value; another has been the promotion of weak ecological definitions of biodiversity; and the third has been the over-focus on local values over global values. In this section, I consider how multi-criteria analyses incorporating regional biodiversity (see e.g. Faith 1995) and related systematic conservation planning (Margules and Pressey 2000) can help reconcile these different values of society.

A key challenge is raised by the idea that an ecosystem (or locality) provides an ecosystem service in protecting some elements of biodiversity that contribute to global option values. Sometimes the local biodiversity is seen as the whole story in addressing biodiversity conservation. For example, Cardinale *et al.* (2012) documented many localized cases where greater biodiversity supported more services, and claimed that these findings help address the biodiversity crisis. However, the study ignored the loss of global biodiversity option values from transforming the land (e.g. to agriculture) and only looked at the issue of more diversity in the transformed system. Put simply, incentives for maintaining local diversity will not guarantee overall regional/global diversity.

Similarly, Polasky *et al.* (2012: 157) reported that "in general, investing in conservation that increases the value of ecosystem services is also beneficial for biodiversity conservation and vice versa." However, Polasky *et al.* recorded only within-place biodiversity, and added up these values over multiple places. Consequently, the study had only a weak link to regional biodiversity and associated option values.

In many cases, such studies also focused on within-ecosystem "biodiversity" that simply reflected ecology of ecosystems. For example, Bateman *et al.* (2013) used a measure that reflected local abundance of different species, with no information on gains or losses in regional biodiversity-as-variation (see Faith 2014).

In discussing complementarity, I noted that a core biodiversity conservation challenge arises from the fact that individual localities typically are the focus of decision-making, but assessment of total biodiversity depends on sets of localities. Thus, every place has elements of biodiversity, but its contribution to the global option values of biodiversity is indicated by its complementarity value, not its total diversity. It is the comparison of the place's current complementarity value to the other values/opportunities in that place that matters when considering trade-offs at a regional scale (Faith 1995; see also Sarkar 2010).

An important scenario is one in which ecosystem services are an added benefit available from intact land which retains its biodiversity (Daily 1997). In this context, the difference between the local within-places and regional among-places perspectives on biodiversity conservation presents a challenge. An apparent win-win scenario is one in which management for ecosystem services within a given place maintains essentially all the biodiversity elements. However, the particular units of biodiversity contributed by that place may not be the ones most needed for biodiversity conservation at the among-places scale. The ecosystem services of a locality may well value exactly what makes that place similar to many others, even though this amounts to redundancy at the regional scale (MEA 2005a, 2005b). A case study (Faith 2014) showed how an increase in ecosystem services conservation, combined with a conservation budget, could mean that the region's capacity for achieving biodiversity conservation collapses.

The local–global problem points to the need for systematic conservation planning (SCP; Margules and Pressey 2000) and related multi-criteria analyses (MCA; Faith 1995, Sarkar 2014a). In SCP, localities have their own specific current benefits, plus some dynamic biodiversity complementarity contribution to a set of, for example, protected areas. Similarly, multi-criteria approaches (Faith 1995) can ensure that both services and biodiversity option values are part of decision-making.

SCP views ecosystem services as a co-benefit or as a cost of conservation (Faith 1995, Margules and Pressey 2000). For a given locality, higher biodiversity complementarity, combined with higher co-benefits and/or lower opportunity costs of conservation, implies greater priority for conservation (see also Faith 2014). Early work, going back to the 1980s, showed how conservation of an area can gain priority based on multiple guidelines: "Any suggested allocation of forest lands between competing uses can be evaluated as to how well it satisfies a range of land allocation guidelines representing the preferences of different stakeholders in relation to the issues that made planning necessary" (Cocks and Ive 1996: 45, and references within). Their suggested protected areas were recognized as satisfying guidelines relating not only to biodiversity, but also to what we would now call ecosystem services, including fresh water, recreation, and wilderness value, plus people's connections to nature such as aesthetics. The suitability score of an area for a land use was the sum of the guidelines' satisfaction ratings for that area related to that use. Thus, various ecosystem services boosted the case for conservation (see also Faith 2014).

The general framework (Table 5.1) points to common calculations for every choice of objects and units, and this common toolbox can extend to SCP. While Sarkar (2014a) described SCP as an "approach to mapping conservation *area* networks," SCP applies to other objects as well. SCP allocates objects to conservation (or "partial protection"; Faith 1995, 2012), maximizing the total biodiversity (count of units) while taking into account other object benefits/dis-benefits, costs, and other constraints. For example, we may apply SCP in a global functional trait space (e.g. Faith 2015). Here, SCP selects objects (species or populations) to maximize biodiversity represented by units (functional traits), while taking into account costs and other factors.

In addition to its complementarity contribution, an object may have other benefits associated with its conservation. Ecosystem services therefore is just one special case (the benefits of ecosystems) of these "object benefits". That special case warns us to avoid what might be called the "ecosystems services error" – where positive gains from conservation that capture high object-benefits and lots of within-object units are seen as a win-win and a pathway for good regional conservation planning. Stated generally, the protection of lots of objects with high current "object benefits" may not be an adequate way to protect overall biodiversity option values. This general problem has been explored for SCP applied to PD, where species are the objects and features are the units. Earlier I referred to the study of Faith and Pollock (2014), who found that a focus on conservation of phylogenetically clumped current-use species (those with high object-benefits) could reduce the capacity to retain high PD option values within a given budget. They concluded that any "sustainable use" of biodiversity must integrate current species uses and conservation of biodiversity option values (preserving both current and possible future evolutionary, or evo-system services; Faith *et al.* 2010).

This highlights again the general "biodiversity conservation paradox": biodiversity reflects potential benefits for humans, and pointing to lots of objects with current benefits is a very good way to advertise and gain support for biodiversity conservation; however, focusing our conservation priorities on those objects with current uses is not an assured pathway to conserve biodiversity. This highlights the need to pursue balanced trade-offs (and synergies) among current benefits and biodiversity option values.

Discussion

The general biodiversity framework outlined here helps to focus the term "biodiversity" on living variation at multiple levels, and on the core values of variation. The shared toolbox spanning all these levels supports many useful calculations – including systematic conservation planning.

Ironically, the ecosystem services movement, in sometimes promoting ecological definitions of biodiversity, in sometimes neglecting biodiversity option values, and in sometimes failing to look beyond the local ecosystem scale, risks demoting, more than promoting, biodiversity conservation for human well-being.

Callicott *et al.*'s (1999: 30) discussion of current normative concepts in conservation lamented that sometimes "partisans of a single normative concept try to make it cannibalise or vanquish all the rest." This recalls the critique (Gasparatos and Willis 2015; see also Sarkar 2014b) that "biodiversity" now has become a vague buzzword within the ecosystem services movement.

This chapter hopefully defuses that problem. A more clearly focused notion of "biodiversity" as variety may help to overcome the current neglect of biodiversity option values, putting them "on the table" alongside other needs of society, and integrated into SCP. The foundations for all this actually are quite old, so progress on implementation is long overdue.

References

Awasthi, A., M. Singh, S. K. Soni, R. Singh, and A. Kalra. 2015. "Biodiversity Acts as Insurance of Productivity of Bacterial Communities Under Abiotic Perturbations." *ISME Journal* 8: 2445–2452.

Bateman, I. J. *et al.* 2013. "Bringing Ecosystem Services into Economic Decision-Making: Land Use in the United Kingdom." *Science* 341: 45–50.

Beier, P., and F. Albuquerque. 2015. "Environmental Diversity is a Reliable Surrogate for Species Representation." *Conservation Biology* 29: 1401–1410.

Bernstein, Aaron S. 2014. "Biological Diversity and Public Health." *Annual Review of Public Health* 35: 153–167.

Callicott, J. B. 1986. "On the Intrinsic Value of Nonhuman Species." In *The Preservation of Species: The Value of Biological Diversity*, ed. B. G. Norton. Princeton, NJ: Princeton University Press.

Callicott, J. B. 1989. *In Defense of the Land Ethic: Essays in Environmental Philosophy*. Albany: State University of New York Press.

Callicott, J. B., L. B. Crowder, and K. Mumford. 1999. "Current Normative Concepts in Conservation." *Conservation Biology* 13: 22–35.

Cardinale, J. B. *et al.* 2012. "Biodiversity Loss and its Impact on Humanity." *Nature* 486: 59–67.

Chao, A., C. Chiu, and L. Jost. 2014. "Unifying Species Diversity, Phylogenetic Diversity, Functional Diversity, and Related Similarity and Differentiation Measures Through Hill Numbers." *Annual Review of Ecology and Evolution Systems* 45: 297–324.

Cocks, K. D., and J. R. Ive. 1996. "Mediation Support for Forest Land Allocation: The SIRO-MED System." *Environmental Management* 20: 41–52.

Daily, G. C., ed. 1997. *Nature's Services: Societal Dependence on Natural Ecosystems*. Washington, DC: Island Press.

Díaz, S., A. Hector, and D. A. Wardle. 2009. "Biodiversity in Forest Carbon Sequestration Initiatives: Not Just a Side Benefit." *Current Opinion in Environmental Sustainability* 1: 55–60.

Díaz, S. *et al.* 2015. "The IPBES Conceptual Framework – Connecting Nature and People." *Current Opinion in Environmental Sustainability* 14: 1–16.

Ehrlich, P., and A. H. Ehrlich. 1992. "Economics of Biodiversity Loss." *Ambio* 21: 219–226.

Faith, D. P. 1992. "Conservation Evaluation and Phylogenetic Diversity." *Biological Conservation* 61: 1–10.

Faith, D. P. 1994. "Phylogenetic Pattern and the Quantification of Organismal Biodiversity." *Philosophical Transactions of the Royal Society of London B* 345: 45–58.

Faith, D. P. 1995. *Regional Sustainability Analysis*. Canberra: CSIRO.

Faith, D. P. 1996. "Conservation Priorities and Phylogenetic Pattern." *Conservation Biology* 10: 1286–1289.

Faith, D. P. 1997. "Biodiversity Biospecifics and Ecological Services." *Trends in Ecology and Evolution* 12: 660.

Faith, D. P. 2003. "Biodiversity." In *The Stanford Encyclopedia of Philosophy* (Fall 2003 edition), ed. Edward N. Zalta, available online at http://plato.stanford.edu/.

Faith, D. P. 2012. "Common Ground for Biodiversity and Ecosystem Services: The 'Partial Protection' Challenge." F1000 Research. Available online at http://f1000research.com/articles/common-ground-for-biodiversity-and-ecosystem-services-the-partial-protection-challenge/.

Faith, D. P. 2013. "Biodiversity and Evolutionary History: Useful Extensions of the PD Phylogenetic Diversity Assessment Framework." *Annals of the New York Academy of Sciences* 1289: 69–89.

Faith, D. P. 2014. "Ecosystem Services Can Promote Conservation Over Conversion and Protect Local Biodiversity, But These Local Win-wins Can Be a Regional Disaster." *Australian Zoologist*, DOI 10.7882/AZ. 2014.0\\.

Faith, D. P. 2015. "The Unimodal Relationship between Species' Functional Traits and Habitat Gradients Provides a Family of Indices Supporting the Conservation of Functional Trait Diversity." *Plant Ecology* 216: 725–740.

Faith, D. P., and L. J. Pollock. 2014. "Phylogenetic Diversity and the Sustainable Use of Biodiversity." In *Applied Ecology and Human Dimensions in Biological Conservation*, ed. L. M. Verdade *et al.*, 35–52. Dordrecht: Springer.

Faith, D. P., and P. A. Walker. 1996. "Environmental Diversity: On the Best-possible Use of Surrogate Data for Assessing the Relative Biodiversity of Sets of Areas." *Biodiversity Conservation* 5: 399–415.

Faith, D. P. *et al.* 2003. "Complementarity, Biodiversity Viability Analysis, and Policy-based Algorithms for Conservation." *Environmental Science & Policy* 6: 311–328.

Faith, D. P., C. A. Lozupone, D. Nipperess, and R. Knight. 2009. "The Cladistic Basis for the Phylogenetic Diversity (PD) Measure Links Evolutionary Features to Environmental Gradients and Supports Broad Applications of Microbial Ecology's 'Phylogenetic Beta Diversity' Framework." *International Journal of Molecular Sciences* 10: 4723–4741.

Faith, D. P. *et al.* 2010. "Evosystem Services: An Evolutionary Perspective on the Links between Biodiversity and Human Well-being." *Current Opinion in Environmental Sustainability* 2: 66–74.

Farnham, T. J. 1997. *Saving Nature's Legacy: Origins of the Idea of Biological Diversity*. New Haven, CT: Yale University Press.

Forest, F. *et al.* 2007. "Preserving the Evolutionary Potential of Floras in Biodiversity Hotspots." *Nature* 445: 757–760.

Francis. 2015. *Encyclical Letter Laudato Si' of the Holy Father Francis: On Care for Our Common Home* [English language version], The Vatican.

Frishkoff, L. O. *et al.* 2014. "Loss of Avian Phylogenetic Diversity in Neotropical Agricultural Systems." *Science* 345: 1343–1346.

Gascon, C. *et al.* 2015. "The Importance and Benefits of Species." *Current Biology* 25: R431–R438.

Gasparatos, A., and K. J. Willis. 2015. *Biodiversity in the Green Economy*. London: Routledge Business & Economics.

Hardy, D. 2008. "Motivating Private Landowner Conservation to Maximize Ecosystem Services." River Basin Center, Odum School of Ecology, University of Georgia.

Haskins, C. 1974. "Scientists Talk of the Need for Conservation and an Ethic of Biotic Diversity to Slow Species Extinction." *Science* 184: 646–647.

Horn, M. E. T., D. P. Faith, and P. A. Walker. 1996. "The Phylogenetic Moment – A New Diversity Measure, with Procedures for Measurement and Optimisation." *Environment and Planning A* 28: 2139–2154.

IUCN. 1980. World Conservation Strategy: Living Resource Conservation for Sustainable Development. International Union for Conservation of Nature and Natural Resources (IUCN). Available online at https://portals.iucn.org/library/efiles/html/WCS-004/cover.html.

Justus, J., and S. Sarkar. 2002. "The Principle of Complementarity in the Design of Reserve Networks to Conserve Biodiversity: A Preliminary History." *Journal of Bioscience* (Suppl. 2) 27: 421–435.

Justus, J., M. Colyvan, H. Regan, and L. Maguire. 2008. "Buying into Conservation: Intrinsic versus Instrumental Value." *Trends in Ecology and Evolution* 24: 187–191.

Kirkpatrick, J. B. 1983. "An Iterative Method for Establishing Priorities for the Selection of Nature Reserves: An Example from Tasmania." *Biological Conservation* 25: 127–134.

Lefcheck J. S. *et al.* 2015. "Biodiversity Enhances Ecosystem Multifunctionality across Trophic Levels and Habitats." *Nature Communications*. DOI: 10.1038/ncomms7936.

Mace, G. M. 2014. "Whose Conservation?" *Science* 245: 1558–1560.

Mace, G. M. *et al.* 2012. "Biodiversity and Ecosystem Services: A Multilayered Relationship." *Trends in Ecology and Evolution* 27: 19–26.

Maclaurin, J., and K. Sterelny. 2008. *What is Biodiversity?* Chicago, IL: University of Chicago Press.

Maguire, L. A., and J. Justus. 2008. "Why Intrinsic Value is a Poor Basis for Conservation Decisions." *Bioscience* 58: 910–911.

Maier, D. S. 2012. *What's So Good About Biodiversity? A Call for Better Reasoning About Nature's Value.* Dordrecht: Springer Science+Business Media.

Margules, C. R., and R. L. Pressey. 2000. "Systematic Conservation Planning." *Nature* 405: 243–253.

McNeely, J. A. 1988. *Economics and Biological Diversity: Developing and Using Economic Incentives to Conserve Biological Resources.* Gland: IUCN.

McShane, K. 2007. "Why Environmental Ethicists Shouldn't Give Up on Intrinsic Value." *Environmental Ethics* 29: 43–61.

Millennium Ecosystem Assessment (MEA). 2005a. *Ecosystems and Human Well-being: Biodiversity Synthesis.* Washington, DC: World Resources Institute.

Millennium Ecosystem Assessment (MEA). 2005b. "Biodiversity." In *Ecosystems and Human Well-being: Current State and Trends.* Washington, DC: World Resources Institute.

Mimura, M., T. Yahara, D. P. Faith, E. Vázquez-Domínguez, R. J. Colautti, H. Araki, F. Javadi, J. Núñez-Farfán, S. Akira, A. S. Mori, S. Zhou, P. M. Hollingsworth, L. E. Neaves, Y. Fukano, G. E. Smith, Y. Sato, H. Tachida, and A. P. Hendry. (accepted) "Understanding and Monitoring the Consequences of Human Impacts on Intraspecific Variation." *Evolutionary Applications.*

Minh, B. Q., S. Klaere, and A. von Haeseler. 2007. *Phylogenetic Diversity on Split Networks.* Technical Report NI07090-PLG, Isaac Newton Institute, Cambridge.

Morar, N., T. Toadvine, and B. J. M. Bohannan. 2015. "Biodiversity at Twenty-Five Years: Revolution Or Red Herring?" *Ethics, Policy & Environment* 18(1): 16–29.

Morgan, G. J. 2010. "Evaluating Maclaurin and Sterelny's Conception of Biodiversity in Cases of Frequent, Promiscuous Lateral Gene Transfer." *Biology & Philosophy* 25: 603–621.

Norton, B. G., ed. 1986. *The Preservation of Species: The Value of Biological Diversity.* Princeton, NJ: Princeton University Press.

Norton, B. G. 2001. "Conservation Biology and Environmental Values: Can There Be a Universal Earth Ethic?" In *Protecting Biological Diversity: Roles and Responsibilities,* ed. C. Potvin *et al.* Montreal: McGill-Queen's University Press.

Noss, R. F. 1990. "Indicators for Monitoring Biodiversity: A Hierarchical Approach." *Conservation Biology* 4: 355–364.

Polasky, S. *et al.* 2012. "Are Investments to Promote Biodiversity Conservation and Ecosystem Services Aligned?" *Oxford Review of Economic Policy* 28: 139–163.

Polasky, S., H. Tallis, and B. Reyers. 2015. "Setting the Bar: Standards for Ecosystem Services." *Proceedings of the National Academy of Sciences* 112: 7356–7361.

Randall, A. 1986. "Human Preferences, Economics, and the Preservation of Species." In *The Preservation of Species: The Value of Biological Diversity,* ed. B. G Norton. Princeton, NJ: Princeton University Press.

Reid, W. V., and K. R. Miller. 1989. *Keeping Options Alive: The Scientific Basis for Conserving Biological Diversity.* Washington, DC: World Resources Institute.

Roush, G. 1977. "Why Save Diversity?" *Nature Conservancy News* 21: 9–12.

Sarkar, S. 2008. "From Ecological Diversity to Biodiversity." In *The Cambridge Companion to the Philosophy of Biology,* ed. D. Hull and M. Ruse. Cambridge: Cambridge University Press.

Sarkar, S. 2010. "Diversity: A Philosophical Perspective." *Diversity* 2: 127–141.

Sarkar, S. 2014a. "Biodiversity and Systematic Conservation Planning for the Twenty-first Century: A Philosophical Perspective." *Conservation Science* 2: 1–11.

Sarkar, S. 2014b. "Environmental Philosophy: Response to Critics." *Studies in History and Philosophy of Biological and Biomedical Sciences* 45: 105–109.

Sarkar, S., and C. R. Margules. 2002. "Operationalizing Biodiversity for Conservation Planning." *Journal of Biosciences* 27: 299–308.

Sober, E. 1986. "Philosophical Problems for Environmentalism." In *The Preservation of Species: The Value of Biological Diversity,* ed. B. G. Norton. Princeton, NJ: Princeton University Press.

Soulé, M. E. 1985. "What is Conservation Biology?" *Bioscience* 35: 727–734.

Turner, R. 1999. "Environmental and Ecological Economics Perspective." In *Handbook of Environmental and Resource Economics,* ed. J. C. J. M. van den Bergh, 1001–1036. Northampton, MA: Edward Elgar.

Vucetich, J. A., J. T. Bruskotter, and M. P. Nelson. 2015. "Evaluating Whether Nature's Intrinsic." *Conservation Biology* 29: 321–332.

Weikard, H. 2002. "Diversity Functions and the Value of Biodiversity." *Land Economics* 78: 20–27.

Weitzman, M. L. 1992. "On Diversity." *The Quarterly Journal of Economics* 107: 363–405.

West, A. 2015. "Core Concept: Ecosystem Services." *Proceedings of the National Academy of Sciences* 112: 7337–7338.

Whittaker, R. J., K. J. Willis, and R. Field. 2001. "Scale and Species Richness: Towards a General, Hierarchical Theory of Species Diversity." *Journal of Biogeography* 28: 453–470.

Wilson, E. O., ed. 1988. *Biodiversity.* Washington, DC: National Academy of Sciences/Smithsonian Institution.

Winter, M., V. Devictor, and O. Schweiger. 2012. "Phylogenetic Diversity and Nature Conservation: Where are We?" *Trends in Ecology and Evolution* 28: 199–204.

6

BIODIVERSITY ELIMINATIVISM

Carlos Santana

In this chapter[1] I defend *biodiversity eliminativism*, the thesis that biodiversity should not be the principal object of conservation. My argument is simple. The principal role played by the concept *biodiversity* in conservation biology is as a comparative measure of value. Biodiversity does not fulfill that role well, in part because it is not a straightforwardly measurable quantity. Moreover, it does not fulfill that role well because it does not closely track *ecological value*, the aggregate of values we place in the environment. Given these failures to fulfill its conceptual role, we should eliminate the biodiversity concept from its central place in conservation biology, in favor of using more direct assessments of ecological value as our primary comparative measures in conservation.

The role of biodiversity

The concept *biodiversity* plays a number of roles in scientific discourse. It is, for instance, a political rallying cry, and the use of the term "biodiversity" in a publication generally signals a commitment to environmental values as well as participation in a public conversation about how best to achieve them. It has also been convincingly argued that the biodiversity concept plays an important role within the sciences by creating an interdisciplinary common ground; it serves as a bridge concept allowing a researcher working in one field to easily locate work from other disciplines relevant to their interests (Meinard *et al.* 2014). These uses for the term "biodiversity" are valuable, and I don't mean to suggest that we should give up on them or eliminate biodiversity talk altogether.[2] I aim only to eliminate biodiversity from its role as the principal target of conservation.

A standard picture of conservation biology is that its goal is to provide systematic methods for prioritizing which areas to protect. These methods prioritize areas on the basis of their biodiversity value, as measured using surrogates (Sarkar and Margules 2002, Margules and Sarkar 2007). In other words, biodiversity plays a central role in conservation biology as the primary measure of value. It is, of course, a comparative measure, used not only to determine the relative worth of the members of a set of areas, but also to measure the extent to which the ecological value found in different sets of areas is complementary. Additionally, it can serve as diachronic measure, comparing a single area across time in order to assess the success or failure of conservation practices. It is this use – as the primary comparative measure in conservation science – for which biodiversity is not well-suited.

Biodiversity is a poor measure

A primary reason why biodiversity is not well-suited to its role is that it has resisted a consistent operationalization, rendering it a poor candidate to be used as a comparative measure. The problem has its root in the standard definitions of biodiversity. Consider, for instance, the definition determined by the 1992 United Nations Convention on Biodiversity: "'*Biological diversity*' means the variability among living organisms from all sources including, *inter alia*, terrestrial, marine, and other aquatic ecosystems and the ecological complexes of which they are part" (Convention on Biological Diversity 1992). A contemporary textbook on conservation biology offers a succinct alternative: biodiversity "is *the variety of life*, in all its many manifestations" (Gaston 2011: 27). Both these definitions are typical, but as Sarkar (2002: 137) observes, such definitions end up saying merely that biodiversity is equivalent to "all of biology."

For many purposes, using the term "biodiversity" as shorthand for all of biology can be useful. A broad, imprecise characterization of biodiversity might be best for the role the concept plays in political activism and interdisciplinary communication. But vagueness and imprecision are poor qualities in a measuring stick. If we are trying to determine whether to prioritize conserving area *A* or area *B*, asking which has a greater portion of "all of biology" won't get us very far. In all but the starkest cases of contrast, there is no obvious sense in which one area has more biology than another. Moreover, conservation biology is tasked with giving us the tools to make triage decisions, and a broad, imprecise measure won't do. We *can't* preserve all of biology, so we need a measure with units of a finer grain.

The conventional response to this problem is to use more precise, quantifiable entities as the actual operational measures in conservation. The most common of these is some variant of *species richness*, the cardinal number of distinct species inhabiting a place. Measures of other levels of organization such as genes or higher phylogenetic taxa are sometimes employed as well. The thought behind using a quantity like species richness as an operational measure is that species richness (or genetic diversity, or phylogenetic diversity, etc.) is both precisely quantifiable and representative of biodiversity in general. To use the standard language, species richness is a *surrogate* for general biodiversity. But as Sarkar and Margules (2002) point out, even a more specific quantity like species richness cannot in practice be directly measured, so we must use methods of estimation. For example, we might count the number of bird species, on the assumption that avian richness is a good proxy for species richness. Sarkar and Margules call this actually measured quantity the *estimator surrogate*, and the quantity for which the estimator serves as proxy for the *true surrogate*. Putting the chain of inference visually, measurement of biodiversity appears as: estimator surrogate → true surrogate → biodiversity. So in practice, biodiversity is at least two steps removed from the empirically observed measure used to make conservation decisions.

This method of using biodiversity surrogates might seem to skirt any issues that might be caused by the vagueness of standard definitions of biodiversity. At the very least, it provides quantities that can be used to make precise comparisons between areas, which is what we expect out of the biodiversity concept. The practice of using surrogates is laden with its own problems, however. One of these is that any standard surrogate will fail to represent some aspects of biodiversity. "No single parameter," Sarkar argues, "whether or not it can be realistically estimated, is likely to capture all biological features that we may find of interest" (2002: 140). The correlation between the true surrogate and biodiversity itself will be imperfect at best, just as there will be some degree of error in estimating the true surrogate from observations of the estimator surrogate. Of course, some uncertainty about the relationship between observed parameters and the entities of interest is just part of the cost of doing science. We can minimize but not eliminate sources of error, and so as long as the sources of potential misrepresentation are ineliminable, we have no reason to question the practice.

My worry, however, is that one major source of potential misrepresentation *is* eliminable. Biodiversity itself is also only a surrogate, because ultimately the goal of conservation is to promote ecological value. By *ecological value* I mean the totality of values we place in the environment. Assume for the sake of argument that biodiversity is a reasonably good proxy for ecological value. Then the chain of representation looks as follows: estimator surrogate → true surrogate → biodiversity → ecological value. Each arrow in that chain is an imperfect relationship, and thus a source of potential error. It's troubling that the parameters we actually use to make conservation decisions are so far removed from the entities of actual interest. More importantly, it's unnecessary. The use of surrogates is ineliminable, but we don't need two layers of vaguely defined objects of value. We could find surrogates for ecological value directly: estimator surrogate → true surrogate → ecological value. By cutting out the middleman – biodiversity – we eliminate one major link in the chain and thus decrease the distance between the operational quantities and the objects we actually want to conserve. Thus, by eliminating biodiversity, we actually improve our ability to conserve ecological value.

My argument is that biodiversity is an extraneous addition to measures of ecological value. The best way to respond to this argument would be to argue that the connection between biodiversity and ecological value is so close that eliminating biodiversity would not eliminate a significant source of error. This response can take at least three forms, each of which I will discuss in detail below. Perhaps the most intuitive argument along these lines appeals to the claim that "all of biology" is intrinsically valuable, so our comparative measure of value must include biodiversity. Another response gives up on intrinsic value, but argues that biodiversity is nevertheless an excellent surrogate for ecological value, because it captures almost all features of interest. Finally, it is commonly argued that biodiversity should be our conservation target not because it necessarily tracks value directly, but because the close link between biodiversity and the stability of ecosystems means that conserving biodiversity is the best way to sustain ecological value. In the following sections, I will explain why each of these responses – intrinsic value, faithful representation, and diversity–stability – falls short.

Biodiversity, ecological value, and intrinsic value

I have defined ecological value as the aggregate of values we place in the environment. This means that ecological value will include a diverse range of values. Many of these will involve practical utility. Healthy fish stocks feed us, extensive forests remove carbon dioxide from the atmosphere, and flourishing vegetation feeds livestock, prevents erosion, and directs the flow of water. Economic values of these sorts are all part of ecological value, but do not exhaust it. Much of ecological value is aesthetic. We value the beauty of healthy ecosystems, of rare animals, and of migratory birds and butterflies not for practical or economic gain, but because we find the experience of them pleasing. This too is part of ecological value. Other ecological values are cultural values. The history of a cultural group is often connected to places, and the natural state of those places can thus be valuable to that group. Similarly, certain organisms, such as the giant panda in China, can take on symbolic meaning for a cultural group, and thus become imbued with ecological value. Some ecological values are none of these, but are merely *existence value*, which is the satisfaction derived merely from knowing that an object persists in a desired form. For example, most of us will never aesthetically experience the environment in the Alaska

National Wildlife Refuge (ANWR), don't benefit economically from the remote ecosystem services it provides, and don't attach cultural significance to it, yet many of us derive satisfaction from the knowledge that ANWR is protected from destructive exploitation. That satisfaction is existence value. Ecological value comprises at least these economic, aesthetic, cultural, and existence values.[3]

All these kinds of value, however, are non-intrinsic[4] values in the sense that they attribute value to a place or an organism only in relation to a valuing agent. This leaves open the possibility that not "all of biology" is valuable, since some parts of biology will not possess that relation to an agent, or will possess it to a comparatively insignificant degree. Because of this, the opponent of biodiversity eliminativism might argue that the focus on non-intrinsic values is misplaced. Using biodiversity as our primary measure of value, they could argue, acknowledges that "all of biology" is intrinsically valuable: that a major portion of the value of places and organisms is not derived from their value to cognitively sophisticated organisms. If this intrinsic value is important, and it is evenly apportioned across all of biology, then biodiversity would turn out to be an ineliminable part of our measurement of ecological value.

I am skeptical that intrinsic value exists, but for the sake of argument, let's assume it does, and that it is significant. Even so, it would not provide a good reason not to eliminate biodiversity from the conservation decision-making process. Why? Because the features of intrinsic value generally aren't amenable to making comparative decisions, so they don't belong in the conservation decision-making process. Some of the problem is technical. If intrinsic value is ubiquitously distributed, or incommensurable with other values, or of infinite worth – all features often attributed to intrinsic environmental value – it will not allow us to use quantitative decision-making tools to prioritize conservation efforts (Colyvan and Steele 2011). In other words, it would not allow us to do conservation biology.

It could be argued that rational conservation decision-making doesn't require the use of these quantitative decision-making tools, and thus intrinsic value could play an important role in environmental planning.[5] The problem with appealing to intrinsic value is not only technical, however, but also political. Even if we can skirt the issues raised by Colyvan and Steele, as Maguire and Justus observe, intrinsic value is "not amenable to the sort of comparative expression needed for conservation decision-making," whether this comparative expression is precisely quantified or not (2008: 910). Moreover, they argue that it lacks the motivational force to "take precedence over competing claims and guarantee conservation" (2008: 910). The problem is twofold. On the one hand, conservation doesn't occur in isolation from social value conflicts, and appeals to intrinsic value may not provide political capital on the same order of magnitude as appeals to non-intrinsic values. On the other, even non-quantitative deliberation requires the ability to assess trade-offs and make compromises, and intrinsic values tend to resist that sort of assessment.

I find Maguire and Justus's reasons why intrinsic value is a poor foundation for conservation planning compelling, but they are controversial.[6] I'll thus buttress them with one of my own. Continue to suppose for the sake of argument that biological units do have intrinsic value, and that this value matters. On top of that intrinsic value those biological units will also have non-intrinsic value. Additionally, if a response to biodiversity eliminativism is going to be grounded in an appeal to intrinsic value, units will not differ in any meaningful way in the intrinsic value they possess. They will, however, differ significantly in the non-intrinsic value attributed to them. Consequently, in situations of conservation triage, decisions between different units must be made on the basis of non-intrinsic value. Because there is no difference in intrinsic value to attend to, intrinsic value would not allow us to make the tough decisions in situations where we can't save everything. Since this is precisely the situation conservation biology finds itself in, intrinsic value cannot be the value conservation biology uses to make comparative deci-sions. Therefore, appeals to the intrinsic value of biodiversity, or to how well the broad nature of biodiversity captures the broad distribution of intrinsic value, don't ground an objection to biodiversity eliminativism.

Biodiversity and non-intrinsic value

The opponent of biodiversity eliminativism can agree that conservation planning shouldn't focus on intrinsic value but maintain that biodiversity is the best measure for conservation on the grounds that biodiversity is a particularly reliable index for ecological value. According to this argument, the broad generality of biodiversity is its strength, as it allows us to capture the varied facets of ecological value. If biodiversity truly is a reliable index for ecological value, then we would have little reason to eliminate it from conservation practice.

But biodiversity is not a reliable index for ecological value. As I'll demonstrate in this section, in many significant cases it fails to capture or even misrepresents important ecological values. Before diving into specific cases, however, we should attend to an in-principle reason why bio-diversity is in fact distant from ecological value: although both are properties of ecosystems, they are, methodologically speaking, properties of very different sorts. Biodiversity is an *endogenous* property of ecosystems. This means that, given a convention about how to operationalize *biodi-versity*, we can measure the biodiversity of a system merely by studying the system itself, much in the same way as we measure other endogenous properties of ecosystems such as biomass, production, and resilience. Ecological value, however, is an *exogenous* property of ecosystems, meaning that it depends on an abundance of facts about entities external to the system as well as some internal-system facts. To determine the ecological value of, for example, ANWR, we have to study both the actual system in Alaska and the states of agents who have an interest of some sort in ANWR. Since ecological value is an exogenous property of ecosystems, it will be determined by a number of facts imperceptible to any endogenous property. Therefore, we should expect in principle that biodiversity, an endogenous property, will frequently fail to track ecological value.

Let's consider this point a little longer, because it is one of the most important reasons to favor biodiversity eliminativism. Conservation biology occupies a particularly value-driven scientific niche. Like any science, it aims in part at creating accurate descriptions of parts of the world, but to a much greater degree than most sciences, conservation biology is driven by non-epistemic normative goals. These normative goals are reflected most clearly in the meas-ures conservation biologists use to make conservation assessments. At issue, then, is where those measures are derived from. My argument is that we should derive those measures in large part from the values of society at large – from states of agents exogenous to the systems in question – and this will require conservation biology to involve much more (admittedly messy) social scientific work in assessing these values.[7] The status quo tends to leave out an assessment of these values in favor of using biodiversity as a catch-all. This approach is understandable, even if not justifiable, because biodiversity, as a property endogenous to ecosystems, is assessable using the tools of biology alone. Conservation biology is, after all, a field of biology, so we should expect it to exhibit a preference for biological methodology. Nevertheless, by elevating biodiversity to its position as primary comparative measure in conservation, conservation biology establishes by scientific fiat what we will take to be valuable in ecosystems. It assumes that the conserva-tion scientists know best what is valuable in nature, and largely excludes the public from the determination of the normative goals of conservation biology. By eliminating biodiversity in favor of ecological value, my aim is to increase the role the public's values play in setting the conservation agenda. The purpose of conservation science is to facilitate the rational management of the environment according to the values of all stakeholders. All stakeholders thus need a say in determining our measures of ecological value.

To see why this matters, let's review a number of examples of how biodiversity in its various forms misrepresents commonly held ecological values. We'll begin with species richness, the most common biodiversity surrogate. The trouble with species richness is twofold: extinction-based measures of value fail to capture a whole range of ecological values, and richness measures implausibly treat species as equally valuable.

If we're using richness as a principle surrogate for biodiversity, and thus ecological value, we will be missing out on a wide variety of gains and losses of value. As Angermeier and Karr have observed, richness-based surrogates can only measure biodiversity loss through species extinction, so they will fail to capture situations such as the "elimination of extensive areas of old growth forest, dramatic declines in hundreds of genetically distinct salmonoid stocks in the Pacific Northwest, and the loss of chemically distinct populations from different portions of a species' range" (1994: 692). Some of these sorts of losses might be captured by expanding our richness measures to additional levels of biological organization, but others cannot, such as the loss of forested land, deviation from a valued historical state, or replacement of one species by an invader. These are significant changes in ecological value, which a richness-based measure can't reliably capture.

Worse, species richness often misrepresents value. Standard prioritization and complementarity algorithms assume that each species "counts for one" (Magurran 2004). This is patently problematic. The extinction of a useful pollinator like the honeybee would be a terrible loss of ecological value, the extinction of a beetle species from a densely packed genus would be a trivial loss of value, and the extinction of a mosquito species which spreads malaria might be a net gain. The right conservation priorities would thus treat the honeybee as a top priority, the beetle as an afterthought, and the mosquito as a priority target for elimination. If, however, our conservation priorities are framed in terms of diversity, we'll frustrate our values by over-prioritizing the less valuable species.

Another example of how richness can misrepresent value comes from the fact that cultural and existence values often attach to the natural, historical state of a place. But if we equate value with biodiversity, and biodiversity with richness, then we can increase the value of an area by introducing new species. It might be objected that the introduction of foreign species is actually prone to decrease richness by driving native species extinct, but such a situation could be avoided through careful experimentation and planning. For example, in a broad-ranging study, Ruesink (2003) found that introduced fish species significantly altered their adopted environments less than a third of the time, which suggests that with careful planning we could easily increase the richness of many aquatic habitats. Presumably, however, if conservation biologists expended their efforts in discovering which foreign species could be safely introduced, they would be failing in their task to promote ecological value (particularly in cases where the historical composition of an ecosystem is valued), despite successfully promoting biodiversity.

Biodiversity in the guise of richness clearly fails to represent ecological value well. But conservation scientists have long been aware that richness is imperfect. It does not, many have pointed out, even fully capture the intuitive notion of diversity. For this reason among others, Maclaurin and Sterelny have argued that species richness should be "supplemented in various ways for various purposes" (2008: 173). One common supplement for richness is measures of *abundance*, the evenness of relative population sizes. The idea behind abundance is intuitive. The folk concept of diversity includes not only the presence of a number of distinct varieties (richness), but also concerns whether each variety is well-represented. A grove consisting of fifty beech and fifty aspen is as species-rich as one consisting of ninety-nine beech and one aspen, but is intuitively much more diverse. Measures of abundance typically use information-theoretic mathematical techniques to quantify how evenly represented each unit is (in conjunction with richness), thus capturing this aspect of diversity.

Supplementing measures of species richness with measures of abundance notably increases our ability to capture intuitive notions of diversity, but does not necessarily improve how well biodiversity measures represent ecological value. Ecological value often has little to do with abundance. Abundance-based measures assume that even population sizes are preferable, but this misses the mark. For some species, a disproportionately small or disproportionately large population size might be the status quo, and a shift in population size to make it more even with other species might impact the system adversely. In fact, in terms of value to humans, rarity itself is sometimes preferable. Booth *et al.* (2011) demonstrate, for instance, that the comparative rarity of a bird species determines how highly prized that species is among birdwatchers. So long as the species has a large enough population size to be out of danger from extinction, then, increasing the abundance of a rare bird species would actually decrease ecological value. In short, abundance faces the same problem as richness – it fails to consider the variety of ways in which species are valued or disvalued, and therefore a poor surrogate for ecological value.

Richness and abundance, of course, don't exhaust the toolkit of surrogates for biodiversity, but they are the most commonly used. Additionally, other measures of biodiversity, focusing on genes, morphology, higher taxa, environment types, etc., tend to either be poor measures of value, or just poorly suited for comparative measurement (Santana 2014). Extant measures of biodiversity thus all exhibit significant departures from non-intrinsic ecological values, which should not be surprising, since most of what is valuable about a particular biological unit isn't due to its contributions to diversity. Consequently, we can't justify using biodiversity as the primary target of conservation by appealing to how well it represents the values we place in the environment.

Biodiversity and ecosystem stability

I have argued that biodiversity is an extraneous addition to our chain of inference in conservation planning, and we have considered two arguments to the contrary. The first emphasized that biodiversity and intrinsic ecological value have similar coverage, thus making biodiversity an important conservation target. In response, I appealed to reasons why intrinsic values should not be part of conservation decisions. The second objection focused on the utility of biodiversity as a faithful surrogate for non-intrinsic ecological values. We saw, however, that it misses out on some values, and misrepresents others, which undermines the force of the objection. One path still remains open to the opponent of biodiversity eliminativism, however. They can contend that although biodiversity frequently falls short as a direct index for ecological value, it succeeds as an index for ecosystem stability. Since all other ecological values rely on some measure of stability, biodiversity remains a useful conservation target despite its failings.

I have two responses to this objection. First, even if there is a close relationship between diversity and stability, general measures of biodiversity are probably not our best way to promote stability. Second, the relationship between stability and ecological value isn't particularly tight.

The *diversity–stability hypothesis* is the claim that there is a close causal relationship between the diversity of an ecosystem and its stability. The hypothesis is controversial, for reasons having to do both with mixed empirical support and with a lack of clarity regarding the concepts *diversity* and *stability* (Justus 2008). Regarding the empirical support, the literature is too large to review here (for entry points to the literature, see Tilman and Downing 1994, Sankaran and

McNaughton 1999, and Gross *et al.* 2014), but the typical finding supports only a carefully caveated version of the hypothesis. For example, a review of the empirical research regarding the diversity–stability hypothesis concludes that biodiversity is at best a rough guide to stability, and that for any particular area there are better methods for maintaining stability than by targeting biodiversity (Johnson *et al.* 1996). If this is the case, the need to maintain stability does not necessarily commit us to employing biodiversity as a conservation target.

On the other hand, it isn't clear that general stability is a guide to the ecological value of the relevant sort at all. "Stability" is taken to mean many different things (Justus 2008), but if, to take one standard definition, "stability" is the persistence of species, populations, or other biological units, then conservation biology by definition aims at stability of some sort. But we cannot preserve all biological units, so aiming for stability of this sort will be a poor guide to conservation for many of the same reasons that aiming for intrinsic values is. More importantly, stability of this sort isn't always valuable to the same degree. As we've seen, biological units vary dramatically in their value, and sometimes we will value the stability of some more than others. In many cases, we will even want to promote change. Ecological value is thus not always closely tethered to stability. In conjunction with the empirical weakness of the diversity–stability hypothesis, this gives us reason to reject the claim that considerations of stability motivate the central role of biodiversity in conservation biology.

Deflationary accounts of biodiversity

Even someone who accepts most of the arguments I've made above might still be hesitant to accept biodiversity eliminativism. Sarkar, for instance, recognizes many of the same problems with the traditional role of biodiversity in conservation science, but opposes eliminativism. "The trouble with eliminativism," he writes, "is its excessive ambition: it would require a complete restructuring of conservation practice with no recourse to 'biodiversity'" (2014: 3). The cost of giving up the term, he thinks, might not be worth the benefit. So instead of eliminativism, he proposes remedying the problems with biodiversity by adopting a deflationary version of the concept. On a deflationary account (Sarkar 2002, 2014), the meaning and operationalization of the term "biodiversity" would be conventionally fixed and adjusted, as needed, to account for the conservation desiderata in particular circumstances. This would allow conservationists to incorporate the interests of all stakeholders into conservation targets.

Weakly deflationary accounts merely allow for flexibility in what aspects of biodiversity to take into account in making conservation decisions. These accounts will be vulnerable to my earlier arguments. But entirely deflationary positions – that is, ones which allow local conventions complete freedom in determining the conservation goals in any particular circumstance – are nearly identical to eliminativism. The only difference is that the deflationary account prefers to retain the word "biodiversity" as a shorthand for ecological value. Given its similarity to eliminativism, I have no major objection to the strong deflationary account, but there is good reason not to retain the word "biodiversity" in its role as representation of value. "Biodiversity" is a historically and theoretically laden term. If we use the term as our primary name for ecological value, we will import that history and theory, and thus skew our perspective of ecological value towards those things which reasonably fall under the concept *biological diversity*. Since much of ecological value does not, it makes sense to push the deflationary account to its logical conclusion, eliminativism.

To reiterate, eliminativism does not propose doing away with biodiversity talk altogether. I acknowledge that the term might continue to play an important role in conservation discussions, both for its political appeal and its utility as a bridge concept. Insofar as Sarkar's objection to eliminativism reflects worries about the loss of these important uses of the term "biodiversity," this acknowledgment should rob the objection of some of its sting. But I do think, given the likelihood that use of the term as shorthand for all of ecological value is likely to elide some important values, that eliminativism is preferable to a deflationary account.

Conclusion

Biodiversity eliminativism is in some ways a drastic proposal. It would require some deep restructuring of conservation practice. I don't think, however, that the cost of adopting eliminativism is disproportionate to the benefit. Wise environmental stewardship is one of the most important items on the human agenda, so it's worth considering drastic options like biodiversity eliminativism if it will help us get conservation done right. Under the biodiversity-focused status quo, we get it right sometimes, but biodiversity isn't optimally sensitive to the varieties of ecological value. Not everything valuable can be adequately labeled "diversity." Nor can we accurately assess ecological values by measuring only endogenous properties of ecosystems. Our best bet for prudentially managing our environment is thus to eliminate biodiversity from its exalted position in conservation planning.

Notes

1 This chapter is a development of ideas found in Santana (2014).
2 I am, however, open to the possibility that we *should* eliminate biodiversity talk from all contexts. It's quite possible that other concepts would fulfill each conceptual role currently played by *biodiversity* better than *biodiversity* currently does.
3 For an illuminating point of comparison, see Sarkar (2012, esp. chapter 2), who outlines a similar taxonomy of what he calls *natural values*: biodiversity, welfare, fidelity, service, and wild nature. His taxonomy of natural values differs from because my taxonomy of ecological value, in part because my taxonomy is meant to be less fine-grained, and in part because Sarkar sees more of a divide between descriptively held and normatively justified values than I do.
4 The literature often contrasts intrinsic values with *instrumental* values. I avoid the ambiguous term "instrumental" here because instrumental values are sometimes understood to comprise merely economic values, but "non-intrinsic" leaves room for aesthetic, cultural, and existence values as well.
5 This objection was brought to my attention by Sahotra Sarkar.
6 Callicott (2006), for instance, argues that we can roughly quantify intrinsic value.
7 See Kareiva and Marvier (2012) for a distinct but compelling argument for the same conclusion.

References

Angermeier, P., and J. Karr. 1994. "Biological Integrity versus Diversity as Policy Directives." *BioScience* 44(10): 690–697.

Booth, J., K. Gaston, K. Evans, and P. Armsworth. 2011. "The Value of Species Rarity in Biodiversity Recreation: A Birdwatching Example." *Biological Conservation* 144(11): 2728–2732.

Callicott, J. 2006. "Explicit and Implicit Values." In *The Endangered Species Act at Thirty: Conserving Biodiversity in Human-Dominated Landscapes*, Vol. 2, ed. J. Michael Scott, Dale D. Goble, and Frank W. Davis, 36–48. Washington, DC: Island Press.

Colyvan, M., and K. Steele. 2011. "Environmental Ethics and Decision Theory: Fellow Travellers or Bitter Enemies?" In *Philosophy of Ecology*, ed. K. deLaplante, B. Brown, and K. Peacock, 285–299. Amsterdam: Elsevier.

Convention on Biological Diversity, June 5, 1992. UN Doc. DPI/1307. 31 I.L.M. 818.

Gaston, K. 2011. "Biodiversity." In *Conservation Biology for All*, ed. N. Sodhi and P. Erlich, 27–44. Oxford: Oxford University Press.

Gross, K., B. Cardinale, J. Fox, A. Gonzalez, M. Loreau, W. Polley, P. Reich, and J. van Ruijven. 2014. "Species Richness and the Temporal Stability of Biomass Production: A New Analysis of Recent Biodiversity Experiments." *The American Naturalist* 183(1): 1–12.

Johnson, K., K. Vogt, H. Clark, O. Schmitz, and D. Vogt. 1996. "Biodiversity and the Productivity and Stability of Ecosystems." *Trends in Ecology and Evolution* 11(9): 372–377.

Justus, J. 2008. "Complexity, Diversity, and Stability." In *A Companion to the Philosophy of Biology*, ed. S. Sarkar and A. Plutynski, 321–350. Oxford: Blackwell.

Kareiva, P., and M. Marvier. 2012. "What is Conservation Science?" *BioScience* 62(11): 962–969.

Maclaurin, J., and K. Sterelny. 2008. *What is Biodiversity?* Chicago, IL: University of Chicago Press.

Maguire, L., and J. Justus. 2008. "Why Intrinsic Value is a Poor Basis for Conservation Decisions." *Bioscience* 58: 910–911.

Magurran, A. 2004. *Measuring Biological Diversity*. Oxford: Blackwell.

Margules, C., and S. Sarkar. 2007. *Systematic Conservation Planning*. Cambridge: Cambridge University Press.

Meinard, Y., S. Coq, and B. Schmid. 2014. "A Constructivist Approach Toward a General Definition of Biodiversity." *Ethics, Policy, & Environment* 17(1): 88–104.

Ruesink, J. 2003. "One Fish, Two Fish, Old Fish, New Fish: Which Invasions Matter?" In *The Importance of Species: Perspectives on Expendability and Triage*, ed. P. Kareiva and S. Levin, 161–178. Princeton, NJ: Princeton University Press.

Sankaran, M., and S. J. McNaughton. 1999. "Determinants of Biodiversity Regulate Compositional Stability of Communities." *Nature* 401: 691–693.

Santana, C. 2014. "Save the Planet: Eliminate Biodiversity." *Biology & Philosophy* 29: 761–780.

Sarkar, S. 2002. "Defining 'Biodiversity'; Assessing Biodiversity." *Monist* 85(1): 131–155.

Sarkar, S. 2012. *Environmental Philosophy: From Theory to Practice*. New York: John Wiley & Sons.

Sarkar, S. 2014. "Biodiversity and Systematic Conservation Planning for the Twenty-first Century: A Philosophical Perspective." *Conservation Science* 2(1): 1–11.

Sarkar, S., and C. Margules. 2002. "Operationalizing Biodiversity for Conservation Planning." *Journal of Biosciences* 27: 299–308.

Tilman, D., and J. A. Downing. 1994. "Biodiversity and Stability in Grasslands." *Nature* 367: 363–365.

7

"BIODIVERSITY" AND BIOLOGICAL DIVERSITIES

Consequences of pluralism between biology and policy

David M. Frank

Introduction: "biodiversity" between science and values

Coined in the 1980s as a portmanteau of "biological diversity" by life scientists reporting to policy-makers about anthropogenic loss of species and ecosystems in the twentieth century, the term "biodiversity" has since taken on positive connotations for many concerned about the fate of life on Earth. However, "biological diversity" as a theoretical term in the life sciences had existed at least since the 1950s, and human interest in life's variety is at least as old as biology. "Biodiversity" has become a term used widely by life scientists, conservation biologists, environmental philosophers, policy-makers, journalists, and activists. The conservation of biological diversity as such, as a more general objective distinct from the conservation of particular species, ecosystems, or landscape features, has become the stated goal of conservation biologists, many conservation organizations, as well as signatory nations to the 1992 Rio Summit's Convention on Biological Diversity.[1]

Against this background, this chapter will present the case for pluralism and contextualism about "biodiversity," where epistemic and non-epistemic values in particular local contexts constrain and guide the development of multiple legitimate definitions and operationalizations of "biodiversity." By *epistemic* goals and values, I mean those characteristically operative in the sciences and rational inquiry more generally, for example the attainment of truth or approximate truth, explanatory and predictive power, etc. (Steel 2010). *Non-epistemic* goals and values include conservation goals like protecting certain ecosystems as well as all other ethical and prudential values, natural values, etc. While both epistemic and non-epistemic values are relevant in scientific practice (consider the practical constraints faced by ecological field researchers, their broader motivations for pursuing particular projects, etc.), scientists of course emphasize the former in the development and testing of models and theories, whereas in public contexts and political discussion, non-epistemic values have more obvious salience.

I will argue that the multiple contexts in which the term "biodiversity" is used, and the epistemic and non-epistemic values at stake in these distinct contexts, should lead us to accept a deep pluralism about "biodiversity," in which only minimal logical constraints limit the concept across all contexts of use. An important consequence of this pluralism is what I will call the problem of *definitional risk*, or risks taken in precise definition or operationalization. These risks may involve epistemic or non-epistemic values, although here I will focus on non-epistemic values. Non-epistemic definitional risks arise when biodiversity is taken as a goal to be pursued, and the concept must be defined and operationalized for conservation prioritization. Risks arise because any particular definition or operationalization might not adequately capture all the goals and values that its users endorse. To take an oft-used example, biodiversity operationalized as species richness (the number of species in an area) may exclude important value judgments about the distinctness of particular species or the importance of rare species in conservation prioritization.[2] The many legitimate ways of defining and operationalizing the term, and its use in practical, decision-making contexts, mean that important judgments about goals and values need to be made in specifying the term's meaning. Which goals and values are relevant is of course a matter of context: we have little reason to believe (and reason to doubt) that ecological theory will employ exactly the same concept of "biodiversity" as practicing conservation biologists, or environmental policy-makers, or environmental ethicists, even if we have good reason to believe that "biodiversity" will never be used to denote, say, geological heterogeneity.

The chapter is organized as follows. In the second section, I survey controversies about broad definitions of "biodiversity," arguing that these controversies suggest that there is no such thing as biodiversity *in general*, only biological diversities as specified in particular contexts. I also lay out very basic constraints for defining "biodiversity" across contexts. In the third section, pluralism is further motivated, and the problem of definitional risk is laid out. In the fourth section, definitional risks are illustrated in the case of conservation planning by examining proposed axioms for diversity functions. In the fifth section I conclude by comparing my contextualist pluralism to other approaches and respond to potential objections.

Before I begin, a brief methodological caveat is necessary: I take the project of this chapter to be partly a descriptive one of conceptual analysis and partly a normative one of conceptual engineering. I hope to provide a clearer understanding of the multiplicity of uses of "biodiversity" across and within contexts, as well as a way to think about conceptual engineering in this case and more generally, by deploying the idea of definitional risk. Decisions about definition and operationalization of terms, especially those in wide use, have important consequences and thus raise risks for science and society.

"Biodiversity" across contexts

"Biodiversity" was coined in the mid-1980s by the botanist Walter Rosen, during the organization of the National Forum on BioDiversity, sponsored by both the National Academy of Sciences and the Smithsonian Institution. A few years later at the 1992 Rio Earth Summit, over 150 nations signed the Convention on Biological Diversity (CBD), which sought to bind nations to conserve biodiversity, which it defined broadly as:

> variability among living organisms from all sources including, inter alia, terrestrial, marine and other aquatic ecosystems and the ecological complexes of which they are part; this includes diversity within species, between species and of ecosystems.
>
> (CBD, Article 2)

The mid-1980s also saw the emergence of conservation biology as an organized field. In 1985, one of the founding documents of conservation biology was published, the article "What is Conservation Biology?" in an issue of *BioScience* devoted to "The Biological Diversity Crisis." Here Soulé included among several "normative postulates" of the discipline that "[d]iversity of organisms is good" (1985: 730). Soulé conceived of conservation biology as an applied, "crisis" field akin to medicine, using insights from multiple disciplines to conserve the diversity of life.

As can be seen from the CBD definition, at its most general, "biodiversity" has been defined as the variety of life at all taxonomic and functional levels of organization, or as the variety of genes, species, and ecosystems (Faith 2008). This broad definition is not limited to conservation and policy contexts. Many ecologists also employ the broad definition to frame their research programs (Loreau *et al.* 2002, Cardinale *et al.* 2012). There is also an established usage, especially in popular, policy, and environmentalist writing, wherein "biodiversity" refers extensionally to an area's biota, while emphasizing its variety, as in "Brazil's biodiversity."

There is significant controversy about these broad definitions that encompass life's variety at all scales. As Sarkar (2005) pointed out, if they entail that biodiversity refers to "all of biology," conservation of biodiversity is infeasible, due to unavoidable evolutionary change, ecological trade-offs wherein preserving one part of the biota necessarily comes at the expense of another, as well as limited resources for biological conservation. Ecologists have also objected to broad definitions on scientific grounds. For example, in the introduction to his monograph on his "unified neutral theory" of biodiversity and biogeography, the ecologist Hubbell (2001) laments what he calls "'biodiversity' in policy parlance," namely,

> the sum total of all biological variation from the gene level to single-species populations of microbes to elephants, and multispecies communities and ecosystems to landscape and global levels … In some usages it also includes all ecological interactions within and among scales of biological organization.
>
> (p. xi)

Hubbell goes on to say that many ecologists would rightly balk at the idea of a unified scientific theory of biodiversity, since at least for "'biodiversity' in policy parlance," developing such a widely applicable theory would be impossible. Thus Hubbell restricts "biodiversity" to the domain of what he calls "classical … ecology" (ibid.), attempting to develop theories that explain the distribution and abundance of macroscopic species in space and time. [The "classical" ecologist Whittaker (1960), for example, distinguished α-diversity, β-diversity, and γ-diversity, referring to species diversity within patches, between patches, and in total, respectively.] Hubbell's point is that "'biodiversity' in policy parlance" covers multiple spatial and temporal scales, for example, genetic heterogeneity within a population and species richness and composition across habitat patches. It does seem exceedingly unlikely that the properties, dynamics, or effects of such different biological diversities at multiple scales would fall under a single explanatory theory.

A related problem with broad definitions, discussed by Sarkar (2002, 2005), Norton (2006), and recently Maier (2012), is that multiple "dimensions" or biological diversities at different scales are incommensurable. This makes it seem impossible to develop an additive "index" of biodiversity to rank areas in terms of their overall diversity. Consider again comparing genetic diversity in two populations and species richness in their respective habitat patches. There is no natural metric of "overall biodiversity" to determine whether, say, an additional species in one area (more species diversity) and a certain loss in heterozygosity in that area (less genetic diversity) would make that area more, less, or equally biodiverse. However, this has not stopped the development of diversity indices. Even across disparate types of diversity such indices can in principle be constructed for conservation decision-making using multi-criteria analysis (Moffett and Sarkar 2006), although the value trade-off judgments necessary to construct such utility functions are difficult, even for biological experts. Since the early statistical treatment of Fisher *et al.* (1943), ecologists have been employing indices of species diversity, for example those based on Shannon's information entropy, that take into account both the number of species in an area and the species' "evenness" or relative abundance.[3] More recently, ecologists interested in the relationship between biological diversity and ecosystem properties have been developing indices of functional diversity, attempting to measure the diversity of traits in eco-systems (Schleuter *et al.* 2010). The incommensurability problem is "solved" in these cases either by convention, as in commonly used indices of species diversity, or by explicit value judgments, as in multi-criteria analysis. Such conventions and judgments are of course revisable and should be responsive to the epistemic and non-epistemic values at stake in these contexts. That there is no "natural" index taking into account multiple types of biodiversity does not mean that it is never useful to *construct* such an index, understanding its limitations.

I take these arguments against broad definitions of "biodiversity" to show that there is no useful general concept of biodiversity. Rather, "biodiversity" in the broad sense (Hubbell's "'bio-diversity' in policy parlance") refers to a plurality of biological diversities, some of which are val-ued in particular contexts, many of which are of interest to biologists. Indeed, the diversities that we see at multiple scales in the biological world are what the life sciences purport to explain (think of Darwin's "tangled bank" and "endless forms"), and there are longstanding research traditions showing that certain types of diversity, and their absence, can have important effects.[4] There have also been proposals that increasing diversity is a kind of "zero-force evolutionary law" of biology, as biological systems left alone will spontaneously differentiate through random mutation, copying errors, etc. (McShea and Brandon 2010). Whether or not this proposal should be accepted, evolutionary biologists have long understood diversity within populations to be the "raw material" of evolutionary change. In conservation contexts, since loss of any species or ecosystem constitutes a loss of diversity, concerns about species and ecosystem loss are eas-ily translated into concerns about declining biodiversity. "Biodiversity" in the broad sense thus groups biological diversities across scientific contexts (evolutionary biology, ecology, etc.) and broader social contexts (conservation biology, environmental policy, etc.) under a single heading, even while our interests in the diversities of life are many and multifaceted.

Thus the only constraints on the use of "biodiversity" across contexts seem to be that (1) the entities, systems, or processes in question are biological; and (2) there is at least implicit reference to variety or heterogeneity.[5] The first constraint rules out, say, geological or chemical heterogeneity from falling under the broad biodiversity concept, although there are borderline cases when one considers systems at the margins of life (Malaterre 2013). It is neutral between biological entities, systems, and processes, as heterogeneity within each of these categories might be important in different contexts. The second condition rules out using "biodiversity" to refer to particular biological entities when some concern for or emphasis on heterogeneity or vari-ety is not present. As I mentioned above, loss of a particular species *ipso facto* counts as a loss of overall species diversity. However, changes in the *composition* of biological communities, the par-ticular species that exist in those communities, may maintain overall species diversity, even while there may be consequences of such changes relevant to other biological diversities, for example genetic diversity or functional diversity. However, the broad extensional usage I mentioned above incorporates biological composition and does not fall outside these constraints: referring to an area's biodiversity denotes the biota there (its composition), while also emphasizing its variety. Similarly, the definition of Groves *et al.* (2002: 500) that includes three "components" of biodiversity, namely "composition, structure and function," explicitly includes composition, as does DeLong's (1996) "consensus definition" that includes the "identity" of species.

These minimal constraints rule out certain overly broad value-based definitions of biodi-versity such as "whatever we value in the biota" or "whatever we value in nature." The former is ruled out insofar as we value some organisms, species, ecosystems, etc. for reasons that have nothing to do with their variety or their contribution to certain kinds of variety. The latter is ruled out insofar as we also value non-living things in nature: all natural values cannot be incorporated into biodiversity. However, these constraints do not rule out value-laden definitions. Certain kinds of biological variety, for example, aesthetic varieties, do not play a role in biological science, even while they presumably supervene on biological differences. These value-laden definitions might be especially relevant to environmental ethics, environmental aesthetics, and conservation policy and planning, where non-epistemic values guide conceptual engineering. Consider Norton's (2006: 15) value-laden definition: "*biodiversity* should refer to those aspects of natural variety that are socially important enough to obligate protection of those aspects for future generations." This is not ruled out as a definition for policy, as long as "natural variety" does not go beyond the biological realm. (Whether it is an adequate definition for policy is another matter and beyond the scope of this chapter.)

These constraints are neutral between what Norton (2006) calls "inventory" and "difference" definitions of biodiversity, where the former involves sets of diverse biological entities and the latter focuses on those differences between entities. Indeed, as Maier (2012) and Maclaurin and Sterelny (2008) emphasize, these logically go hand in hand, since adding an additional entity to a set means it is not identical to the other members of the set (it is different in some way), and any set of differences can only be understood as differences *between* distinct things. Thus Maclaurin and Sterelny (2008) offered a general "units and differences" approach to understanding biodiversity across contexts in the life sciences and conservation biology, where the "units" may be at multiple scales and the "differences" specified in biological terms relevant to that scale and unit.

Norton presents what he calls a "strategic dilemma" (2006, 12) for defining "biodiversity" between the contexts of biological science and conservation policy:

> ... should we start with a scientifically accurate definition before identifying values that would accrue from systems that are diverse in the biologists' sense of the term, or should we ask what we value about biological diversity and then seek a definition that captures those values?

On the view I am presenting here, the "dilemma" is one of negotiating multiple epistemic and non-epistemic values within particular contexts, but there is no single correct resolution across contexts. In some contexts, a "biologist's sense of the term" should take precedence, since the goals and values, especially epistemic goals and values, of a particular sub-field of biology are most relevant. Evolutionary biologists or agroecologists interested in the causes and consequences of genetic diversity within a population will not define biodiversity in the same way as ecologists interested in the causes and consequences of species diversity within or between habitat patches. In policy or conservation decision-making contexts, non-epistemic values and goals like conserving valued species or ecosystems will take precedence, and biodiversity will be defined and operationalized according to these goals involving particular ecosystems or taxa. Similarly, definitions of "biodiversity" in environmental ethics may emphasize aspects of biological diversity that are particularly salient to conceptualizing our ethical obligations.

Biological diversities: contextualist pluralism

Biodiversity pluralism is the view that there are multiple, incompatible ways of defining "biodiversity" and measuring biological diversities within the minimal constraints given above. *Contextualism* is the view that the values relevant to assessing a definition of "biodiversity" or engineering a biodiversity concept within these constraints are given by the context of use. The relevant values in some contexts are mainly epistemic, as in the life sciences, and in other cases non-epistemic, as in environmental ethics, conservation planning, etc. To further motivate pluralist contextualism, consider the following "decision points" for more precisely defining and/or operationalizing "biodiversity":

1. *Spatio-temporal scale of interest*: diversity exists within and between genes, genomes, organisms (in morphologies and phenotypic traits), populations, assemblages or ecological communities, ecosystems, etc. Choice of temporal scale allows consideration of the dynamics of diversity or diversity of biological processes occurring over time.

2. *Biological taxonomy*: even within particular spatio-temporal scales there exist multiple concepts and strategies of classification, for example in taxonomy. Maclaurin and Sterelny (2008: 32–33) helpfully enumerate the multiple species concepts from the biology and philosophy of biology literature. Kitcher (1984) defends pluralism about species concepts by distinguishing historical from structural approaches, arguing that both are useful for generating certain kinds of interesting biological explanations. Dupré's (1993) more "promiscuous" pluralism goes further, allowing non-epistemic values (for example, culinary values) to influence biological classification. Contextualism about the epistemic and non-epistemic values relevant to assessing a definition of "biodiversity" can encompass Kitcher's taxonomic pluralism within biological science and Dupré's in broader social contexts.

3. *Diversity concepts and measures*: there are multiple distinct concepts and measures of variety, difference, and the diversity of a particular area. The technical surveys of biodiversity measurement found in Gaston (1996), Magurran (2004), and Magurran and McGill (2011) reveal a wide variety of measures and indices, including simple species counts (species richness), relative abundance or evenness metrics, measures of commonness and rarity, indices of compositional differences between areas (β-diversity), and measures of functional, trait, and phylogenetic disparity. I explain these concepts in more detail below.

4. *Mathematical operationalizations or formalizations of diversity concepts*: within these conceptual classes, many mathematical and statistical frameworks have been proposed. For example, genetic diversity is often measured using the so-called F-statistics, which measure expected heterozygosity, but there exist other measures, like Nei's D, which measures the genetic "distance" or divergence between populations or sub-populations. As mentioned above, Fisher *et al.* (1943) initiated research constructing statistical indices of ecological diversity. Their index related the number of individuals to the number of species in a community sample, while later indices developed by Simpson (1949) and others were constructed out of the frequency of types (e.g. species) in a community.[6] In a recent study of quantitative concepts of ecological diversity, which incorporate both the richness and relative abundance or evenness of species in an ecological community, Justus (2010) evaluated eleven distinct indices that take species richness and evenness into account. In their recent review of compositional similarity and β-diversity, Jost *et al.* (2011) list two incidence-based and eleven abundance-based similarity indices, which measure the similarity of two or more species assemblages. Similarly, Velland *et al.* (2011) review at least nine distinct indices of phylogenetic diversity.

By making decisions at each of these points it is possible to construct many distinct biodiversity concepts and measures. Some decisions may not be necessary for particular contexts; for example, in environmental ethics one might reasonably discuss the value of species diversity while remaining relatively open or pluralistic on taxonomic issues. In scientific contexts, of course, precise definition and operationalization are crucial for formulating testable hypotheses and generating data.

To further motivate pluralism, it is worth focusing more closely on the central concepts used by conservation biologists to define and operationalize diversity (decision points 3 and 4 above). As discussed above, conservation biology as a discipline was formed in the decade in which "biodiversity" was coined, and its early expositors took maintaining biodiversity to be its central goal. Sarkar (2002) went so far as to take the practices of conservation biologists to *implicitly* define "biodiversity," with special attention to the ways in which they prioritize areas for conservation management. Given a particular taxonomic or functional unit of analysis (e.g. species or ecosystem), conservation biologists may measure the biodiversity of a particular area by taking into account at least one or more of the following criteria:

1. *Richness*: the number of units. Other things being equal, an area with more units (e.g. more distinct species) is more diverse than an area with fewer units.
2. *Disparity*: the differences between the units. Imagine two areas with the same species richness, but where the first area has many species that are closely related phylogenetically, while the second area has many species that are more phylogenetically disparate. In certain contexts we may take the second area to be more biodiverse. Faith (1992) and others advocate phylogenetic measures of disparity. Measures used include disparity of DNA sequence and morphological disparity, especially within a clade where a local "morphospace" may be constructed (Raup 1966, Maclaurin and Sterelny 2008, ch. 4). Morphological and genetic disparity can come apart from each other and, in theory, from phylogenetic disparity, although genetic disparity is usually taken as a rough measure of phylogenetic disparity.
3. *Complementarity*: the number of new (distinct) units of an area relative to a background set. Other things being equal, an area with more distinct units adds more diversity to the total set than an area with fewer distinct units (Vane-Wright *et al.* 1991, Sarkar 2012a). Here related measures of the compositional difference between areas (β–diversity) are appropriate.
4. *Evenness*: uniformity of the relative abundance of units. Imagine two areas with the same number of species, but in one area a single species dominates the ecosystem. According to the measures used by ecologists, other things being equal, the area with a more *even* distribution (e.g. 40% species *A*, 30% species *B*, 30% species *C*) is more diverse than the area with a more skewed distribution (e.g. 90% species *A*, 9% species *B*, 1% species *C*).
5. *Rarity*: how rare the units are. Different kinds of rarity include abundance rarity (when there are few organisms of a species left), geographical or habitat rarity (endemics with limited or specific range), and temporal rarity (a biological event that only happens rarely). An area with rare or endemic species is more diverse than one without, other things being equal. The abundance rarity criterion is of course directly in conflict with the evenness criterion, since communities with more rare species will be by definition less even.

Combined with the decision points presented above, this list of criteria suggests that all biological diversities cannot be fully captured by a single measure, although single measures and indices are often convenient and useful in practice. As noted, some of the criteria above trade off as a logical matter, for example, evenness and abundance rarity. But the problem is not just a conceptual one, as there is also empirical evidence that measures of diversity can be non-concordant. For example, Hughes *et al.* (2002) show that there is low concordance between species richness and endemic fish and coral species in Indo-Pacific coral reefs. Another example is provided by the data set from Neige's (2003) study of the biogeography of Old World Sepiids or cuttlefish (Cephalopoda). Neige's data show no simple relationship between cuttlefish species richness and his constructed measure of morphological disparity: richness does not predict disparity.

I take the multiple conceptually and empirically incompatible ways of defining and operationalizing "biodiversity" given by the options above to establish a solid preliminary case for pluralism and at least shift the burden of proof to the anti-pluralist, for example, the eliminativist, an option canvassed below. Some of the possible ways of more carefully defining and operationalizing "biodiversity" given the plethora of options above will no doubt prove useless in some contexts. But it is the salient values in particular contexts that determine how we ought to evaluate a definition of "biodiversity." What works for conservation planning, for example, may be useless to an ecologist. Consider the systematic conservation planning framework, which defines biodiversity for prioritization by identifying "constituents" or components of biodiversity: favored alleles, organisms, populations, species, or communities whose existence and persistence across space and time may be tracked for the purposes of conservation and management, with an eye toward "maintaining … complexity" (Margules and Pressey 2000: 245).[7] It should not be controversial that the selection of biodiversity constituents in this sense depends on our broader goals and values: we would not target disease or pest organisms for conservation management, for example. While an ecologist may be interested in particular taxa due to the value or disvalue society places on them, a definition of "biodiversity" that referred to all valued taxa would not be scientifically salutary for the purposes of developing and testing ecological models.

An important consequence of this pluralism is that in any given context, the choices made in defining "biodiversity" raise risks for their users. I call these risks "definitional risks." Definitional risks arise for conservation planning when biodiversity is taken as a goal to be pursued, and the concept must be defined and operationalized for prioritization. In systematic conservation planning, for example, any particular set of constituents that are taken to define "biodiversity" might not adequately capture conservation priorities that decision-makers, on reflection, would actually endorse. This can be particularly problematic in contexts in which conservation biologists have different values than the people who live in areas under conservation management.[8] Definitional risks also arise in relatively "pure" scientific contexts, but the relevant values at stake are epistemic. Consider an ecologist who operationalizes "biodiversity" as species richness, while it is actually another kind of biological heterogeneity (for example, trait diversity), which crosscuts species richness, that has relevant effects on the variable the ecologist is interested in tracking. I discuss a case of definitional risk in more detail in the next section, focusing on the development of axioms for diversity functions that take sets of pairwise distance values and return a diversity value.

Consequences of pluralism: definitional risks in conservation optimization

Consider characterizing biodiversity formally for optimization purposes, as has been attempted by economists, beginning with Weitzman (1992). Gerber (2011), following Weitzman, takes the axioms below to be desiderata for diversity functions $V(.)$ that take sets Q of pairwise distance values $d(i, j)$, representing the difference between units i and j, and return a diversity value for the set. Gerber and Weitzman take the units to be species, where disparity or distance values $d(i, j)$ could be generated via DNA hybridization experiments, comparison of DNA sequences, comparison of location in a theoretical morphospace, etc. Here I informally present the most important axioms (leaving out only their continuity condition) using the more general term "units":

1. *Monotonicity in units*: when a new unit is added to the set, diversity should increase.
2. *Twin property*: if the added unit is identical to one already in the set, diversity should not increase.

3. *Monotonicity in distances*: when distances between the units increase, diversity should increase.
4. *Favor the most distantly related units*: diversity increases more when an added unit is more distant.

Acceptance of these plausible axioms entails definitional risk. The first axiom states that diversity should increase when a new unit is added to the set. We may want this axiom to fail our measure depending on the unit (e.g. species) under consideration: in an applied context, not all new species will lead to greater diversity. For example, if the added species is extremely common, or if we have some evidence that the added species may lead to a decline in the species diversity of the region over the long term (consider the introduction of domesticated cats or an invasive plant), our judgments may fail this axiom. If monotonicity in units fails, then monotonicity in distances will also fail, since the introduction of a new (distinct) unit i entails that there exists a j such that the distance between i and j is positive $(d(i,j) > 0)$.

The second axiom entails definitional risk insofar as it formally rules out taking relative abundance data into account, since it states that adding the occurrence of an already-occurring species should not change the overall diversity value of the set. Common measures of ecological diversity (for example, those based on Shannon's entropy measure) that take richness and relative abundance into account are maximized as evenness is maximized.

According to Axiom 4, disparity maximizes diversity. Gerber (2011: 2279–2280) notes that this axiom "suggests a value judgment pertaining to the optimization problem at hand." More than a suggestion, it *is* a value judgment when this measure is used in an applied context, namely the judgment that disparity is valuable. It is worth stressing again that disparity itself may be defined and measured in multiple ways, and the use of a particular disparity metric (e.g. phylogenetic disparity) will carry definitional risk, since it may or may not capture the property or properties decision-makers care about. In general, since these axioms apply to diversity functions over sets of pairwise differences, the use of other types of data (relative abundance or evenness data, or data on rarity, etc.) are formally ruled out from this type of analysis, entailing definitional risk when we have some valuation over variation in these properties.

Because the order that units are input into the function can change a set's overall diversity value, Weitzman's function that he proves satisfies these axioms is recursively defined as a simple additive function. While Weitzman's function is plausible, *any* function (linear, exponential, logarithmic, hyperbolic, etc.) that is monotonic, continuous, and increasing in the addition of species and differences will satisfy these axioms. Since new units added to a set will have positive difference values paired with units already in the set, richness is also taken into account, albeit indirectly. These various functions will represent different trade-offs between adding new units and the differences between that new unit and the set. Insofar as a function satisfying these axioms does not satisfy other properties we may desire, for example decreasing marginal diversity of additional units added to a set, its use will entail definitional risk.

Objections to contextualist pluralism and the eliminativist option

The pluralist, contextualist view presented here is contestable. I begin by dismissing what I take to be the least plausible objections from the point of view of reductionists or monists. A naïve genetic reductionist might argue that all biological diversities can be explained by genetic diversity. This genetic reductionist would hold that the differences between species can be explained by genetic differences, the differences between ecosystems can be explained by these differences in component species which in turn can be explained by genetic differences, and so on. This reductionist view can be dismissed on biological grounds due to phenotypic plasticity and properties of biological diversities that have abiotic or environmental causes. For example, types of phenotypic or morphological diversity may be studied in populations of genetically identical organisms exposed to different environments. Other forms of monism that focus narrowly on, say, species diversity, to the detriment of all other biological diversities, conflict with the practices of conservation biologists, ecologists, and other users of the concept. Even if taxonomic monism were true, for example if cladistics used the only legitimate species concept, other biological diversities (e.g. morphological diversity, genetic diversity within species, etc.) would be left out of such an account. I conclude that monistic approaches should be rejected on empirical and conceptual grounds.

Other versions of pluralism in the literature include Sarkar's (2002, 2012b) and Maclaurin and Sterelny's (2008). The difference between the view presented here and Sarkar's (2012b) is that while his focuses on the practices of conservation biologists, particularly conservation planners' choice of constituents, the view presented here attempts to take other contexts and uses into account. Thus of his four conditions on the selection of constituents (that they be biotic, that *variability* and *taxonomic spread* must be incorporated, and that concern "not be limited to material resource use," p. 116), I retain only the first two for constraints on "biodiversity" across contexts. While I agree that in most conservation planning contexts his latter two constraints are also plausible, there are some contexts even within biological conservation in which we might drop or relax these constraints. For example, consider focusing on maintaining genetic diversity within a particular species population, a kind of biodiversity conservation that is limited to a single taxon. In practice this might involve introducing or translocating individuals (e.g. wolves) to maintain genetic diversity in a metapopulation. This might be best understood as genetic biodiversity conservation in the service of species conservation. The last constraint would rule out, for example, measures of agricultural biodiversity (e.g. genetic diversity) where the focus is on increasing yields or yield potential. While this certainly does not fall within the field of conservation biology, I claim that genetic diversity even within resource exploitation contexts is still a kind of biodiversity.

Maclaurin and Sterelny offer a kind of pluralism that focuses on the life sciences and conservation biology, with a special role for species richness:

> We [do] … not find much reason to accept the idea that diversity is essentially captured by species and their phylogeny. But we shall see that a somewhat more modest view deserves to be taken seriously: that a phylogenetically informed species count is a good … indicator or surrogate for total biodiversity.
>
> (p. 7)

I accept their rejection of monism here. The problem with their "more modest" proposal is that, for reasons given above, it is doubtful that there is any such thing as "total biodiversity." If they mean that species richness is a good indicator of other biological diversities, this is an empirical claim that may no doubt hold true for *some* biological diversities, many of which they convincingly consider explicitly, but not all. Within conservation planning this is known as the surrogacy problem (Sarkar and Margules 2002), and there remain many open empirical questions about the relationship between different types of biological diversities.

Furthermore, their view is insufficiently pluralistic, as it is limited to certain scientific contexts. While I agree with them that "the gastronomic or medico-herbal biodiversity of a rainforest [does not have] the same status as an account of its species richness" (p. 8), this is only because the contexts of use for such distinct measures are so different, not because (cladistic) species richness has some independent special status.

Maclaurin and Sterelny might respond that the pluralist contextualist view is too permissive: it advocates a kind of "anything goes" methodological anarchism bordering on relativism. But this would be to misapprehend the importance of context. While "biodiversity" brings multiple kinds of biological diversities under a single concept (including, e.g. "medico-herbal biodiversity"), the relevant values at stake in the context of use constrain definitions of "biodiversity" for reasons that have to do with epistemic values in scientific contexts and other values elsewhere.

Finally, consider the challenge from the eliminativist (Santana 2014), who argues that the reasons for pluralism given above are really reasons to *eliminate* the concept from use, particularly in conservation science. One version of the argument may be put like this: if there is no such thing as biodiversity in general, and biological diversities are conceptually and empirically distinct, then there is no reason to retain a single overarching concept. A distinct argument focusing on non-epistemic values and conservation planning goes like this: all biological values (Santana calls this "ecological value") cannot be captured by concern for biological diversity, so it is misleading to use "biodiversity" as a stand-in for conservation value.

Against the first argument, Sarkar (2014) has argued that linguistic legislation on such a vast, multi-disciplinary scale given the concept's popularity is quite unlikely to succeed. Thus eliminativism may fail merely due to "ought implies can" considerations. But much progress has been made in conservation biology, ecology, and elsewhere in clarifying and operationalizing concepts of "biodiversity," and in understanding biological diversities. As long as researchers and practitioners are clear about their usage, eliminating "biodiversity" across fields, or even just in conservation biology, seems an unnecessarily radical change of conceptual practice.

Regarding the second argument, for the reasons I discuss in the second section, I agree that "biodiversity" should not stand in for "ecological value" or, more broadly, "everything we care about in the biological world." However, as long as some conservation values are captured by diversity concepts, those uses of diversity concepts tell against eliminativism. The broad use of "biodiversity" in conservation planning, which identifies "biodiversity" with the set of diverse constituents deemed worth conserving on a particular landscape is obviously capturing multiple values not related to diversity. For example, many aesthetic, economic, and ecological values are not related to diversity itself, but rather to the properties of particular species, communities, or ecosystems. (Similarly, conservationists sometimes focus resources on particular species populations for multiple reasons, including reasons that may not have much to do with that species, for example, concerns about habitat for other, perhaps less charismatic species, or concern with broader biodiversity, etc.) But as long as conservationists' use of "biodiversity" is also capturing important diversity-related values, and planners are clear about the many goals and values at stake, there is justification for retaining the concept there. Conservationists value life's diversities for many reasons, from the effects of diversity on ecosystem properties to the sheer wonder and aesthetic appreciation diversity inspires. Use of the term "biodiversity" emphasizes these values of diversity itself while also capturing other values. That its use as a goal of conservation might leave out other values is the problem of definitional risk, a consequence of pluralism that should be acknowledged. Since maintaining biodiversity is not the only goal of conservationists, other values need to be explicitly considered as well. Thus I agree with Santana that these other values should not be brought under the heading of "biodiversity," but rather under distinct evaluative criteria used in conservation decision-making. Thus conceptual confusion can be avoided while avoiding the costs associated with discarding "biodiversity."

Acknowledgments

Thanks to Sahotra Sarkar, Justin Garson, Dale Jamieson, Laura Franklin-Hall, and audiences at the University of Texas at Dallas, the Environmental Studies Department at New York University, and the Philosophy & Theory in Biology Symposium at CUNY-Lehman College for comments on versions of this chapter.

Notes

1 The first book to use the neologism "biodiversity" in its title was Wilson (1988). For an early sociological history of the idea of biodiversity, see Takacs (1996). Sarkar (2002, 2005) initiated discussion of biodiversity from a philosophical point of view, continued in, for example, Okansen (2004), Maclaurin and Sterelny (2008), and Maier (2012). On the use of biodiversity concepts in ecology, see Magurran (2004) and Magurran and McGill (2011). Sarkar (2007) connects the early discussions of ecological diversity with the emergence of "biodiversity" in conservation biology.

2 Similar problems may arise in defining and operationalizing "health" in medical contexts or "welfare" in economics. Since health and welfare may be taken as goals of medicine and prescriptive economics, their definition and measurement imply value judgments that carry risks of implicitly excluding important values.

3 For a discussion of the history of ecological diversity concepts and its relation to biodiversity, see Sarkar (2007). Justus (2010) discusses several types of indices that attempt to take richness and abundance into account.

4 See, for example, Cardinale *et al.* (2012) and Loreau *et al.* (2002). For critiques of this work, see Sarkar (2005) and Maier (2012).

5 These are closely related to the first two of Sarkar's (2012b) proposed four constraints on sets of biodiversity constituents for conservation biology. In the fifth section I explain why I reject Sarkar's final two constraints, which are that "taxonomic spread is important" and "concern should not be limited to material resource use" (Sarkar 2012b: 116).

6 For the beginnings of a detailed history, see Sarkar (2007).

7 Biodiversity *constituents* or "true surrogates" should be distinguished from *surrogates*, which are biotic or abiotic measures correlated with units of conservation concern. Margules and Sarkar (2007) originally used the terminology "true surrogate."

8 For these reasons, the systematic conservation planning framework explicitly emphasizes stakeholder engagement. See Sarkar and Illoldi-Rangel (2010).

References

Cardinale, B. J. *et al.* 2012. "Biodiversity Loss and Its Impact on Humanity." *Nature* 486: 59–67.

Convention on Biological Diversity, available online at www.cbd.int/convention/.

DeLong, D. C. 1996. "Defining Biodiversity." *Wildlife Society Bulletin* 24: 738–749.

Dupré, J. 1993. *The Disorder of Things: Metaphysical Foundations of the Disunity of Science*. Cambridge, MA: Harvard University Press.

Faith, D. 1992. "Conservation Evaluation and Phylogenetic Diversity." *Biological Conservation* 61: 1–10.

Faith, D. 2008. "Biodiversity." In *Stanford Encyclopedia of Philosophy*. Available online at http://plato.stanford.edu/entries/biodiversity/.

Fisher, R. A., A. S. Corbet, and C. B. Williams. 1943. "The Relation between the Number of Species and the Number of Individuals in a Random Sample of an Animal Population." *Journal of Animal Ecology* 12: 42–58.

Gaston, K. 1996. *Biodiversity: A Biology of Numbers and Difference*. Oxford: Blackwell.

Gerber, N. 2011. "Biodiversity Measures Based on Species-level Dissimilarities: A Methodology for Assessment." *Ecological Economics* 70: 2275–2281.

Groves, C. *et al.* 2002. "Planning for Biodiversity Conservation: Putting Conservation Science into Practice." *BioScience* 52(5): 499–512.

Hubbell, S. 2001. *The Unified Neutral Theory of Biodiversity and Biogeography*. Princeton, NJ: Princeton University Press.

Hughes, T. P., D. R. Bellwood, and S. R. Connolly. 2002. "Biodiversity Hotspots, Centres of Endemicity, and the Conservation of Coral Reefs." *Ecology Letters* 5(6): 775–784.

Jost, L., A. Chao, and R. L. Chazdon. 2011. "Compositional Similarity and β (Beta) Diversity." In *Biological Diversity: Frontiers in Measurement and Assessment*, ed. A. E. Magurran and B. J. McGill, 66–84. New York: Oxford University Press.

Justus, J. 2010. "A Case Study in Concept Determination: Ecological Diversity." In *Handbook of the Philosophy of Science. Volume 11, Philosophy of Ecology*, ed. K. deLaplante, B. Brown, and K. A. Peacock. Amsterdam: Elsevier.

Kitcher, P. 1984. "Species." *Philosophy of Science* 51(2): 308–333.

Loreau, M., S. Naeem, and P. Ichausti, eds. 2002. *Biodiversity and Ecosystem Functioning: Synthesis and Perspectives*. New York: Oxford University Press.

Maclaurin, J., and K. Sterelny. 2008. *What is Biodiversity?* Chicago, IL: University of Chicago Press.

Magurran, A. E. 2004. *Measuring Biological Diversity*. Oxford: Blackwell Science.

Magurran, A. E., and B. J. McGill. 2011. *Biological Diversity: Frontiers in Measurement and Assessment*. New York: Oxford University Press.

Maier, D. 2012. *What's So Good about Biodiversity?* New York: Springer.

Malaterre, C. 2013. "Microbial Diversity and the 'Lower Limit' Problem of Biodiversity." *Biology and Philosophy* 28(2): 219–239.

Margules, C., and R. Pressey. 2000. "Systematic Conservation Planning." *Nature* 405: 243–253.

Margules, C., and S. Sarkar. 2007. *Systematic Conservation Planning*. New York: Cambridge University Press.

McShea, D., and R. N. Brandon. 2010. *Biology's First Law: The Tendency for Diversity and Complexity to Increase in Evolutionary Systems*. Chicago, IL: University of Chicago Press.

Moffett, A., and S. Sarkar. 2006. "Incorporating Multiple Criteria into the Design of Conservation Area Networks: A Minireview with Recommendations." *Diversity and Distributions* 12: 125–137.

Neige, P. 2003. "Spatial Patterns of Disparity and Diversity of the Recent Cuttlefishes (Cephalopoda) across the Old World." *Journal of Biogeography* 30: 1125–1137.

Norton, B. 2006. "Toward a Policy-relevant Definition of Biodiversity." In *The Endangered Species Act at Thirty, Volume 2*, ed. J. Michael Scott, Dale D. Goble, and Frank W. Davis. Washington, DC: Island Press.

Okansen, M. 2004. "Biodiversity Considered Philosophically: An Introduction." In *Philosophy and Biodiversity*, ed. M. Okansen and J. Pietarinen, 1–26. New York: Cambridge University Press.

Raup, D. 1966. "Geometric Analysis of Shell Coiling: General Problems." *Journal of Paleontology* 40: 1178–1190.

Santana, C. 2014. "Save the Planet: Eliminate Biodiversity." *Biology and Philosophy* 29: 761–780.

Sarkar, S. 2002. "Defining 'Biodiversity'; Assessing Biodiversity." *The Monist* 85(1): 131–155.

Sarkar, S. 2005. *Biodiversity and Environmental Philosophy: An Introduction*. New York: Cambridge University Press.

Sarkar, S. 2007. "From Ecological Diversity to Biodiversity." In *The Cambridge Companion to the Philosophy of Biology*, ed. D. L. Hull and M. Ruse, 388–409. New York: Cambridge University Press.

Sarkar, S. 2012a. "Complementarity and the Selection of Nature Reserves: Algorithms and the Origins of Conservation Planning, 1980–1995." *Archive for the History of Exact Sciences* 66: 397–426.

Sarkar, S. 2012b. *Environmental Philosophy: From Theory to Practice*. Malden, MA: Wiley-Blackwell.

Sarkar, S. 2014. "Biodiversity and Systematic Conservation Planning for the Twenty-First Century: A Philosophical Perspective." *Conservation Science* 2(1): 1–11.

Sarkar, S., and C. Margules. 2002. "Operationalizing Biodiversity for Conservation Planning." *Journal of Biosciences* 27: 299–308.

Sarkar, S., and P. Illoldi-Rangel. 2010. "Systematic Conservation Planning: An Updated Protocol." *Natureza & Conservação* 8: 19–26.

Schleuter, D., M. Daufresne, F. Massol, and C. Argillier. 2010. "A User's Guide to Functional Diversity Indices." *Ecological Monographs* 80: 469–484.

Simpson, E. H. 1949. "Measurement of Diversity." *Nature* 163: 688.

Soulé, M. 1985. "What is Conservation Biology?" *BioScience* 35: 727–734.

Steel, D. 2010. "Epistemic Values and the Argument from Inductive Risk." *Philosophy of Science* 77(1): 14–34.

Takacs, D. 1996. *The Idea of Biodiversity: Philosophies of Paradise*. Baltimore, MD: Johns Hopkins University Press.

Vane-Wright, R. I., C. J. Humphries, and P. H. Williams. 1991. "What to Protect? Systematics and the Agony of Choice." *Biological Conservation* 55: 235–254.

Velland, M., W. K. Cornwell, K. Magnuson-Ford, and A. Mooers. 2011. "Measuring Phylogenetic Biodiversity." In *Biological Diversity: Frontiers in Measurement and Assessment*, ed. A. E. Magurran and B. J. McGill, 194–207. New York: Oxford University Press.

Weitzman, M. L. 1992. "On Diversity." *Quarterly Journal of Economics* 107(2): 363–406.

Whittaker, R. H. 1960. "Vegetation of the Siskiyou Mountains, Oregon and California." *Ecological Monographs* 30: 279–338.

Wilson, E. O., ed. 1988. *Biodiversity*. Washington, DC: National Academy Press.

8

ECOLOGICAL HIERARCHY AND BIODIVERSITY

Christopher Lean and Kim Sterelny

A *prima facie* challenge

Valentine's wonderfully rich though somewhat quirky *Origins of Phyla* includes a discussion of the cellular diversity of the various metazoan phyla, cells after all being the essential building blocks from which organisms are constructed. Table 8.1 lists his estimate of the cell-level diversity of some of the better-known phyla (Valentine 2004).

This table shows that (at least on one important conception of biodiversity) we would not measure diversity well by counting cell types; that would not, for example, capture the extraordinary exuberance of arthropod evolution. This poses no deep metaphysical mystery: different systems can be built out of a common set of basic elements, just as different sentences can be built from the same words. Development and evolution have exploited this combinatorial and structural freedom in building the incredible variety of arthropods. In general, when structured ensembles can be built from components with many degrees of freedom, and thus many different ensembles can be built from the same components, we do not track ensemble-level characteristics just by tracking the characteristics of the components.

This poses a potentially very serious problem for a theory of biodiversity. Biology appears to be hierarchically organized; with higher-level structures built out of lower-level constituents. On one very standard formulation of this idea, organisms comprise populations, populations comprise communities, and communities comprise ecosystems. Yet our tools for thinking about

Table 8.1 Cell diversity of metazoan clades

Phylum	Estimated basal cell diversity	Estimated crown cell diversity (if different)
Brachiopods	34	
Arthropods	37	90
Molluscs	37	60
Chordates	60	215
Echinoderms	40	
Annelid worms	40	

biodiversity typically focus on the lower-level components out of which biological systems are built: populations of species. If the relationship between ecological systems and the species (or, more exactly, the populations of species) from which they are built is like the relationship between morphological diversity and cell type diversity, in focusing on populations of species, we will fail to capture biologically important differences between different communities and ecosystems. Thus when we compare the biodiversity of one region to that of another (say, in making conservation decisions), our method might be seriously flawed were we to rely on one level of analysis, as it would if we regarded the brachiopod and arthropod clades as similarly diverse, by counting cell types.

Not all theories of biodiversity conceive of biodiversity as an objective and causally important quantity of biological systems. In a series of well-known and important publications, Sahotra Sarkar has defended a *constrained conventionalism* about the concept of biodiversity (Sarkar 2005, 2011). In his view, we cannot count just anything as biodiversity: any conception of biodiversity has to be measurable, comparable, and have something to do with the local biota. But within those broad limits, biodiversity measures are for the concerned parties to decide, reflecting their values, interests, preferences, and compromises. The contribution of ecological organization to biodiversity poses no special problem to this view of biodiversity. If local groups have a particular attachment to patchworks of burnt grassland – the result of a culturally salient and deeply valued foraging tradition – then counting patchwork structure would be part of the relevant biodiversity measure. If not, then we ought not count patchwork structure. However, following Wilson's original plea for conservation, Maclaurin and Sterelny aimed for something more ambitious: well-designed biodiversity measures should map onto an explanatorily important quantity of local biological systems (Wilson 1992, Maclaurin and Sterelny 2008). Differences in biodiversity should make a difference; in particular to stability, productivity, and ecosystem services of various kinds.[1] If we can characterize and measure such a quantity, and if increased biodiversity would contribute positively to ecological function, communities would have good prudential reasons to value higher levels of biodiversity, both to buffer their current access to critical resources, and to hedge their bets against future contingencies. These contingencies might include nature's unexpected surprises, but also changes in the values of resources that biological systems provide, for example, changes in the terms of trade between food and other commodities.

Such "causal relevance" accounts of biodiversity face serious challenges. The most obvious is empirical. The evidence we have for the causal relevance of biodiversity is patchy and weakly compelling at best. Experimental studies (even in the field) of, for example, biodiversity–stability relations are constrained by problems of scale. The temporal depth and spatial extent of experimental plots do not match the spatial and temporal scales at which these effects, if they are real, will act (Tilman *et al.* 2001, Tilman and Snell-Roode 2014). The evidential value of natural experiments is eroded by the usual worries about unconstrained variables, and the less usual worry that the supposed key variable – biodiversity – can normally be measured only via proxies of doubtful reliability. A second challenge is decision-theoretic. It is hard to convincingly crunch the numbers so that prudence recommends foregoing a current benefit for a supposedly greater future benefit. For such prudence requires rational confidence in access to future benefits, and regions in which biodiversity hotspots are under threat are typically also regions of socio-political instability, and agents rationally have little confidence in institutional commitments to just futures. Moreover, prudential investment in biodiversity requires confidence that local and regional initiatives will not be swamped by the negative impact of larger-scale processes, most obviously climate change and sea-level shifts. Biodiversity is another commons, another tragedy.

The problems of demonstrating the causal importance and prudential value of biodiversity are important and unsolved. However, our focus in this chapter is a third challenge, the interface between species richness and ecology. The main argument of this chapter is to recognize the complexity, power, and importance of ecological interactions between populations, but to argue that ecological assemblages[2] are typically quasi-systems. They have some enduring structural properties, but they do not interact or act as wholes. Thus we resist the idea that ecological organization is an independent vector of biodiversity, one that needs to be tracked in addition to species richness (or similar species-centric measure). That problem is challenging enough. The most systematic conceptual and theoretical work on biodiversity has been on the evolved components of biological systems: on populations and species. There is an array of sophisticated formal measures of the species richness of habitat patches; measures which combine information about the sheer number of species present with information about their abundance; information about the species profile of the focal patch, and the extent to which it contrasts with the profiles of neighbouring patches. In addition, there are measures of the phylogenetic distinctiveness of the species in the patch: ways of assessing the extent to which a focal species (say, Albert's lyrebird) has closely related species – in that patch; nearby; nowhere. Theory development has resulted in measures that combine information about the species richness of a local patch with information about their relative abundance and their phylogenetic distinctiveness. Table 8.2 describes some of the standard measurements of "biodiversity"; they are mostly focused on species, or the attributes of species. That said, empirical data on the ecological importance of increased diversity typically rely just on species richness.

However, these individual organisms, and the populations they compose, seem to be components of larger, relatively stable, relatively organized ecological systems. So, for example, in urging the importance of niche construction ideas for ecology, Baker and Odling-Smee write: "organisms and their environments are in reciprocal causal relationships capable of generating feedback

Table 8.2 Common "biodiversity" measures

Types of "biodiversity" measurement	Description	Examples of the methodology / measures
Functional diversity	The role that a population trait plays in maintaining an ecological system	Convex Hull measures Dendrogram measures
Trait diversity (phenetic diversity)	Morphological features	Same as functional diversity but unconstrained by describing a use for the trait
Phylogenetic diversity	Measures the differentiation of population lineages and quantifies over the branching pattern of life	Node-based measures, i.e. taxonomic distinctness Distance-based measures, i.e. phylogenetic distinctness
Genetic diversity	The identification of alleles and their abundance in populations	Genetic barcoding
Ecosystem function	Local nutrient retention	Onsite monitoring of nutrients and GIS studies
Species diversity	Combines species richness with other variables, usually relative abundance	Shannon Evenness Indices Simpson Evenness Indices

effects; that organisms figure as agents of change rather than merely as passive objects of selection; and that *organisms and their local environments must be considered as integrated systems that evolve together*"[3] (Barker and Odling-Smee 2014: 201). A crucial question is whether these ecological assemblages really are such systems, in some rich sense of "system". For if they were, that would suggest that we should incorporate their system-level properties into our measures of biodiversity.

A paradigm of an organized system is a mechanism, for with a mechanism, we cannot explain the behaviour of the system as a whole just through information about its components (and their numbers). The packing slip telling you what is in the box does not double as an explanation of machine function. We need to understand the spatial organization of the components, and the specific interactions scaffolded by that organization. Designed mechanisms are not ensembles of autonomous individuals, but of organized, interconnected, and often quite distinct components. Typically, the behaviour of the mechanism depends on the presence and placement of all or most of these components. These systems behave predictably, for the components are reliable (given the stresses they are typically under) and so are the connections between the components. Systems built by natural selection have many of these characteristics too. Populations are quite different from mechanisms, in that they are typically composed from a large number of individuals but with relatively few types, and they are not organized: what happens to the population rarely depends on the precise identity and placement of specific individuals. In deciding whether you are likely to be bogged while driving across a sand dune, the precise location and identity of any specific grain of sand is rarely salient. Population behaviour is often predictable, but only because population-level effects are aggregate outcomes of individual operations, no one of which matters. So populations are not very system-like (Godfrey-Smith 2009: 147–150). Are ecological associations more like heaps of sand, or more like a village, with its division of labour, specialization, and mutual dependence?

Our example of a village is no accident: it reflects Elton's original model of a community organized through a set of complementary biological roles (Elton 1927). Elton's niche concept has been superseded, but ecologists still study the structure of these compositional systems: their food webs and other aspects of their trophic structure. Ecosystem ecologists investigate the ways material flows through these systems; for example, the ways detritivores recycle crucial nutrients back into the soil. Ecologists and natural historians track both the relative stability of the species composition of these local systems, and the predictable changes in that composition in response to major disturbance. More recently, the role of ecological engineering in these local systems has come into focus: the ways populations modify not just their own physical and biological environment but that of other organisms there too. Thus a particular stand of eucalypts as they grow will affect the soil chemistry, moderate the effects of storms by acting as windbreaks and as (very leaky) umbrellas; provide numerous nesting cavities and retreats as hollows form; provide food for honeyeaters and other pollinators; more reluctantly, food for appropriately specialized herbivores as well; shelter for spiders and the like under their bark, both on and off the tree. In addition, and depending on the species, they make the site more fire-prone. This recent turn is especially relevant to the idea that the contribution of community organization to ecological processes does not reduce to the contribution of its member populations, for if niche construction effects are important, history is important too, for these effects accumulate and ramify over time (Jones *et al.* 1997, Pearce 2011, Barker *et al.* 2014). History becomes even more important if, as is quite plausible, the accumulation of change is path-dependent. It may well matter which species establishes first: tree species have different profiles as ecosystems engineers, and once a stand establishes, its members can be present for a very long time. One of us has in the past defended this view of local communities, arguing that the stability of species associations, and of their relative abundance, is hard to explain unless they collectively organize and filter their local patch. For physical features of habitat patches vary quite significantly from year to year, yet local species lists often remain current for decades (though obviously with some changes on the margins), with common species remaining common, and rare species remaining rare (Sterelny 2006).

Phenomenologically, then, these interconnected, interacting, co-located networks of populations seem to be real systems. They seem to be (in Rob Cummins's sense) functionally organized, hierarchically composed local systems. There is no doubt that there has been a strong tradition within population and community ecology of treating these co-located interacting collectives as real systems (Cooper 2003), supposing that these interactions in a local patch constrain one another's abundance, impose order effects on the formation of new communities after major disturbance, and filter potential immigrants (Agrawal *et al.* 2007). Ricklefs calls this the assumption of "local determinism" in community ecology (Ricklefs 2004, 2005, 2006, 2008). If they are real, these system-level properties, with stabilized associations between local populations, are plausibly relevant to the ecosystem services that supposedly make biodiversity management prudentially important. For example, if local species composition is determined by these local interactions, that will determine the extent to which, say, pollination is buffered by redundancy. If these local interactions permit a rich guild of pollinators to be present, pollination will be buffered against chance fluctuations in the number of any specific pollinator. The community, partially co-constructing one another's niche, stabilizes the system in the face of disturbance, excludes many potential exotic invaders; in general, it increases the robustness and predictability of the local ecological dynamics. Conservation decisions, one might suggest, should reflect the value of these stabilized associations, especially to the extent that such decisions involve trading patches of the conservation estate for land to be restored after exploitation.

A sceptical response

It is arguable, however, that this appearance of genuinely organized and structured ecological systems is an illusion. Angela Potochnik and Brian McGill have recently argued against the standard version of the view that ecological interaction is organized into real systems; a stratified conception of nested structure in ecology, with organisms comprising populations, populations comprising communities, and communities comprising ecosystems. They do not think that ecological interactions conform to the model of a system organised into discrete and well-defined levels (Potochnik and McGill 2012). They argue against the view that ecological interactions are organized into hierarchical structures on the following grounds:

- *Metaphysical significance*: compositional relationships are not always in the form of one level being built exclusively from elements at the next level down. A termite mound, for example, is composed from non-living but organized matrix, termites, plant materials, fungi, bacterial colonies, and no doubt assorted fellow travellers and parasites. Likewise, properties can have a complex, multi-level structural basis. The camouflage of a nest, for example, is not only structurally complex (with the outer layer often sourced from many places); its being camouflaged depends on its placement, and on the perceptual profile of the nest predators.
- *Explanation and evidence*: metaphysical supervenience relations do not indicate a direction of explanation. Higher-level theories can be explanatory without direct reference to their lower-level constituents. For example, the principles of island biogeography do not depend on the specific taxa on the islands.

- *Causal*: objects can play a causal role at more than one level; they can "cut" between "levels" (Guttman 1976). For example, waste molecules can act directly on organisms, making one organism move away from a particularly smelly deposit, serving as a signal to another, while at the same time playing a causal role in the ecosystem, moving nitrogen through the system.

We think that Potochnik and McGill interpret the idea of hierarchical organization in too simple and rigid a way (though it is true that ecologists sometimes write in ways that encourage this interpretation of their views). Consider an uncontroversial example of a hierarchical, nested organization: the morphology of a metazoan, with cells organized into tissues; tissues into organs; organs into organ systems (like a mammal circulatory system). No-one supposes organs interact only with other organs. The lungs interact with gasses and particles direct from the atmosphere; with blood; with hormones and other signalling molecules; with muscles and nerves. There is plenty of cross-cutting causal interaction in a metazoan body, despite the fact that it is clearly a hierarchically organized system. Likewise, facts about that system as a whole often explain features of its components. The mass of an elephant explains the size and strength of much of its skeleton. As Potochnik and McGill note, the same is true of ecological interactions. Echidnas interact with termite colonies rather than individual termites. Springtails and other tiny arthropods in the leaf litter interact with bacterial colonies and with biofilms rather than individual bacteria. Likewise, in many cases, it is probably best to conceptualize phytophageous insects as interacting with a system that includes the tree and its associated symbiotic fungi, rather than with the tree alone. These causal interactions cut from organism to population and organism to community and yet are typical of what ecologists study. The importance of these cross-species associations, both for the partners themselves and for third-party interactions, is among the phenomena that make the local community perspective plausible. It is certainly no reason to reject the view that local communities are real, hierarchically organized systems. That rejection, as the comparison with morphology shows, seems to depend on saddling the classic, nested hierarchy conception of ecological organization with extraneous metaphysical and causal commitments.

The reasons for scepticism about the standard conception of ecological organization are empirical rather than metaphysical. There has long been an individualist, "Gleasonian" voice in ecology that has regarded communities as no more than ephemeral associations of organisms that happen, for now, to tolerate a similar range of conditions. The distribution and abundance of organisms is essentially controlled by large-scale environmental factors: moisture, temperature, seasonality, and the like. A tree cares how much it rains, perhaps how far away the next favourable patch is, but not about the specific identity of its next-door neighbour. This view of ecology regards historical evidence of the existence of very different associations revealed as the glaciers retreat and advance, as decisive evidence that so-called communities are merely unstructured multi-species associations. Colinvaux (2007) is part travelogue, part triumphalist assertion of this argument. If this individualist view of ecological pseudo-organization is right, then the apparently structured, compositional, and hierarchical organization of ecological systems poses no special extra problem for the project of giving a realist account of biodiversity: system-level behaviour will be some form of a relatively simple statistical reflection of the properties and the numbers of the components. But the individualist perspective is at best controversial, as we shall see, even among those who reject the idea that community organization is under local control.

Ricklefs has long rejected the view that local communities are genuine biological systems, but not from a Gleasonian perspective. He argues, first, that in the typical case, local communities are not composed from genuine biological populations. We both live and work in Canberra, near a bush reserve, Black Mountain. Black Mountain has a healthy population of brushtail possums. But the Black Mountain brushtails are not a population; they are an arbitrary and transient segment of a population, for there is a continuous population of brushtails that includes Black Mountain, the nearby O'Connor Ridge, and most of the suburban gardens of inner North Canberra. That population is real, for it has semi-permeable boundaries formed by a lake, dense urban infrastructure, and farmed open grassland. So the North Canberra brushtails influence one another's fate, in ways they do not influence other brushtails. A virus, for example, could spread through this group without affecting others. But this genuine demographic unit is not nested in Black Mountain; it is not a component of a Black Mountain community. Second, the demographic units are not typically spatially congruent. There are echidnas on Black Mountain too. But the North Canberra echidna population is not congruent with the brushtails. Echidnas do not mind open grassland; they are the most broadly distributed of the Australian native mammals. But they do not penetrate suburbs gardens with the ease of a brushtail. This line of argument – the fact that populations are not congruent, and hence there is no local system into which they can all be nested as components – can be repeatedly recycled for other species.

In one of ecology's landmark publications, in 2001 Stephen Hubbell proposed a neutral model of local diversity and distribution, denying that the composition of tropical forests was structured by local competition, or by other fine-grained selective forces (the theory was general, but Hubbell's empirical research was on these forests; Hubbell 2001). Ricklefs shares Hubbell's intuition that tropical forests have the wrong composition for local forest communities to be at equilibrium as the result of interspecies interactions: they have too many species, and too few exemplars of any one species (Ricklefs 2005: 595–597). A hectare of tropical forest sometimes supports 300 or so species of tree, but often with only one or a few individuals per species, and almost never with patches of single-species stands. So it is extremely implausible to suppose that its diversity and richness is the effect of niche differentiation and interspecific competition, with each tree finding its way to the five square metres where it is competitively superior to 299 rivals for that same spot. This intuition matters. If the most rich and diverse biological assemblages on Earth are not structured by local interactions, then, at the very best, the local community concept has very limited application. Ricklefs concludes that the Black Mountain community and similar ensembles are not structured out of component populations. Such communities are merely interaction zones, spaces where many distributions overlap.

In our introductory section, we suggested that a causal-relevance conception of biodiversity might have to be a two-factor model, with one factor focusing on genealogical units, reflecting the fact that species play roles (often different ones[4]) in many different ecological systems. The other derives from ecological organization, reflecting the fact – if it is a fact – that these are organized, enduring systems, with causally important properties that are not simple reflections of the properties of the genealogical units from which they are composed (Hutchinson 1965, Hull 1989, ch. 7). The Gleasonian suggestion, though, is that ecological interaction is not organized into systems at all, so the "second factor" disappears. That is not quite Ricklefs's view, as we shall see, despite his scepticism about local determinism. Rather, his positive suggestion is to increase the spatial scale of our analysis. We should think of regional systems – landscapes – as our bounded and organized ecological units. At this point, we need to introduce some conceptual machinery from Bill Wimsatt.

Flies, stones, and territories

In his "Complexity and Organisation", Bill Wimsatt compares a granite pebble and a fly to distinguish between two different forms of compositional organization. Flies and granite pebbles are uncontroversial examples of real structures; they are discrete, bounded, can move independently of other objects; they have important collective physical properties (and in the case of the fly, biological properties too). But the fly is complexly organized in a way the granite cobble is not, for that cobble has a simple and privileged internal organization. To a first approximation, whatever drives our scientific interest in the granite and its composition – its crystal structure, chemical composition, mass distribution, electrical and thermal conductivity – we will decompose it into parts in the same places. Its crystal organization, chemical organization, variance in mass, in electrical and thermal conductivity vary with one another. The boundary where one crystal gives way to another is also a boundary where tensile strength or thermal conductivity changes too. That is not true of the fly; a map of its cell types will look very different from a map of its anatomical parts, which in turn looks different from a map of the circulation of fluids or of its gas exchange with its environment (Wimsatt 2007: 183). Each of the maps is robust. We can, for example, investigate cell types through a number of different experimental techniques: light and electron microscopy; different staining techniques to reveal cell structures. Robustness is important: when multiple streams of evidence reveal the same structures in the same places, we can be much more confident that we have identified real features of the world (Wimsatt 1981, Hacking 1983, Calcott 2011). Though each of the maps are real, they are not congruent.

As Wimsatt sees it, multiple decomposition reflects an objective feature of the world, and thus an inescapable feature of scientific practice. Different sub-disciplines describe their target explananda through their local theoretical perspectives, and these guide the identification of systems and their salient parts. Two different perspectives will result in different profiles of the parts of a system, and as we have seen in considering the fly, these need not be congruent. Generalizing from the fly, Wimsatt thinks of "multiple decomposition" as a process in which different theoretical perspectives are overlayed onto the system under investigation. This provides information on the system's complexity and on the commensurability of differing perspectives. Wimsatt describes multiple decomposition as follows:

(1) Systems can be understood given different theoretical perspectives.
(2) Different theoretical perspectives give different characterizations of the parts of the system. That is, they use different criteria, and different empirical techniques, to identify the parts of the system, and the boundaries of those parts.
(3) Once two different perspectives of the one system have been developed, we can attempt to spatially align the parts identified via one decomposition with those identified through other decompositions. In the case of the granite cobble they align quite well. Not so, the fly.

In Wimsatt's terminology, the granite cobble is descriptively simple, because its parts are spatially coincident over different perspectives. If not, as with the fly, the system is descriptively complex. Wimsatt's conceptual machinery helps us see the limits of the Potochnik-McGill critique of hierarchy in ecology: their tacit model is of a descriptively simple system of hierarchical organization. Ecological systems are not descriptively simple; the components specified from one perspective (say, locating the different guilds in the system) do not match up with those from another (say, modelling the key factors in response to fire). But flies have genuine compositional organization, even though they are not descriptively simple.

We read Ricklefs's suggestions that ecological stability depends on regional rather than local processes through the Wimsattian lens: landscapes (or territories) are real, but descriptively complex, hierarchically organized ecological systems. Landscapes are descriptively complex, first, because as we noted above, demographically connected local populations rarely have congruent populations in a territory. So, for example, the Atlas of Living Australia maintains an

Figure 8.1 Kangaroo Island species distribution map. This map shows little pygmy possum (black) and western pygmy possum (grey) sightings.

Figure 8.2 Kangaroo Island species distribution map. This map shows coast ground berry (black) and wiry ground berry (grey) sightings.

online database of species records. If one looks, say, at Kangaroo Island,[5] and checks the records of the little pygmy possum, we see that the records are clustered heavily at the western end. By contrast, the western pygmy possum is clustered at both ends (with a few records in the middle); it is reasonably congruent with the southern brown bandicoot. To shift from animals to plants, the coast ground berry is heavily clustered on the south coast of the island; the wiry ground berry is more evenly spread, but heavily clustered towards the eastern end.[6]

Community ecologists and population ecologists are often interested in explaining the distribution and abundance of specific species, especially when these are vulnerable, and those focusing on different populations will decompose Kangaroo Island into different interacting components. The little pygmy possum is "near-threatened", so a possum ecologist would need to identify the distribution of this species, and those other species with which it had important interactions (predators, host trees with hollows where it can shelter and nest; competition for those hollows; food sources). But she could probably afford to ignore the echidna distribution. The same is true when we consider the orthogonal explanatory agendas of community and ecosystem ecology. Ecosystem ecologists are primarily interested in explaining the cycling of materials through their target systems, and so the physical geography of a landscape is central to their explanatory projects; different aspects of that geography, for different materials. For example, in understanding the flow of water through the system, relief is very important, so capturing the fact that the western end is much hillier than the east is critical. In considering nutrient flows, the base geology, the ground cover, and the direction of the prevailing winds will all matter. Notice that these decompositional descriptions are all robust: there are many techniques for censusing population distributions; for identifying and assessing ecologically relevant physical features of an environment; and for measuring the flow of materials through a system. In brief, there are multiple ecological perspectives on Kangaroo Island, and the components the different perspectives identify will often not be spatially congruent.

We also think that there is a quite persuasive case for thinking of landscapes or regions as objective features of the biological world, structuring ecological interactions. There are four considerations that favour taking landscapes seriously. First, they are bounded: the edges of landscapes or territories are defined by physical boundaries or by physical gradients which reach thresholds [of salt levels; night temperatures (frost or snow), aridity] which influence the movement or viability of many species of organisms. Obviously, these boundaries are not absolute: some plants are salt-tolerant; some animals can do without surface water. But the skin is not an absolute boundary either. Humans (like most animals) harbour huge populations of microorganisms, and some migrate in and out despite that barrier. So these territories are the arenas in which demographically real units – demographically connected populations – interact with one another in zones of overlap, and with the abiotic environment.

Second, landscapes are the spatial scale at which ecological and evolutionary processes connect. One problem with the focus on local communities is that it makes it difficult to see how to integrate ecological and evolutionary thinking (Sterelny 2001). Evolutionary change takes place in populations and in ensembles of populations. Local communities and the interactions therein – our Black Mountain – are too spatially localized to be of much evolutionary significance. Obviously, an important mutation might occur through a Black Mountain reproductive episode, but in the typical case the new variant cannot go to equilibrium on Black Mountain, if as Ricklefs argues, the Black Mountain animals are an arbitrary and ephemeral fragment of a population. Likewise, local communities are often too short-lived to generate significant evolutionary change; grasslands turn into forest or bake to clay; ponds dry out; silt up. The shift to landscapes takes us to the right temporal and spatial scale to link ecology and evolution. The evolutionary mechanisms that build diversity seem mostly to operate on a regional scale; the more boundaries filter movement, the more free populations are to diverge from their siblings. Ecological change, both fast and slow, takes place on all spatial scales, from the very local to the global. But disturbances – a major storm system, for example – will often have region-wide effects, and the same is true of slower environmental changes. So if we take regions or landscapes to be the most salient level of ecological organization, its scale matches the spatial and demographic scale of microevolutionary change. These evolutionary responses include responses to the other populations in the landscape, and as John Thompson has shown, these coevolutionary responses can be marked, even when the populations in question only overlap, and even when the interactions are between multiple populations. This will be the typical situation in territories, as Ricklefs repeatedly notes. Coevolution does not require congruent, tightly coupled populations (Thompson 1994, 2005).

Third, Ricklefs argues that this regional turn enables us to capture the genuine insights derived from thinking about local communities. As Ricklefs sees it, the local community paradigm is committed to making two strong predictions: (i) community richness correlates with the physical heterogeneity and productivity of the local patch; (ii) local richness is independent of regional richness: physically similar local patches embedded in different biological regions should have similar levels of diversity. While he thinks there is a reasonable case to be made for the first of these predictions, the second fails. European plant communities, for example, are impoverished (in tree species) compared to East Asian ones, and that is because the regional diversity of European trees has not recovered from glacial extinctions.[7] But he is open to the possibility that local interactions filter regional diversity. Regional diversity presents a list of potential community members to local patches, and these are filtered by habitat selection (tolerance for physical conditions, as they are originally, and as they become modified by niche construction effects); competitive interactions; mutualisms; the effects of predation and disease; and of course chance (Ricklefs 2005). In principle, local diversity might be very strongly shaped by these local interactions, but they are interactions between population fragments whose presence and abundance is explained by events at larger spatial scales and longer temporal scales. There are echidnas on the Australian National University campus, but that might well be a consequence of source–sink dynamics; an overflow from the echidnas of Black Mountain.

Finally, it is worth mentioning that hypothesis formation and testing on regional scales is more tractable than it once was. The recent development of Geographical Information Systems (GIS) has rapidly increased the ability of scientists to test hypotheses on these larger scales. There are, however, important limits to this. While GIS provides precise detail on co-variation in population distributions it does not directly represent the local causal interaction of the individuals that constitute these population distributions (Kozak *et al.* 2008, González-Orozco *et al.* 2011).

Seeing the fresco in the ecological mosaic

In the previous sections, we developed a case for taking the apparently organized, structured character of ecological associations seriously, but on spatial scales of landscapes rather than as local communities. The broad-brush stability of local habitat patches genuinely needs explanation. But we also saw that there were powerful objections to seeing local patches as organized systems. In contrast, there is a persuasive case to be made for taking seriously regional organization. However, that case has two limitations. First, there may well be large stretches of continental plains which are not, in the relevant sense, regionalized. It is an open empirical question whether populations are always, or typically, in bounded territories. If we consider large continental expanses without major physical barriers – for example, the western slopes and plains running west from Australia's Great Dividing Range – it is conceivable that populations reach the limits of their physical tolerances in ways that are not at all coordinated. The less heat and arid-adapted populations drop off, and the more desert-adapted organisms drop in, but there is no zone where the less hardy hit the wall more or less together. Phenomenologically, that does not look plausible: Australian natural historians write of red and yellow box woodlands; the mallee belt; Mitchell grass country; saltbush–spinifex plains as if these named large stretches of country with a fairly stable and predictable character. But for most species, historical distribution data are patchy. We simply do not know. Populations nested in Kangaroo Island-like bounded territories may be more the exception than the rule.

Second, even if regions do have the structural and organizational features we have noted, in other important respects they are not system-like. Unlike flies and rocks, they do not interact with their environment, including other flies and rocks, as a single integrated entity. Our granite cobble, swept up in a flood, bumps and bangs into other rocks, bits of wood, and the like, and its global properties determine the effects of these collisions. We see no case for thinking that territories or regions interact with other territories or regions as a single system. Nor do local communities.

So we suggest that ecological assemblages – perhaps local communities, perhaps spatially larger, bounded territories – are somewhat system-like. They are not mere aggregates, like a heap of sand. Within a territory, there are many populations, and their specific character and their spatial locations both matter to the overall ecological and evolutionary dynamics of a region. But at least in most cases, they are not tightly integrated and interdependent; despite Vermeij's metaphor (Vermeij 2009), the communities interacting in a region are not like a modern economy; the connections are much looser. Very likely, most species on Kangaroo Island would not notice if our possums and berries were to vanish. They are quasi-systems. As a consequence, our best bet is that a ground-level, species richness-based account of biodiversity is all we need (if indeed we can get even that).[8] Ecological organization is not machine-like enough for us to need to count machine types as well as the parts from which they are made. Moreover, conservation triage decisions almost invariably involve comparisons within landscapes or territories, not between landscapes or territories; which parts of Kangaroo Island should be in the conservation estate, not whether Kangaroo Island as a whole is more biodiverse than (say) Groote Eylandt, let alone more biodiverse in virtue of its landscape-level properties.

This chapter has explored one of the challenges to an ambitious, realist concept of biodiversity: a line of thought that suggests that such a view needs to develop ways of conceptualizing the differences between ecological systems, and ways of testing for their causal importance. We have argued that while the realist project has plenty of problems, that is not one of them. The Convention on Biological Diversity (CBD) defined biodiversity as: "Diversity between species, within species and of ecosystems." We suggest dropping the "of ecosystems". Biodiversity should be defined in terms of biological taxonomy, though we have not addressed the specifics of that project in this chapter. The specifics matter: for example, the quoll and the feral cat are both meso-predators of the Australian bushland. In our view, though not one we have defended in this chapter, the quolls's phylogenetic distinctiveness gives us *prima facie* reasons to privilege it in conservation decisions. The domestic cat has a global distribution, is closely related to many felids, and is strongly suspected of being implicated in the defence of many small Australian endemics. No similar charges are made against the quoll.

In this chapter, we have used conceptual machinery developed by Bill Wimsatt to identify the targets of conservation decisions; to zero-in on how to set conservation priorities, and to argue that while ecological aggregates have some causally important structure and organization, they are marginal rather than paradigm cases of organized systems. We shall end by discussing the consequences of this view of ecological hierarchy for conservation biology, and in particular, the unresolved tension between local and regional perspectives. The project of conservation biology is to stabilize important aspects of our biota, but if we accept Ricklefs's line of argument as developed it in the third section, to the extent to which there are equilibrium processes in ecological systems, these seem mostly to be on regional rather than local scales. Conservation biologists need to think regionally, in part because stability seems more regional than local. For example, since the introduction of cane toads, Australian snake species have increased in body size (making toad poison less likely to be fatal), but the head gape size has reduced (making it less likely that they will eat big poisonous toads) (Phillips and Shine 2004). There can be little doubt about the form of this interaction: it is a stable, aggregate outcome of probably quite varied and fluctuating interactions across many local patches.

However, this regional perspective has to connect to more localized and taxon-specific descriptions of the causal interactions which drive change in local populations (or population fragments). For these are the typical sites of conservation interventions. These taxon-specific and local phenomena include the genetic diversity of local populations (for example, whether inbreeding depression is a threat), as well as their size, spatial distribution, age structure, gender balance. Yet local patches and the populations they support are not inherently stable. The temporal beta-diversity in local areas often appears to be extremely high.[9] If this is indeed typical, the species composition in local communities rapidly changes. In one study of 100 biomes across Earth, 75 percent of these systems had at least one in ten species disappear locally per decade (Dornelas *et al*. 2014). This is often coupled with little change in regional diversity, which is more stable. Populations simply shift their distribution across the larger landscape (Thuiller *et al*. 2007). As populations change in their local abundance, so the interactions between them also change. When the populations of two different species overlap, both the strength and the type of interaction can vary over their shared range, depending on their relative abundance and the local abiotic factors (Poisot *et al*. 2015).

Local communities are the stuff out of which landscapes or territories are composed. But if the studies we have just cited are typical of the behaviour of local communities, these highly local interactions and population fragments are often ephemeral. In many cases, there is no sense in which a stable set of population, regulated around an equilibrium number, is their natural state. That in turn implies that there is a problem in treating local communities and their boundaries as the right area for preserving species. As a consequence of local patch dynamism, conservation of species involves not just a focus on where the population is currently found but where the population can be locally sustained. Conservation biologists have to think locally, in part for economic reasons. Very often, conservation decisions are about small patches. Sometimes quite large chunks of territory are part of the conservation estate, but active intervention tends to be on much smaller spatial scales. New Zealand's Kapiti Island is still one of the largest islands from which all rats have been removed; it is somewhat less than half of 1 percent of the area of Kangaroo Island.[10] But conservation biologists also have to think locally because regions are indeed ensembles of patches, and so are aggregated from patch-specific interactions. They have to act locally but think regionally.

Notes

1 Thus we think of ecosystem services as a product of biodiverse systems, and as a reason for conserving such systems. We do not think of such services as part of the diversity of a community or region. We think there is some confusion in the literature about this: see Lean and Maclaurin (forthcoming).
2 We use "ecological assemblage" as a neutral term, to capture the natural history truth that there are spatial patches where populations are found together, and interact, and that we can project from one population census to the next with some reliability, but without committing ourselves to any claim about the causal basis of these fairly stable groups of co-occurring organisms.
3 Emphasis added.
4 For example, coyotes are top predators in some regions, but not when wolves are present.
5 We choose this island as an uncontroversial example of a region; its length is about 80 km east–west; about 20 km north–south.
6 The examples are arbitrary, except in that we have chosen taxa where there are enough records for the recorded distribution to be some guide to where the organisms actually are.
7 In East Asia and North America, but not Europe, tree populations could shift south in glacial cycles, as the mountain ranges run north–south rather than east–west.
8 Both of us support taxonomic accounts of biodiversity, but one of us believes biodiversity is better understood in reference to phylogenetic structure rather than species richness.
9 Beta-diversity is $\beta = \gamma/\alpha$ where α (alpha-diversity) represents species richness in a local assemblage and γ (gamma-diversity) represents the species richness of the region comprised by all the local assemblages being analysed. Temporal beta-diversity assesses species diversity at single local assemblage over multiple time slices.
10 19.65 square kilometres to 4,416 square kilometres.

References

Agrawal, A. A., D. D. Ackerly, F. Adler, E. Arnold, C. Cáceres, D. Doak, E. Post, P. Hudson, J. Maron, K. Mooney, M. Power, D. Schemske, J. Stachowicz, S. Strauss, M. Turner, and E. Werner. 2007. "Filling Key Gaps in Population and Community Ecology." *Frontiers in Ecology and the Environment* 5(3): 145–152.

Barker, G., and J. Odling-Smee. 2014. "Integrating Ecology and Evolution: Niche Construction and Ecological Engineering." In *Entangled Life: Organism and Environment in the Biological and Social Sciences*, ed. G. Barker, E. Desjardins, and T. Pearce, 187–211. Dordrecht: Springer.

Barker, G., E. Desjardins, and T. Pearce, eds. 2014. *Entangled Life: Organism and Environment in the Biological and Social Sciences*. Dordrecht: Springer.

Calcott, B. 2011. "Wimsatt and the Robustness Family." *Biology & Philosophy* 26(2): 281–293.

Colinvaux, P. 2007. *Amazon Expeditions: My Quest for the Ice-age Equator*. New Haven, CT: Yale University Press.

Cooper, G. 2003. *The Science of the Struggle for Existence*. Cambridge: Cambridge University Press.

Dornelas, M., N. J. Gotelli, B. McGill, H. Shimadzu, F. Moyes, C. Sievers, and A. E. Magurran. 2014. "Assemblage Time Series Reveal Biodiversity Change but Not Systematic Loss." *Science* 344(6181): 296–299.

Elton, C. 1927. *Animal Ecology*. New York: Macmillan.

Godfrey-Smith, P. 2009. *Darwinian Populations and Natural Selection*. Oxford: Oxford University Press.

González-Orozco, C. E., S. W. Laffan, and J. T. Miller. 2011. "Spatial Distribution of Species Richness and Endemism of the Genus *Acacia* in Australia." *Australian Journal of Botany* 59(7): 601–609.

Guttman, Burton S. 1976. "Is 'Levels of Organization' a Useful Biological Concept?" *BioScience* 26(2): 112–113.

Hacking, I. 1983. *Representing and Intervening: Introductory Topics in the Philosophy of Science*. Cambridge: Cambridge University Press.

Hubbell, S. P. 2001. *The Unified Neutral Theory of Biodiversity and Biogeography*. Princeton, NJ: Princeton University Press.

Hull, D. 1989. *The Metaphysics of Evolution*. Albany: State University of New York Press.

Hutchinson, G. E. 1965. *The Ecological Theater and the Evolutionary Play*. New Haven, CT: Yale University Press.

Jones, C., J. Lawton, and M. Shachak. 1997. "Positive and Negative Effects of Organisms as Physical Ecosystems Engineers." *Ecology* 78: 1946–1957.

Kozak, K. H., C. H. Graham, and J. J. Wiens. 2008. "Integrating GIS-based Environmental Data into Evolutionary Biology." *Trends in Ecology & Evolution* 23(3): 141–148.

Maclaurin, J., and K. Sterelny. 2008. *What is Biodiversity?* Chicago, IL: University of Chicago Press.

Pearce, T. 2011. "Ecosystem Engineering, Experiment and Evolution." *Biology & Philosophy* 26(6): 795–812.

Phillips, B. L., and R. Shine. 2004. "Adapting to an Invasive Species: Toxic Cane Toads Induce Morphological Change in Australian Snakes." *Proceedings of the National Academy of Sciences of the United States of America* 101(49): 17150–17155.

Poisot, T., D. B. Stouffer, and D. Gravel. 2015. "Beyond Species: Why Ecological Interaction Networks Vary Through Space and Time." *Oikos* 124: 243–251.

Potochnik, A., and B. McGill. 2012. "The Limitations of Hierarchical Organization." *Philosophy of Science* 79(1): 120–140.

Ricklefs, R. E. 2004. "A Comprehensive Framework for Global Patterns in Biodiversity." *Ecology Letters* 7(1): 1–15.

Ricklefs, R. E. 2005. "Historical and Ecological Dimensions of Global Patterns in Plant Diversity." *Biologiske Skrifter* 55(3): 583–603.

Ricklefs, R. E. 2006. "Evolutionary Diversification and the Origins of the Diversity–Environment Relationship." *Ecology* 57(7 Supplement): S3–S13.

Ricklefs, R. E. 2008. "Disintergration of the Ecological Community." *American Naturalist* 272(6): 741–750.

Sarkar, S. 2005. *Biodiversity and Environmental Philosophy*. Cambridge: Cambridge University Press.

Sarkar, S. 2011. *Environmental Philosophy: From Theory to Practice*. New York: John Wiley & Sons.

Sterelny, K. 2001. "Darwin's Tangled Bank." In *The Evolution of Agency and Other Essays*, 152–178. Cambridge: Cambridge University Press.

Sterelny, K. 2006. "Local Ecological Communities." *Philosophy of Science* 73(2): 215–231.

Thompson, J. N. 1994. *The Coevolutionary Process*. Chicago, IL: University of Chicago Press.

Thompson, J. N. 2005. *The Geographic Mosaic of Coevolution*. Chicago, IL: University of Chicago Press.

Thuiller, W., J. A. Slingsby, S. D. Privett, and R. M. Cowling. 2007. "Stochastic Species Turnover and Stable Coexistence in a Species-rich, Fire-prone Plant Community." *PLoS ONE* 2(9): e938.

Tilman, D., and E. Snell-Roode. 2014. "Ecology: Diversity Breeds Complementarity." *Nature* 515: 44–45.

Tilman, D., P. Reich, J. Knops, D. Wedin, T. Mieke, and C. Lehman. 2001. "Diversity and Productivity in a Long-term Grassland Experiment." *Science* 294(5543): 843–845.

Valentine, J. 2004. *On the Origin of Phyla*. Chicago, IL: University of Chicago Press.

Vermeij, G. 2009. *Nature: An Economic History*. Princeton, NJ: Princeton University Press.

Wilson, E. O. 1992. *The Diversity of Life*. New York: W.W. Norton.

Wimsatt, W. C. 1981. "Robustness, Reliability, and Overdetermination." In *Scientific Inquiry and the Social Sciences*, ed. M. Brewer and B. Collins, 124–163. San Francisco, CA: Jossey-Bass.

Wimsatt, W. C. 2007. "Complexity and Organisation." In *Re-Engineering Philosophy for Limited Beings: Piecewise Approximations to Reality*, ed. W. C. Wimsatt, 179–192. Cambridge, MA: Harvard University Press.

9

UNNATURAL KINDS

Biodiversity and human-modified entities

Helena Siipi

Introduction

Human beings have an enormous capacity to modify the living world on Earth. The terms "genetic modification," "cis-genesis,"[1] "synthetic biology,"[2] "ecosystem restoration,"[3] "assisted migration,"[4] "de-extinction,"[5] and "domestication," for example, refer only to a handful of different ways of conceptualizing intentional human impacts on living entities. From the point of view of biodiversity, how should these human influences be understood? Should the term "biodiversity" be taken to include everything living from the deep sea fish to genetically modified crop plants and from primeval forests to stocks of industrial broiler farms? What about living entities unintentionally influenced by human beings, for example, through pollutants and species introductions?

The questions presented do not concern the possible effects of the use of genetically modified, cis-genic, or synthetic organisms in agriculture and industry. Nor is the interest in the risks assisted migration, de-extinction, and species introductions may pose to the receiving ecosystems. Rather, it is asked whether the outcomes of these procedures themselves, regardless of their effects, should be understood as constituents of biodiversity.

The academic world is far from agreeing. Many biodiversity definitions and descriptions omit to take an explicit stand on the issue, although it is common to suggest the term "biodiversity" to refer to *all* biotic variety (see e.g. Thomas Eisner, Terry Erwin, and Reed Noss in interviews of Takacs 1996: 47–48). Daniel P. Faith (2008), for example, describes biodiversity as "the variety of all forms of life, from genes to species, through to the broad scale of ecosystems." However, despite the commonness of the "all" indicator in definitions, some scholars are keen to associate biodiversity with "wilderness" (for criticism, see Guha 1997, Sarkar 1999) or with "nature" (see e.g. Reed Noss, Michael Soulé, and David Woodfuff in interviews of Takacs 1996: 76–79, Lanzerath 2014). Some go as far as explicitly stating their willingness to exclude at least some anthropogenic entities from the sphere of biodiversity:

> Our definition [of biodiversity] excludes exotic organisms that have been introduced …
>
> (Sala *et al.* 2000)

Other misuses of the term [biodiversity] stem from inclusion of human-generated elements in assessments of an area's biodiversity.

(Angermeier and Karr 1994: 692)

I consider artificial diversity to be generated by any addition of biotic elements to wild systems through direct manipulation by humans. Diversity may be enhanced artificially at any organizational level, including genome (by gene transfer), assemblage (by exotic species), and landscape (by fragmentation) levels. … Artificial diversity should be explicitly excluded from conceptions of biodiversity …

(Angermeier 1994: 600, 602)

Contrasting views can also be found. Some writers define biodiversity to include any living entity regardless of its relation to human beings:

Biodiversity … might be wherever you find it – including genetic labs, inner cities, zoos, the halls of office buildings, artificially constructed "landscapes" … Certainly, human design and intervention of the sort that destroys wilderness does not necessarily have a biodiversity-diminishing effect. Nor does the lack of discernible human design and intervention now and in the past have a biodiversity-enhancing one. In fact, human influence might just as well increase biodiversity, while human restraint might just as well decrease it.

(Maier 2012: 114)

The term biodiversity refers to the totality of species, populations, communities, and ecosystems, both wild and domesticated, that constitute the life of any one area or the entire planet … it specifically includes cultural modifications of natural world.

(Dasmann 1991: 8)

Limiting the scope of biodiversity to that which is native because of the value judgment about the importance of native biodiversity would allow a number of other biases to be built into definition of the term.

(DeLong 1996: 743)

Whether "unnatural living entities" should be included in the sphere of biodiversity partly depends on what is understood by (un)naturalness. The terms "natural" and "unnatural" are highly ambiguous (Mill 1969, Vogel 2011, Van Haperen *et al.* 2012, Heller and Hobbs 2014) and they are used in variety of different ways in biodiversity conservation and research (Willis and Birks 2006, Aplet and Cole 2010, Hobbs *et al.* 2010). The crucial question is whether the term "natural" has a meaning that pertains to all and only those entities that we are, in the name of morality and consistency, willing to include in the sphere of biodiversity.

In this chapter, six different understandings of (un)naturalness are analyzed with respect to their implications regarding genetic modification, cis-genesis, synthetic biology, ecosystem restoration, assisted migration, de-extinction, domestication, and unintentional human influences. The analysis does not take a stand on which, if any, of the outcomes of these human influences should be accepted to the sphere of biodiversity. Rather, the goal is to reveal what it would mean to limit the sphere of biodiversity to living entities that are natural in the presented six senses. If implications of all of the presented understandings of (un)naturalness turn out to be problematic, this may offer one reason to include also unnatural entities to the sphere of the biodiversity.

If a tolerable way of limiting the sphere of biodiversity to natural entities can be found, this opens the door for a naturalness-based understanding of biodiversity but does not, alone, imply that such understanding should be adopted.

Types of (un)naturalness

The different understandings of (un)naturalness can be divided into two groups depending on whether (un)naturalness is understood as a continuous gradient or as an all-or-nothing affair (Varner 1998: 125–126, Attfield 1999: 15–16, Siipi 2008: 77). When (un)naturalness is understood as a continuous gradient it is seen to vary between extremes of entirely natural and entirely unnatural. Often naturalness is associated with absence of human influence. The more human-dependent and the more affected by humans a living entity is, the less natural it is taken to be (Rolston 2001: 272, Verhoog *et al.* 2003, Van Haperen *et al.* 2012: 806–807). Conceptualizing (un)naturalness as a continuum enables one to judge human-influenced beings as differing in degree with respect to their (un)naturalness. One may think, for example, that outcomes of cis-genesis are more natural than other transgenic organisms, or that outcomes of ecosystem restoration are more natural than gardens (even though both are less natural than primeval forests). (Un)naturalness as a continuous gradient does not offer much help in determining whether a particular human-influenced entity should be taken to belong to the sphere of biodiversity or not. This goal requires dividing living entities into two classes – to those that belong and those that do not belong to the sphere of biodiversity – and a continuum cannot easily provide a tool for that. This is not to say that naturalness as a continuous gradient cannot offer any help for conservation decisions.

(Un)naturalness as an all-or-nothing affair is a conceptual dichotomy with no degrees. In it, entities are divided into natural and unnatural ones, and everything is taken to be either natural or unnatural (or to escape the definition) (Varner 1998: 125–126, Attfield 1999: 15–16, Siipi 2008: 77). (Un)naturalness is understood as an all-or-nothing affair, for example, when unnaturalness is associated with acting against God's will (for this interpretation, see Mill 1969, Sagoff 2001, Cooley and Goreham 2004) or with being supernatural (for this interpretation, see Mill 1969, Vogel 2002: 26). Even though we may be unaware of God's will or whether a happening is supernatural, in both cases the understanding of (un)naturalness rests on the idea of dichotomy of two mutually excluding groups. Happenings either are supernatural or not and actions either are against God's will or not. Even though a murder, for example, may be more a serious violation of God's will than a pickpocketing, they both belong to the group of actions that are against it. When (un)naturalness is understood as an all-or-nothing affair, the crucial question concerns the way of distinguishing between natural and unnatural entities: what is it that sets natural living entities apart from the unnatural ones?

There are three main ways that commentators have tried to justify this distinction: by appealing to the history of an entity, its current properties, or its relationships to other entities. Strictly speaking, all of the three ways concern relations an entity has to other entities. History-based naturalness is about its past causal relations, property-based concerns similarity to other entities, and relation-based is taken to cover other relations that offer a reason for taking something to be natural or unnatural. Thus, the first two ways can be taken as instances of the third. Yet, centrality of history-based and property-based understandings in environmental philosophy and conservation justifies discussing them separately from other relation-based understandings of (un)naturalness.

The idea of history-based understanding of (un)naturalness is that the way an entity came into being and the kinds of modifications it has gone through determine whether it should be considered natural or not (Varner 1998: 125, Vogel 2003: 160, Siipi 2008). The interest is in causal relations behind the existence and features of entities, in their history and genesis. For example, animal clones and transgenic plants are often understood to be unnatural in this sense for they are the outcomes of intensive high-technology-based human activities (see e.g. Lee 2003, Verhoog *et al.* 2003, Cooley and Goreham 2004). In an ecosystem context, naturalness is often strongly associated with non-manipulation, lack of human influence or at least lack of certain kinds of human influences (Hobbs *et al.* 2010, Yung *et al.* 2010, Boldt 2013).

In the property-based understandings of (un)naturalness, the interest is not in the entities' history or genesis, but in their current properties – for example, in species composition and functions of an ecosystem or in morphological, genetic, and behavioral features of an organism (Varner 1998: 125, Vogel 2003: 160, Siipi 2008). Naturalness is seen as a question of similarity to properties of non-human-influenced entities. Due to their extraordinary properties, many genetically modified organisms also can be taken to be unnatural in this sense. However, because of their similarity to non-modified entities, perfect clones of wild organisms are natural with respect to their properties. Analogously, property-based naturalness of an ecosystem implies that it is approximately similar to some human-independent ecosystem. According to this line of thought, even when the ecosystem is human-produced (and thus unnatural in the history-based sense), it may be natural with respect to its properties provided that it appears and functions similarly to some human-independent ecosystem (Aplet and Cole 2010: 23, Hobbs *et al.* 2010: 485).

There are other relations besides similarity (property-based (un)naturalness) and causal origin (history-based (un)naturalness) can offer a basis for understandings regarding (un)naturalness. This is the case when naturalness is taken to mean an entity's familiarity to someone (for this understanding, see Mill 1969, Richards 1984) or that an entity is suitable to someone or something. The expressions "fish is natural food for human beings" and "savanna is a natural environment for antelopes," for example, may be interpreted to refer to nutrition's suitability for humans or an environment's suitability for antelopes (Siipi 2011). Of the six understandings of (un)naturalness to be discussed in what follows, the first three are instances of history-based (un)naturalness, the two after them of property-based (un)naturalness, and the last concerns relation-based (un)naturalness distinct from the other two.

History-based (un)naturalness: naturalness as independence from humans

Sometimes naturalness is understood as absence of (almost) all human influence and as independence from human beings. Natural entities are then an opposition to everything that is human-produced or affected by humans (Lee 1999: 82, Delaney 2003: 34, Lanzerath 2014: 2). In his *End of Nature*, Bill McKibben understands (un)naturalness this way:

> [N]ow that we have changed the most basic forces around us, the noise of [… the] chain saw will always be in the woods. We have changed the atmosphere, and that will change the weather. The temperature and rainfall are no longer to be entirely the work of some separate, uncivilizable force, but instead in part a product of our habits, our economics, our way of life.
>
> (McKibben 1989: 47)

With some rare exceptions (such as subglacial lakes of Antarctica), every place on Earth and every living being on Earth is at least indirectly affected by humans. As a result, we have, according to McKibben (1989), literally destroyed nature and nothing is natural anymore. Since unnaturalness is seen to follow from any human influence, all human-influenced living entities, from transgenic cotton to penguins, and from unexplored rainforests to industrial corn fields, are equally unnatural.

Even though understanding naturalness as total independence from human beings may be fruitful in other contexts, this sense of (un)naturalness is useless for the task of distinguishing between entities that belong and those that do not belong to the sphere of biodiversity. Restricting the concept of biodiversity to living entities unaffected by humans would leave the sphere of biodiversity (almost) empty. As a result, there would not be biodiversity to conserve and conservation of biodiversity could never be a sensible goal. Conserving biodiversity, the sphere of which is limited to entities that are natural in this sense, would be as reasonable as defending the Holy Roman Empire or protecting *Tyrannosaurus rex* (Vogel 2011).

However, history-based (un)naturalness as an all-or-nothing affair can be interpreted in other ways also. Naturalness does not need to indicate absence of all human influences. Rather, naturalness may refer to the lack of certain kinds of human influences, for example, ones that rely on advanced technologies (for this understanding, see Angermeier 2000: 374, Lee 2003: 56, Verhoog *et al.* 2003) or ones that stem from human culture (or from the human mind) and not from the biological side of human beings (or human bodies) (for this understanding, see Brennan 1988: 88, Fukuyama 2002: 130, Vogel 2011: 93). According to these lines of thought, only certain kinds of human influences in the history of a living entity make it unnatural. Other human influences are compatible with naturalness (Lee 1999: 83, Siipi 2008). This kind of understanding of (un)naturalness, of course, raises the question about the kinds of human influences that are (and are not) relevant for (un)naturalness. Two alternatives are discussed next: naturalness as a lack of intentional human influences and naturalness in contrast to being an artifact.

History-based (un)naturalness: intentional control as a source of unnaturalness

Some living entities – most insects of inhabited areas, for example – are only indirectly and unintentionally affected by human beings. At the same time, other living entities are under continuous, direct, and intentional human control, for example, in farms, laboratories, and homes. Following these differences, Richard J. Hobbs *et al.* (2010: 484) define naturalness as "absence of intentional human control". Along the same lines, Gregory H. Aplet and David N. Cole (2010: 13) identify one meaning of the term "natural" with the "freedom from intentional human control." They further clarify that a natural ecosystem may well bear marks of human presence and it may be influenced by human beings. An ecosystem may retain its naturalness even when it suffers from pollutants or invasive species. What is crucial to naturalness is that the area is "not subject to intentional manipulation and human intervention."

The idea behind this understanding is that natural entities do not follow human-set goals or targets. They do not have, lose, or retain their properties because human beings have urged them to be of a certain kind. As Holmes Rolston III (2001: 274) puts it, "intentional, ideological construction is exactly what natural entities do not have." Naturalness does not follow from human-set endpoints or goals of being of a certain kind (for further discussion, see Heller and Hobbs 2014); rather, it follows from the absence of them. A river ecosystem affected by pollutants, for example, is not unnatural in the way that a cultivated corn (as a species) and a corn field (as an ecosystem) are. The polluted river ecosystem differs from them in being left to its own devices, to react to human inputs according to its own tendencies, and in that way being "free from the constraints of human intentionality" (Aplet and Cole 2010: 13).

If (un)naturalness in the presented sense is used as a criterion for belonging to the sphere of biodiversity, outcomes of domestication, genetic modification, cis-genesis, synthetic biology, de-extinction, and intentional species introductions fall outside the sphere of biodiversity. Unnaturalness of outcomes of intentional species introductions does not hold only with respect to the creation of gardens and commercial fields, but also with respect to environmentally motivated species introductions. Outcomes of ecosystem restoration, assisted migration, and projects aiming to protect ecosystems by introducing members of keystone species (or populations of a closely related species[6]) are unnatural in this sense and cannot contribute to conservation of biodiversity.

According to the presented sense of (un)naturalness, not just uninfluenced, but also unintentionally influenced living entities are natural. Thus, using it as a criterion for belonging to the sphere of biodiversity implies that, contrary to intentionally introduced species, unintentionally introduced species are constituents of biodiversity. This is not to say that unintentionally introduced species cannot be *causes* of extinctions or other losses of biodiversity. The point is that, apart from their consequences, the outcomes of unintentional species introductions are not unnatural and thus not excluded from the sphere of biodiversity.

Aplet and Cole's (2010) example of (un)naturalness as lack of human control is the Chesapeake & Ohio (C&O) Canal. It is a human-created waterway parallel to the Potomac River. The canal was used for transportation of coal and other products during the industrial revolution. It fell into commercial disuse in the beginning of twentieth century and since then its ecosystem has been left to develop in its own order (National Park Service 2014). Even though the ecosystem of the canal is now natural in not being intentionally controlled by humans, it seems clear that the canal itself, as a waterway, is far from being natural. It is a human-made artifact. What are the implications of using the distinction between natural entities and artifacts as a basis for determining which entities belong to the sphere of biodiversity?

History-based (un)naturalness: natural in contrast to an artifact

Artifacts are outcomes of intentional human labor. Yet, all intentional human influences on living entities do not turn them into artifacts and not all living entities under "intentional human control" should be considered as artifacts. Many game animal populations, for example, are controlled by feeding and hunting and this is not sufficient for them to be artifacts. Thus, as a criterion for belonging to the sphere of biodiversity the natural–artifact distinction implies a broader sphere of biodiversity than the lack of intentional control criterion.

What is crucial for being an artifact is that an artifact is brought into existence by human beings. As Keekok Lee (2003: 4) puts it, an artifact "does not exist in the absence of human manipulation and intervention, but is deliberately created by humans." Risto Hilpinen (1995: 138) argues along the same lines: "an artifact necessarily has a maker or author, or several authors, who are responsible for its existence. … Artifacts are products of intentional making."

Neither Hilpinen nor Lee is suggesting that human production is sufficient for an entity to be an artifact. Even though human babies, interpersonal relationships, and great physical fitness may be the intended products of human action, they cannot be considered as artifacts (Katz 1997: 122; Vogel 2011: 92). Thus, further requirements for being an artifact are set. Hilpinen (1992: 65) considers artifacts to be distinct from their raw materials; different sortal descriptions distinguish them from each other. Artifacts also differ from other human-influenced entities in having designed functions (Dipert 1986: 402; Lee 1999). One possibility is to consider the union of modification and creation as necessary for being an artifact and claim that artifacts are brought into existence by intentionally causing them to have certain properties (Siipi 2003).

All in all, distinguishing between artifacts and natural entities has proven to be a complicated issue and no academic consensus prevails regarding it (for further discussion, see e.g. Brennan 1984, Lee 2003, Vogel 2003). Despite the conceptual disagreements, some implications of using the artifact–natural distinction as a criterion for belonging to the sphere of biodiversity seem clear. Excluding artifacts and only artifacts from the sphere of biodiversity implies that biotic entities that are (even to a great extent) modified, controlled, and influenced by human beings may be constituents of biodiversity as long as they are not created or brought into existence by humans. This means that species introductions that do not cause the receiving ecosystem or the organism translocated to lose its identity (as a certain kind of being) may, as such, enhance biodiversity. Assisted migration as well as other environmentally oriented species introductions might well, under this understanding, be used for biodiversity conservation.

The other side of the coin is that many ecosystems brought into existence by human beings are artifacts and, thus, fall outside the sphere of biodiversity. Eric Katz (1992, 2012) and Robert Elliot (1982) famously argue that restored ecosystems are artifacts (and thus less valuable than the original ones). When restoration means rebuilding a destroyed ecosystem, the outcome of the restoration activities certainly should be considered as an artifact (Siipi 2003). Analogous cases can be found from the level of species and organisms. Animals born from de-extinction procedures have, despite their similarity to members of the natural species that once died out, been brought into existence by human beings and fulfill all mainstream criteria set for being an artifact. Along the same lines, the outcomes of synthetic biology seem to be artifacts (Oksanen 2014: 161).

Domestic plants and animals, genetically modified organisms, as well as cis-genic organisms, are less clear cases. According to J. Baird Callicott, "[d]omestic animals are creations of man. They are living artifacts, but artifacts nevertheless …" (Callicott 1980: 330; see also Katz 1997: 128–129, Lee 2003, 2004). Yet, with respect to some species, it seems unclear whether the process of domestication should be seen as a form of creating a new type of being or merely as a modification of already existing species. Even though domestic horses, for example, differ from wild ones in many respects, it is not clear whether they should be considered as a distinct human-created species (Gibbons 2014). The same kind of lack of clarity seems to concern genetically modified organisms. As indicated by their name, they are usually considered merely as varieties of already existing types of organisms. However, this may not be self-evident: "The other [possibility] is to say that it differs so fundamentally from a normal non-transgenetic … tomato plant that it would be misleading to say simpliciter that it is a common or garden variety. One could perhaps call it … a *Tg* tomato plant" (Lee 2003: 154).

To conclude, the above discussion indicates that limiting the sphere of biodiversity to any sense of history-based (un)naturalness may not be simple. If naturalness is understood as the absence of all human influences, the sphere of biodiversity becomes empty. For some, naturalness as lack of intentional human control and naturalness as a contrast to artifacts may seem to be too tight (over-exclusive) and too loose (over-inclusive) at the same time. It is likely that this is the case also regarding many other history-based understandings of (un)naturalness.

Property-based (un)naturalness: naturalness as similarity to human-independent entities

In property-based understandings of (un)naturalness the interest is in the current (non-historical) properties and features of entities, not in the influence and modifications that they have gone through (Vogel 2003, Cooley and Goreham 2004, Siipi 2008, 2011, Heller and Hobbs 2014). Property-based (un)naturalness is always a question of comparison and similarity. In order to find out whether something is natural in this sense, its current properties and features need to be compared with properties and features of some ideally natural entity. Entities that are similar to the ideally natural entity are taken to be natural (Cooley and Goreham 2004, Siipi 2008). Thus, it is possible for a totally artificial entity, an outcome of synthetic biology for example, to be natural in this sense. The crucial question concerns the choice of comparative models: which entities should be used as the ideally natural entities, the properties of which the other entities are compared?

The idea of property-based naturalness is that an ecosystem is natural if, and only if, its current properties (species composition, functions, etc.) are similar to those of an ecosystem that is natural with respect to its history. In other words, if an ecosystem is similar to an eco-system that has not been (at least not to a great extent) modified by human beings, then that ecosystem is natural – regardless of the way it received its properties. For example, outcomes of a "historically accurate but highly manipulated prairie restoration project" (Aplet and Cole 2010: 21) may be natural in this sense. Restored ecosystems are controlled and strongly modi-fied by humans. Some of them can even be considered human artifacts. Yet, their properties are not due to human invention but similar to (and copies of) the ones of a human-independent ecosystem that was in the place before destruction (Yung *et al.* 2010, Heller and Hobbs 2014). Thus, accepting the proposed sense of property-based (un)naturalness as a criterion for belong-ing to the sphere of biodiversity allows one to consider ecosystem restoration as an important contribution to biodiversity conservation.

Using this sense of naturalness as a criterion further implies that organisms translocated in (intentional and unintentional) species introductions may well retain their naturalness and be constituents of biodiversity – provided that their properties do not change in this process. Naturalness of the receiving ecosystems, however, is more questionable. With an introduced spe-cies in it, an ecosystem usually cannot be considered natural with respect to its species composi-tion. Yet, an ecosystem with an introduced species may, depending on what kind of changes the introduced species does and does not bring about in it, retain its naturalness with respect to its functions. As a matter of fact, retaining these functions may be the goal of some species intro-ductions (see e.g. Bowen 1999) and in such cases the introductions may be taken to conserve the functional aspects of biodiversity.

Transgenic organisms, cis-genic organisms, members of most domestic species, and those outcomes of synthetic biology that are not copies of already existing organisms are clearly not natural in this sense and do not, according to the criterion set, belong to the sphere of biodi-versity. Outcomes of successful de-extinction projects, however, differ from them in this respect. The goal of de-extinction is exactly to bring into existence organisms (or even populations) with properties similar to ones of members of a naturally evolved species (Oksanen and Siipi 2014, Revive and Restore 2015). Along similar lines, those outcomes of synthetic biology that are copies of organisms that have come into being through natural evolution are natural with respect to their properties and can, thus, be taken as constituents of biodiversity.

Property-based (un)naturalness: naturalness as similarity to entities that could have existed

Sometimes property-based naturalness is understood as similarity not to living entities that exist or have existed without human influence, but to living entities that *would* or *could* have existed under certain counterfactual circumstances.

The goal of ecosystem restoration is not always to bring about an ecosystem similar to one that existed in the site before human-caused destruction. Since ecosystems are changing and evolving entities, the goal of restoration is sometimes to create an ecosystem similar to one that *would* have existed in the site had the human-caused destruction not taken place (Aplet and Cole 2010: 13, Heller and Hobbs 2014). Outcomes of this kind of restoration project are considered natural because of their similarity to an ecosystem that has never existed but *would* have existed had a certain human destruction not taken place. Hällfors *et al.*'s definition of assisted migration rests on a similar idea of (un)naturalness:

> Assisted migration means *safeguarding biological diversity* through the translocation of representatives of a species or population harmed by climate change to an area outside the indigenous range of that unit where it *would* be predicted to move as climate changes, *were it not* for anthropogenic dispersal barriers or lack of time.
>
> <div align="right">(Hällfors <i>et al.</i> 2014, italics added)</div>

Hällfors *et al.* (2014) consider species translocations to predicted ranges to be more natural and to make greater contributions to biodiversity than species introductions to other areas. The translocations to predicted ranges are seen as more natural, since were there plenty of time and no human-created obstacles (such as cities or canals) the species would migrate to the predicted range by itself. This counterfactual understanding of (un)naturalness is strongly connected to naturalness as similarity to historically natural entities. In both understandings, naturalness is associated with similarity to something uninfluenced by human beings. The difference lies in whether that human-independent something ever existed. That an entity never came into being is not an obstacle for using it as the ideal comparative model. Its failure to come into being was, after all, caused by humans. In practice, these counterfactual entities used as ideal comparative models probably have to be quite similar to historically natural entities.

The often presented view that outcomes of cis-genesis are more natural than outcomes of other forms of genetic modification (see e.g. Holme *et al.* 2013, Eriksson *et al.* 2014) seems to rest on another kind of a counterfactual. With respect to their history, outcomes of cis-genesis are quite similar to ones of GMOs. Neither GMOs nor cis-genic organisms are similar to beings that exist, have existed, or would exist without human-made obstacles. The outcomes of cis-genesis are considered natural (or at least more natural than other transgenic organisms) because they *could* or *might* have come into being without human assistance (or at least without use of modern genetechnologies) (Schouten *et al.* 2006, Jacobsen and Schouten 2009: 240, Eriksson *et al.* 2014). Accepting this understanding of (un)naturalness as a criterion for belonging to the sphere of biodiversity opens that sphere to various beings that could have come into existence (without human assistance) under different counterfactual circumstances.

Relation-based (un)naturalness: naturalness as suitability and belonging

In addition to similarity (property-based (un)naturalness) and causal relations (history-based (un)naturalness), other relations have been suggested to offer a basis for (un)naturalness. One might, for example, be tempted to judge a palm tree in Lapland to be unnatural, even when one is unaware of its genesis and properties, if one thinks that palm trees simply do not *belong* to the nature of Lapland. (Un)naturalness is then taken to mean that Lapland is not a *suitable* place for palm trees.

Naturalness as suitability can refer to something being beneficial or, at least, harmless to something else. Certain ecosystems can be said to be natural for an organism, for example, when the organism in question can succeed in it. In other words, an ecosystem *x* is suitable for an organism y, if organism y has adapted or can easily adapt to the ecosystem x (Brennan 1988, Siipi 2011). To put it in another way, an organism can be (un)natural for an ecosystem in this sense. (Un)naturalness then refers to the influence that the organism has or would have to the ecosystem in question: an invasive species driving others to extinction is unnatural for them. Along these lines, Nicole Heller and Richard J. Hobbs (2014: 6) suggest understanding (un)-naturalness as a relationship between humans and other living beings: "recognizing naturalness as a process and a relationship between organisms, not an intrinsic identity; human behavior may be more or less natural."

Accepting this sense of naturalness as suitability to be the criterion for belonging to the sphere of biodiversity rests on an understanding of what counts as success, harm, and benefit in the natural world. Yet, it is not certain whether the language of harms and benefits can sensibly be used for collective natural entities (for discussion, see e.g. Sober 1988, O'Neill *et al.* 2008). Moreover, the capability to enhance biodiversity cannot offer a sensible criterion for belonging to the sphere of biodiversity. Such a criterion is circular, since, in order to say whether something enhances biodiversity, we already need to know which entities belong to its sphere.

The relations of suitability and belonging may justify (un)naturalness statements even when an organism (e.g. the palm tree in Lapland) is not causing any changes in an ecosystem. The idea then is that living entities have an essence or "nature". Naturalness means that something is authentic and true to its essence or nature (Sagoff 2001, Bergin 2009). A natural entity has all its necessary constituents and it is not spoiled by materials not belonging to it. Materials that do not belong to an entity and that are not suitable for it make it unnatural.

Along these lines, some have seen the unnaturalness of transgenic organisms to follow from the foreign genes that do not really belong to these organisms (for discussion, see e.g. Cooley and Goreham 2004, Bergin 2009). Since genes transferred in cis-genesis originate from the same (or closely related) species to which they are moved, cis-genic organisms may escape this kind of unnaturalness. When it comes to ecosystems, presumably the most oft-cited example of assisted migration concerns Torreya Guardians who transplanted seedlings of Florida torreya (*Torreya taxifolia*) in North Carolina. The translocation was motivated by the endangered status of Florida torreya and by their view that the species belongs to the Appalachian Mountains as it is thought to have lived there before (Torreya Guardians 2015). As these examples show, naturalness as belonging leaves room for different kinds of interpretations of the relation of "belonging". One might see this as a weakness that hinders the possibility of using this sense of naturalness as a criterion for something being a constituent of biodiversity. Yet, Gill Aitken (2004) sees the openness of the understanding as its strength. For her, belonging can offer guidance for biodiversity conservation for "it succeeds in taking into account the human/cultural perspective" and, thus, is not indifferent to human desires regarding environments.

Conclusions and suggestions

The above-presented six understandings of (un)naturalness differ widely with respect to their implications. Many living entities are natural in one sense of the term and unnatural in another sense. As a result, in order to offer any guidance for biodiversity conservation and research, all attempts to restrict the sphere of biodiversity to natural living entities must contain a description of the sense of (un)naturalness meant.

Can any of the presented understandings of (un)naturalness offer an acceptable criterion for what belongs to the sphere of biodiversity? One cannot realistically expect everyone to be ready to accept any single one of them. More problematically, implications of the understandings may also seem intolerable to those who, at the outset, would be willing to restrict the sphere of biodiversity to natural entities. If that is the case, the first alternative is to claim that the most significant understanding of (un)naturalness has not been presented here and formulate a better interpretation. The second alternative is to state that (un)naturalness should be understood in a multidimensional way that consists of bringing together some of the different senses of (un)-naturalness. Instead of merely pronouncing the term "natural" to be ambiguous, one should bring some of the different understandings of (un)naturalness together and take each of them as a necessary condition for (un)naturalness. The downside of these kinds of multidimensional understandings of (un)naturalness is that they easily become over-restrictive and, in practice, leave the sphere of biodiversity empty.

It may seem tempting to conclude that the implications of the above-presented understandings of (un)naturalness are enough to show that the sphere of biodiversity cannot consistently be limited to natural living entities. It is better to understand biodiversity to include, literally, *all* diversity of life – including the anthropogenic ones. However, this solution is not unproblematic. Many authors have worried that including unnatural living entities in the sphere of biodiversity might imply a loss of value of natural biodiversity and even the natural biodiversity itself. Angermeier, for example, famously writes:

> Through genetic engineering, species introduction, and environmental modification, we conceivably could manufacture a world even more biologically "diverse" than the one derived through evolutionary processes. Moreover, none of that variety needs to consist of native elements or share any evolutionary history. Such "management" strategy is arguably consistent with some conceptions of biodiversity. Nevertheless, to most conservation biologists – including me – such a world would present a straggling loss of biodiversity, and such management strategy would be unconscionable.
>
> (Angermeier 1994: 601)

Dieter Birnbacher (2014: 41) calls Angermeier's worry "a substitution problem" and connects it to biodiversity's indifference to individuals. The problem is that, if unnatural living entities are accepted in the sphere of biodiversity, then, in principle, every natural living entity can be replaced by an unnatural living entity without loss of biodiversity. If a substitute differs more from other living entities than the replaced entity, such substitutions might even increase the overall diversity. One might then go as far as suggesting that we should increase biodiversity by creating new lifeforms and ecosystems (for discussion, see Boldt 2013). Very few are ready to accept that this kind of substitution of natural entities with unnatural ones can even in principle enhance biodiversity. The common solution to the problem is, following Angermeier, to exclude unnatural diversity from the sphere of biodiversity or to argue that unnatural diversity is less valuable than natural diversity. However, these are not the only possible solutions.

One solution to the substitution problem is to accept the different senses of (un)naturalness as factors of biodiversity. The idea is that, just like a world with many plant species is more diverse than a world with only a few plant species, a world in which all living entities are similar with respect to their (un)naturalness (e.g. all are artifacts) is less diverse than a world in which living entities differ with respect to their (un)naturalness (e.g. some are artifacts and others are not). Thus, substituting all natural entities with unnatural ones would imply a great loss of biodiversity. (Analogously, of course, losing all unnatural living entities would diminish biodiversity.) If different senses of (un)naturalness are accepted as factors of biodiversity in the presented sense, conserving natural living entities (in different senses of the term) becomes an important goal – and it becomes especially important in the current world in which naturalness in a certain sense of the term has already been lost.

Notes

1 Cis-genesis is a form of plant breeding in which genes from a plant species itself or from crossable species are introduced to it by GM technology. Cis-genesis is comparable to the way existing genetic variation is used in traditional methods of plant breeding: the genes transferred by it could also have been transferred by traditional breeding techniques (Schouten *et al.* 2006, Jacobsen and Schouten 2009: 240, Eriksson *et al.* 2014).

2 The goal of synthetic biology is to create artificial life from scratch. In practice, two distinct lines of research can be distinguished. In its first sense, the term "synthetic biology" refers to "the design and fabrication of biological components and systems that do not already exist in the natural world" and in its second sense to "the re-design and fabrication of existing biological systems" (Synthetic Biology Community 2014a). Synthetic biologists working for the first goal use natural or artificial molecules and assemble them to create novel types of biological systems. The ones working for the second goal use artificial molecules to mimic natural molecules with the goal of creating artificial life (Synthetic Biology Community 2014a, 2014b).

3 The term "ecosystem restoration" refers to different kinds of procedures aiming to diminish human disturbance or destruction in an ecosystem. At one extreme, cleaning trash away from a stream can be considered as restoration. At the other end of the continuum are cases in which a human-destroyed ecosystem is rebuilt (Katz 2012: 69).

4 Assisted migration is defined by Hällfors *et al.* (2014) as "safeguarding biological diversity through the translocation of representatives of a species or population harmed by climate change to an area outside the indigenous range of that unit where it would be predicted to move as climate changes, were it not for anthropogenic dispersal barriers or lack of time."

5 "De-extinction" means bringing extinct species back to life, or to put it more precisely, creating an organism which is similar to members of an extinct species. De-extinction has not been carried out yet, but biotechnology offers various promising alternatives for achieving this purpose, and various research groups around the world are working towards the goal of de-extinction (Oksanen and Siipi 2014, Revive and Restore 2015).

6 For example, in the 1990s Texas panthers were introduced into southern Florida to supplement the native stock of Florida panthers. The justification for the introduction was to save the southern Florida ecosystem. Without a top predator the ecosystem would have disappeared (Bowen 1999).

References

Aitken, G. 2004. *New Approach to Conservation: The Importance of the Individual through Wildlife Rehabilitation.* Aldershot: Ashgate.

Angermeier, P. 1994. "Does Biodiversity Include Artificial Diversity?" *Conservation Biology* 8: 600–602.

Angermeier, P. 2000. "The Natural Imperative for Biological Conservation." *Conservation Biology* 14: 373–381.

Angermeier, P., and J. R. Karr. 1994. "Biological Integrity versus Biological Diversity as Policy Directives." *BioScience* 44: 690–697.

Aplet, G. H., and D. N. Cole. 2010. "The Trouble with Naturalness: Rethinking Park and Wilderness Goals." In *Beyond Naturalness: Rethinking Park and Wilderness Stewardship in an Era of Rapid Change*, ed. D. N. Cole and L. Yung. Washington, DC: Island Press.

Attfield, R. 1999. *The Ethics of the Global Environment.* Edinburgh: Edinburgh University Press.

Bergin, L. 2009. "Latina Feminist Metaphysics and Genetically Engineered Foods." *Journal of Agricultural and Environmental Ethics* 22: 257–71.

Birnbacher, D. 2014. "Biodiversity and the 'Substitution Problem'." In *Concepts and Values in Biodiversity*, ed. D. Lanzerath and M. Friele. Abingdon: Routledge.

Boldt, J. 2013. "Do We Have a Moral Obligation to Synthesize Organisms to Increase Biodiversity? On Kinship, Awe, and the Value of Life's Diversity." *Bioethics* 27: 411–418.

Bowen, B. W. 1999. "Preserving Genes, Species, or Ecosystems? Healing the Fractured Foundations of Conservation Policy." *Molecular Ecology* 8: 5–10.

Brennan, A. 1984. "The Moral Standing of Natural Objects." *Environmental Ethics* 6: 35–56.

Brennan, A. 1988. *Thinking About Nature: An Investigation of Nature, Value and Ecology*. London: Routledge.

Callicott, J. B. 1980. "Animal Liberation: A Triangular Affair." *Environmental Ethics* 2: 322–338.

Cooley, D. R., and G. A. Goreham. 2004. "Are Transgenic Organisms Unnatural?" *Ethics & the Environment* 9: 46–55.

Dasmann, R. F. 1991. "The Importance of Cultural and Biological Diversity." In *Biodiversity, Culture, Conservation, and Development*, ed. M. L. Oldfield and J. B. Alcorn. Boulder, CO: Westview Press.

DeLong, D. C. 1996. "Defining Biodiversity." *Wildlife Society Bulletin* 23: 738–749.

Delaney, D. 2003. *Law and Nature*. Cambridge: Cambridge University Press.

Dipert. R. 1986. "Art, Artifacts, and Regarded Intentions." *American Philosophical Quarterly* 23: 401–408.

Elliot, R. 1982. "Faking Nature." *Inquiry* 25: 81–93.

Eriksson, D., S. Stymne, and J. K. Schjoeering. 2014. "The Slippery Slope of Cisgenesis." *Nature Biotechnology* 32: 727.

Faith, D. P. 2008. "Biodiversity." In *The Stanford Encyclopedia of Philosophy*, ed. E. N. Zelta, available online at http://plato.stanford.edu/archives/fall2008/entries/biodiversity/.

Fukuyama, F. 2002. *Our Posthuman Future: Consequences of the Biotechnology Revolution*. New York: Farrar, Straus and Giroux.

Gibbons, A. 2014. "The Thoroughly Bred Horse." *Science* 346(6216): 1439.

Guha, R. 1997. "The Authoritarian Biologist and the Arrogance of Anti-humanism: Wildlife Conservation in the Third World." *The Ecologist* 27: 14–20.

Heller, N. E., and R. J. Hobbs. 2014. "Development of a Natural Practice to Adapt Conservation Goals to Global Change." *Conservation Biology* 28: 696–704.

Hilpinen, R. 1992. "On Artifacts and Works of Art." *Theoria* 58: 58–82.

Hilpinen, R. 1995. "Belief Systems as Artifacts." *The Monist* 78: 136–155.

Hobbs, R. J., D. N. Cole, L. Yung, E. S. Zavaleta, G. H. Aplet, F. S. Chapin III, P. B. Landers, D. J. Parsons, N. L. Stephenson, P. S. White, D. M. Graber, E. S. Higgs, C. I. Millar, J. M. Randal, K. A. Tonnessen, and S. Woodley. 2010. "Guiding Concepts for Park and Wilderness Stewardship in an Era of Global Environmental Change." *Frontiers in Ecology and the Environment* 8: 483–490.

Holme, I. B., T. Wendt, and P. B. Holm. 2013. "Current Developments of Intragenic and Cisgenic Crops." *ISB News Report*, available online at www.isb.vt.edu/news/2013/Jul/HolmeWendtHolm.pdf.

Hällfors, M. H., E. Vaara, M. Hyvärinen, M. Oksanen, L. Schulman, H. Siipi, and S. Lehvävirta. 2014. "Coming to Terms with the Concept of Moving Species Threatened by Climate Change – A Systematic Review of Terminology and Definitions." *PLoS ONE*: e102979. doi:10.1371/journal.pone.0102979.

Jacobsen, E., and H. J. Schouten. 2009. "Cisgenesis: An Important Subinvention for Traditional Plant Breeding Companies." *Euphytica* 170: 235–247.

Katz, E. 1992. "The Big Lie: Human Restoration of Nature." *Research in Philosophy and Technology* 12: 231–241.

Katz, E. 1997. "Artifacts and Functions: A Note on the Value of Nature." In *Nature as Subject: Human Obligation and Natural Community*, ed. E. Katz. Lanham, MD: Rowman & Littlefield.

Katz, E. 2012. "Further Adventures in the Case against Restoration." *Environmental Ethics* 34: 67–97.

Lanzerath, D. 2014. "Biodiversity as an Ethical Concept: An Introduction." In *Concepts and Values in Biodiversity*, ed. D. Lanzerath and M. Friele. Abingdon: Routledge.

Lee, K. 1999. *The Natural and The Artefactual: The Implications of Deep Science and Deep Technology for Environmental Philosophy*. Lanham, MD: Lexington Books.

Lee, K. 2003. *Philosophy and Revolutions in Genetics: Deep Science and Deep Technology*. Basingstoke: Palgrave Macmillan.

Lee, K. 2004. "There is Biodiversity and Biodiversity: Implications for Environmental Philosophy." In *Philosophy and Biodiversity*, ed. M. Oksanen and J. Pietarinen. New York: Cambridge University Press.

Maier, D. S. 2012. *What's So Good About Biodiversity? A Call for Better Reasoning About Nature's Value*. New York: Springer.

McKibben, B. 1989. *The End of Nature*. New York: Random House.

Mill, J. S. 1969. "Essays on Ethics, Religion and Society." In *Collected Works of John Stuart Mill*, part 10, ed. J. M. Robson. Toronto: Toronto University Press.

National Park Service. 2014. "Chesapeake & Ohio Canal," available online at www.nps.gov/choh/index.htm.

Oksanen, M. 2014. "Biodiversity and the Value of Human Involvement." In *The Ethics of Animal Re-creation and Modification: Reviving, Rewilding, Restoring*, ed. M. Oksanen and H. Siipi. London: Palgrave Macmillan.

Oksanen, M., and H. Siipi (2014) "Introduction: Towards the Philosophy of Resurrections Science." In *The Ethics of Animal Re-creation and Modification*, ed. M. Oksanen and H. Siipi. London: Palgrave Macmillan.

O'Neill, J., A. Holland, and A. Light. 2008. *Environmental Values*. Abingdon: Routledge.

Revive and Restore. 2015. "Genetic Rescue for Endangered and Extinct Species," available online at http://longnow.org/revive/.

Richards, J. R. 1984. *The Sceptical Feminist: A Philosophical Enquiry*. Harmondsworth: Penguin Books.

Rolston III, H. 2001. "Natural and Unnatural; Wild and Cultural." *Western North American Naturalist* 61: 267–276.

Sagoff, M. 2001. "Genetic Engineering and the Concept of Natural." *Philosophy and Public Policy Quarterly* 25: 2–10.

Sala, O. E., F. S. Chapin III, J. J. Armesto, E. Berlow, J. Bloomfeld, R. Dirzo, E. Huber-Sanwald, L. F. Huenneke, R. B. Jackson, A. Kinzig, R. Leemans, D. M. Lodge, H. A. Mooney, M. Oesterheld, N. L. Poff, M. T. Sykes, B. H. Walker, M. Walker, and D. H. Wall. 2010. "Global Biodiversity Scenarios for the Year 2100." *Science* 287(5459): 1770–1774.

Sarkar, S. 1999. "Wilderness Preservation and Biodiversity Conservation – Keeping Divergent Goals Distinct." *BioScience* 49: 405–412.

Schouten, H. J., F. A. Krens, and E. Jacobsen. 2006. "Cisgenic Plants are Similar to Traditionally Bred Plants: International Regulations for Genetically Modified Organisms Should Be Altered to Exempt Cisgenesis." *EMBO Reports* 7: 750–753.

Siipi, H. 2003. "Artefacts and Living Artefacts." *Environmental Values* 12: 413–330.

Siipi, H. 2008. "Dimensions of Naturalness." *Ethics & The Environment* 13: 71–103.

Siipi, H. 2011. "Non-backward-looking Naturalness as an Environmental Value." *Ethics, Policy and Environment* 14: 329–344.

Sober, E. 1988. "Philosophical Problems for Environmentalism. In *The Preservation of Species: The Value of Biological Diversity*, ed. B. Norton. Princeton, NJ: Princeton University Press.

Synthetic Biology Community. 2014a. "What is Synthetic Biology?," available online at http://syntheticbiology.org/FAQ.html.

Synthetic Biology Project. 2014b. "What is Synthetic Biology? Defining the Concept," available online at www.synbioproject.org/topics/synbio101/definition/.

Takacs, D. 1996. *The Idea of Biodiversity: Philosophies of Paradise*. Baltimore, MD: The Johns Hopkins University Press.

Torreya Guardians. 2015. "American Conifer Tree Endangered by Climate Change," available online at www.torreyaguardians.org/.

Varner, G. 1998. *In Nature's Interests? Interests, Animal Rights and Environmental Ethics*. New York: Oxford University Press.

Van Haperen, P. F., B. Gremmen, and J. Jacobs. 2012. "Reconstruction of the Ethical Debate on Naturalness in Discussions about Plant-Biotechnology." *Journal of Agricultural and Environmental Ethics* 25: 797–812.

Verhoog, H., M. Matze, E. L. van Bueren, and T. Baars. 2003. "The Role of the Concept of Natural (Naturalness) in Organic Farming." *Journal of Agricultural and Environmental Ethics* 16: 29–49.

Vogel, S. 2002. "Environmental Philosophy after the End of Nature." *Environmental Ethics* 24: 23–39.

Vogel, S. 2003. "The Nature of Artifacts." *Environmental Ethics* 25: 149–168.

Vogel, S. 2011. "Why 'Nature' Has No Place in Environmental Philosophy." In *The Ideal of Nature: Debates about Biotechnology and the Environment*, ed. G.E. Kaebnick. Baltimore, MD: Johns Hopkins University Press.

Willis, H. J., and Birks, H. J. B. 2006. "What is Natural? The Need for a Long-Term Perspective in Biodiversity Conservation." *Science* 314(5803): 1261–1265.

Yung, L., D. N. Cole, D. M. Graber, D. J. Parsons, and K. A. Tonnessen. 2010. "Changing Policies and Practices: The Challenge of Managing Naturalness." In *Beyond Naturalness: Rethinking Park and Wilderness Stewardship in an Era of Rapid Change*, ed. D. N. Cole and L. Yung. Washington, DC: Island Press.

10

GOING SMALL

The challenges of microbial diversity

Christophe Malaterre

Questions of biodiversity are so much more likely to be associated with the fate of larger plants and animals that one may wonder whether microorganisms matter at all in this debate. Nevertheless, microbial diversity has become the focus of intense research in the past decades, owing much to technical advances that now greatly facilitate the identification of microorganisms and their study. Despite being largely invisible to the naked eye, microorganisms account for a significant proportion of Earth's biomass, species abundance, and richness. Even more importantly, it is now well-established that microorganisms drive massive biogeochemical cycles that affect the entire planet. Microbial diversity matters, and this chapter will be largely about that. Along the way, my objective will also be to point to specific philosophical questions that the study of microbial diversity raises. In the first section, I first provide a brief account of the concept of microorganisms and how the field of microbial diversity studies has recently developed in relationship to both microbiology and ecology. I then review, in the second section, a number of reasons why microorganisms are interesting to look at from a diversity point of view, in particular their quantity, ubiquity, and ecological significance. In the third section, I address the question of defining microbial diversity, and focus on the units of diversity problem in the context of microorganisms, with a special attention to taxonomic and functional perspectives. In the fourth section, I review key reasons why microbial diversity matters, including a set of applied and purely theoretical reasons. Because diversity studies venture into smaller and smaller entities, such as viruses, this raises the question of extending microbial diversity to non-cellular micro-entities, as I show in the fifth section. And in the sixth section, I review how conservation questions also arise in the context of microbial diversity.

Microorganisms?

Microbial diversity is nothing other than the diversity of microorganisms, also called "microbes". The term "microorganisms" refers to a very broad range of minute unicellular organisms that are too small to be seen with the naked eye (Madigan *et al.* 2014). Because the smallest things that the human eye can see are about 0.1 mm in length, that makes microorganisms include any organism smaller than 0.1 mm. This characteristic is shared by many different organisms. So, it should not come as a surprise that the term "microorganism" should cut across different phylogenetic groups. The term is usually taken to denote members of the domains Bacteria and

Archaea (all of which are unicellular), as well as microscopic unicellular members of the domain Eucarya (for instance, unicellular algae, some fungi, and protists). Bacteria, for instance, are typically 0.5–5.0 micrometers in length (i.e. 0.0005–0.005 mm), less than one-twentieth of what can be seen with the unaided human eye. Microscopic eukaryotes can be bigger: phytoplanktonic algae, for instance, are typically 20–200 micrometers long. The group of microorganisms as defined by size is very heterogeneous from a phylogenetic point of view.

Despite not occupying the central place in biodiversity studies in the past decades, especially in the eyes of the public who have been more concerned by the fate of larger plants and animals, microorganisms have been objects of diversity studies ever since their discovery by van Leeuwenhoek at the end of the seventeenth century. Since the 1990s, microbial diversity has also been investigated through molecular phylogenetic perspectives (Moreira and López-García 2002). Philosophical attention to microorganisms and their impact on biodiversity is more recent (O'Malley and Dupré 2007, Morgan 2010, Bapteste and Dupré 2013, Malaterre 2013, O'Malley 2014). In any case, microbial diversity has not received so far the type of attention that larger organisms and ecosystems have.

The field of microbial biodiversity has often fallen within the scope of microbiology and molecular biology, and it has developed – at least up until quite recently – largely independently from general biodiversity studies (Øvreås and Curtis 2011). Size matters. And the small size of microorganisms requires specific instruments and taxonomic methodologies that have come from microbiology, and not from ecology. Very few microorganisms can be reliably classified upon visual inspection of their morphology. Historically, the classification of microorganisms often relied on indirect biochemical methods. For instance, when their habitats and nutritive environments can be reproduced in the laboratory, microorganisms can be cultivated *in vitro* and then tested against specific biochemical traits (Alain and Querellou 2009). Another common approach is to use gene sequences as molecular taxonomic markers. This can be done through DNA–DNA binding experiments that assess the overall genetic similarity among microorganisms by looking at the degree to which their genomes hybridize with each other, or through the sequencing of well-known genes such as 16S ribosomal RNA genes and comparisons with gene databases of known organisms (Pace 1997). The fact that such tools are typical of the disciplines of microbiology and molecular biology can explain why microbial diversity studies have been more strongly linked to these disciplines rather than to general biodiversity studies and ecology.

As a consequence of this hiatus between microbial diversity studies and general biodiversity studies, the former are often granted to have developed tools and methodologies to probe diversity at the micro-scale but are criticized for their relative lack of theoretical foundations in terms of diversity modeling (Prosser *et al.* 2007). On the other hand, biodiversity studies are granted to have developed much theoretical knowledge, yet criticized for largely ignoring the microbial world whose ecosystemic impact, however, can be extremely significant (Horner-Devine *et al.* 2004).

Why look at microbes?

There are many reasons to investigate the microbial world. Microbes are not just the oldest forms of life, they are also the most numerous, the most diverse, and the most ecologically significant. Microbial entities have been around for several billions of years. Fossilized microbes have been found that are over 3.2 billion years old, possibly even up to 3.5 billion years old (Javaux *et al.* 2010, Sforna *et al.* 2014), and there are good reasons to think that life on Earth was exclusively unicellular, and therefore microbial, for over two billion years.

One of the most salient characteristics of microorganisms is their ability to thrive in a range of habitats that go well beyond the usual habitats of larger plants and animals. Microorganisms are almost everywhere. They are found in many temperate climates and nutrient-rich environments: they colonize soil, lakes, and oceans, as well as man-made artifacts and buildings, and the inside of larger organisms. They have also evolved adaptations that enable them to survive in numerous otherwise deadly environments. Such microorganisms, often bacteria or archaea, are typically called "extremophiles." Some of them ("hyperthermophiles") can withstand temperatures above 100°C as in hot springs and submarine hydrothermal vents. Others ("psychrophiles") thrive in the Antarctic ice, way below the freezing point of water. Still other microorganisms have been found that know how to cope with the extreme pressures encountered in the deepest layers of the oceans ("barophiles"), with very high concentrations of salt ("halophiles") or with very acidic environments ("acidophiles") (López-García 2005). The study of microorganisms is therefore key to charting the boundaries of habitable space on Earth, as well as the ubiquity of life and identifying the diversity of adaptive life-sustaining responses to many extreme environments.

Another salient characteristic of microorganisms is their numerical abundance and taxonomic diversity. It is estimated that they are, by far, the most numerous living entities on Earth. To give an order of magnitude, over 10 billion bacteria live in a single cubic centimeter of soil (Torsvik *et al.* 2002). Worldwide, the number of all bacterial and archaeal microorganisms is likely to be over 10^{29}–10^{30} individuals. To these numbers should be added microbial eukaryotes counts, which are estimated to be extremely numerous as well: they are, for instance, estimated to account for 10–30 percent of total cell counts in the deep oceans (Moreira and López-García 2002). This results in an overall microbial biomass that is about a third of Earth's living biomass (e.g. Whitman *et al.* 1998, Kallmeyer *et al.* 2012).

This abundance of microbial life goes hand in hand with an amazing diversity of species (but see the third section below for a discussion of the concept of species in the context of microbial diversity). For instance, estimates of bacterial species alone run as high as 10^7–10^{12} species (Dykhuizen 1998), to which should be added species of archaea and of microbial eukaryotes, which are taken to be very high as well. These numbers are one to several orders of magnitude higher than the estimated number of species of larger animals and plants, which is in the vicinity of 10^6 species (Staley and Gosink 1999).

This abundance and diversity of microorganisms enables them to play a broad range of chemical transformation roles whose aggregated effects can even become tangible on a global planetary scale. Microorganisms have proven to be capable of engendering some of the greatest ecological changes on our planet; for instance, they are responsible for the significant rise in atmospheric oxygen some 2.4–2.2 billion years ago (Catling *et al.* 2001). Microorganisms also play a unique chemical role in the fixation of atmospheric nitrogen (Postgate 1998), and are capable of transforming organic waste into useful compounds to be reused by other living organisms. Their metabolic role is even more pressing when microorganisms happen to live in symbiosis with larger organisms (e.g. Tannock 1990). An important component of microbial diversity is its diversity of functions. This functional diversity of microbes plays key classificatory and theoretical roles in ecology, for instance when it comes to modeling the functions of ecosystems in which microorganisms partake, or in explaining how specific ecosystem services are realized.

What is microbial diversity?

One simple way to approach microbial diversity is simply to define it as biological diversity applied to microorganisms. Like general biological diversity, microbial diversity may be phrased in terms of different *units* of diversity: genes, populations, species, functions, and so forth. Assessing microbial diversity therefore encounters the classic problem of choosing the items of diversity that are going to be measured, a problem sometimes referred to as the "units-and-difference" problem (Maclaurin and Sterelny 2008). In principle, microbial diversity could be understood in a large number of ways, depending on the units that are chosen, possibly leading to a conceptual pluralism in this regard.

In practice, however, and because not all such units can be easily accessed, microbial diversity will often be assessed in terms of *taxonomic units*, such as species, very much like general biodiversity is. Indeed, instrumental concepts in the study of biodiversity – such as detectability, sampling, measurement, abundance in space and time, or density – are all concepts that are very often defined in terms of number of species (e.g. Magurran and McGill 2011). On this basis, different diversity indices can be calculated that generally take into account species richness (i.e. number of species) and evenness (i.e. relative abundance of species). Of course, such a construal of diversity runs into the problem of defining "species," a problem that is even more pressing in the case of microorganisms (Franklin-Hall 2007, Ereshefsky 2010). The microscopic size of these organisms very often does not make it possible to identify species on the basis of directly observable morphologic traits. Species identification is indirect, often based on approaches that combine phenotypic (e.g. results of biochemical tests, fatty-acid composition), genotypic (e.g. degree of DNA–DNA hybridization) and phylogenetic properties (e.g. 16S rRNA gene-sequence identity). Unlike higher organisms for which evolutionary and ecological processes underpin the concept of species, microorganism species demarcation tends to be more arbitrary. For instance, one of the major criteria often used to delineate species is a criterion of minimum 70 percent DNA–DNA binding combined with a minimum of 97 percent of 16s rRNA gene-sequence identity, these threshold values just having been calibrated empirically to yield many of the previously phenotypically recognized species (Gevers *et al.* 2005). However, such criteria now appear insufficient in light of recent advances in sequencing technology. The complexity of the problem is also increased by the high replication rates and high adaptability of microorganisms, as well as by the influence of "lateral gene transfer" (LGT) through which microorganisms exchange genetic material not only within species but also across species or even domains, thereby blurring genealogical relationships and phylogenies (Gogarten *et al.* 2002). In sum, delineations between microbial species are extremely delicate. This has resulted in a debate over the relevance of a natural classification view of microorganisms as compared to a more pragmatic view based on operational taxonomic units (OTUs) such as those used in phylogenetic studies for grouping organisms on percent similarity thresholds of various genes or sets of genes (e.g. multilocus sequence analyses; see Gevers *et al.* 2005). The critical question is whether different OTUs can be compared across different studies, as there are several ways of defining OTUs.

An alternative approach to microbial diversity is to switch away from taxonomic units such as "species" and frame diversity in terms of *functions*. One of the main motivations to do so is the fact that microorganisms carry out processes that deserve specific attention. So construed, microbial diversity becomes a diversity whose units consist of microbial functional traits or "functions". Such functions can be extremely diverse. For instance, in the case of soil microorganisms, functions include a vast array of activities such as nutrient transformation, decomposition, plant growth promotion or suppression, or modification of soil physical processes (e.g. Waldrop *et al.* 2000). Such activities are often investigated at an even finer level: for instance, nutrient transformation can be analyzed in terms of the number of substrates metabolized by the soil microbial community and in terms of the variability in substrate use. Because functions are linked to genes, assessing functional diversity becomes a more tractable objective (though this link between function and genes is far from straightforward). Genomic and proteomic techniques now make it possible to identify key molecular units – protein coding genes, enzymes, metabolites – that can be associated with specific microbial functional traits that are in turn regarded as key functional components of an overall ecosystem (Green *et al.* 2008). Such approaches to microbial diversity no longer focus on identifying and sorting organisms into species or other taxonomic units, but employ metagenomic tools to assess the presence or absence of specific functions within heterogeneous communities of microorganisms. Because functions can be realized by different species of microorganisms, functional approaches to microbial diversity make it possible to directly focus on functions, without being trapped by the problem of sorting organisms into species first. And, by targeting function-bearing genes and not microorganism species, such approaches also elude the problem raised, for instance, by lateral gene transfer that spreads functions by transferring genes from species to species. Furthermore, functional approaches could provide a common currency with which to develop ecological theories bridging macroorganisms and microorganisms (e.g. Kembel *et al.* 2014). This is not to say that taxonomic perspectives on biodiversity should be dropped – since there are still good reasons to endorse such perspectives, as mentioned above – but rather that a multifaceted approach to biodiversity that comprises both functional and taxonomic perspectives could hold the key to promising advances in the field.

Why study microbial diversity?

There are two broad sets of reasons to study microbial diversity. First, microbial diversity studies have often been linked to directly *applied objectives*, related to public health, agriculture, or the food-processing industry. Human pathogens, for instance, have received considerable attention, and their diversity matters to understand, predict, and control the spread and evolutionary dynamics of diseases. For instance, it has been shown that taking into account microorganisms' diversity – and not solely individual taxonomic groups – impacts the way one understands a broad range of phenomena, from antibiotic resistance to air quality management in hospitals (Cho *et al.* 2012, Kembel *et al.* 2012). In addition, microorganisms can be put to work and help in the production of a broad range of goods and services. This is the case, for instance, in biotechnology and pharmaceuticals, but also in food-transformation processes – from beer brewing to yoghurt fermentation and wine production – in which microorganisms play key transformation roles, and for which their functional diversity is carefully studied. Other direct applications of microbial diversity concern the improvement of raw sewage treatment, the generation of bioenergy and biofuels, the extraction of metals through microbial leaching, or the implementation of environmental remediation services targeting heavy metals, uranium or oil spills. Indicators of microbial community diversity and function could serve as early warning systems for environmental changes, somehow using bacteria as climate change sentinels (Ogunseitan 2005, Madigan *et al.* 2014).

Second, there is also a certain number of more *theoretical reasons* to study microbial diversity. One such reason is simply to assess the extent of microbial diversity, to characterize its units and quantify how diverse these units are within a specific location and timeframe. Such place- and time-specific descriptive assessments are often a first step towards gathering a more comprehensive understanding of microbial diversity in terms of spatial and temporal patterns. Obviously, identifying patterns of microbial diversity leads to the question why such patterns exist. Hence another theoretical objective of microbial diversity studies consists in explaining the origin of those patterns: why is microbial diversity different here and there, or now and then? What are its drivers of change, both from a spatial and a temporal point of view? This includes studying how microbial diversity is influenced by the type of habitat and by its heterogeneity, as well as assessing the extent to which this diversity follows particular geographical distributions and responds to specific environmental changes (e.g. Horner-Devine *et al.* 2004). In this context, an interesting question is whether microbial diversity patterns are better explained by natural selection or by chance dispersal. At stake is a better understanding of the rate and range of global dispersal, evolution, and extinction of microorganisms.

These theoretical objectives also connect to those of ecology. For instance, a significant debate in microbial diversity is whether its models of geographical distribution patterns are different or not from those of larger plants and animals (Martiny *et al.* 2006). Whereas some argue that microbial diversity corroborates classical ecological models, for instance when it comes to explaining the absence of specific geographic patterns by the presence of high dispersion rates, others argue that there are geographic barriers to dispersion of microorganisms, at least for certain species of extremophiles. One of the questions is whether geographic patterns are as tightly linked to phenotypic traits in the case of microorganisms as they may be in the case of larger plants and animals (Green *et al.* 2008). To answer these questions, one must compare microbial ecological models to those of general ecology. A further observation that bears on this question is the frequent tight functional interactions between microorganisms and larger plants and animals cannot be ignored. These interactions make it necessary to study microbial diversity in light of plants and animals diversity, and viceversa, for instance showing how each type of diversity influences the other according to specific spatio-temporal patterns (e.g. Kembel *et al.* 2014).

Microbial diversity is also epistemically valuable when it comes to understanding large environmental dynamics, and in particular those that are linked to the cycling of key chemical elements such as carbon, nitrogen, or sulfur. Microorganisms play an immense geochemical transformation role thanks to their metabolic diversity and their abundance in the soils and oceans (e.g. Falkowski *et al.* 2008). For instance, numerous phototrophic microorganisms fix CO_2 in aquatic environments (photosynthesis), whereas other microorganisms produce CO_2 through the degradation of organic compounds (respiration). Microorganisms also contribute to the carbon cycle through the production or consumption of methane (CH_4), respectively by metanogens and by metanotrophs. Still other microorganisms play key roles in the cycling of nitrogen, for instance by fixing atmospheric N_2 or by transforming nitrogen compounds such as as ammonia (NH_3) or nitrate (NO_3^-) ultimately back into N_2. And still other microorganisms take part in the sulfur, iron, phosphorus, calcium, or silica cycles. Because all these nutrient cycles are tightly coupled with each other, minor changes in one cycle may strongly affect the functioning of the other cycles. Ultimately, this is how microbial diversity happens to play a major role in the balance of key planetary environmental parameters, such as the quantity of atmospheric CO_2 and CH_4 (that are both greenhouse gases) or of ozone-depleting gases (such as NO). The incorporation of microbial diversity into ecosystem models is therefore key to the prediction of carbon dynamics under global warming or the forecast of precipitation regimes (Treseder *et al.* 2012).

Extending microbial diversity to non-cellular micro-entities?

Microbial diversity is sometimes understood as referring not just to the diversity of microorganisms as classically construed – bacteria, archaea, microbial eukaryotes – but also to the diversity of even tinier entities that, despite their non-cellular state, nonetheless do play significant ecosystemic roles. In such contexts, the definition of microorganisms is enlarged to include "all prokaryotes (archaea and bacteria), some eukaryotic organisms (fungi, yeasts, algae, protozoa), *non-cellular entities (e.g. viruses), their replicable parts and other derived materials e.g. genomes, plasmids, cDNA"* (OECD 2007: 59, italics added). With such a broad definition of microorganisms, microbial diversity encompasses the diversity of many – if not all – non-cellular entities of biological interest, be they extracted from unicellular organisms – as in the case of genomes, plasmids, or complementary DNA – or freely floating entities such as viruses, viroids (viral particles composed of a short stretch of circular single-stranded RNA), satellites (viral particles that can only reproduce if their host cells are also simultaneously infected with another specific helper-virus), virophages (viruses that infect other larger viruses) or even prions (proteins considered to be in a misfolded form).

Enabled by the development of novel sequencing techniques and biochemical tools, the identification of these types of micro-entities has, so far, been mostly driven by research on pathogens. It is therefore reasonable to expect that many more such non-cellular entities will be identified in the near future, and not just those pathogenic ones, thereby drawing considerable attention to another dimension of microbial diversity. Viroid diversity, for instance, is the topic of nascent research studies (e.g. Wang *et al.* 2008), while virus diversity is already well researched and could very well serve as a model for other non-cellular diversity studies.

The study of virus diversity has revealed that these microbial non-cellular entities – similarly to cellular entities – are not only abundant, but also highly diverse and tightly ecologically integrated with other living entities. As a matter of fact, the abundance of viruses is the focus of intense studies. As of today, almost 3,700 virus species have been identified and sorted into a specific taxonomic system (International Committee on Taxonomy of Viruses 2015). But these known species – most of which are health- or agriculture-related – are thought to be but an extremely small proportion of all species of viruses out there in the natural world, in constant interaction with the rest of microbial and macrobial diversity. Some estimate virus diversity to be in the range of 10^5 to 10^{13} species, possibly making virus diversity an order of magnitude larger than the diversity of unicellular organisms (Rohwer 2003). Viruses, like unicellular organisms, are extremely common and found in a very broad range of habitats. Because of their very small size, their diversity and abundance can be very high at a very small spatial scale: for instance, it has been estimated that 200 liters of ocean water may possibly contain over 5,000 different virus species (Breitbart *et al.* 2002). It has also been estimated that a single cubic centimeter of ocean water may contain as many as 10^8 viruses (10^9 for a cubic centimeter of nearshore surface sediment), adding up to some 10^{30} viruses worldwide, thereby making viruses a most significant component of biomass (Suttle 2005).

Viruses also display an interesting range of genotypic and morphological diversity. Viruses can vary a lot in size, starting from a dozen nanometers (10^{-3} micrometers) up to several hundreds of nanometers in the case of giant viruses like the "mimivirus," a size comparable to that of a small bacterium (La Scola *et al.* 2003). Viruses can also vary a lot in terms of morphotypes. One finds, for instance, "complex viruses" (such as the classical head-and-tail viruses that have a capsid and a protein tail acting as a syringe), "helical viruses" (that can be rod-shaped or filamentous), "icosahedral viruses" (that possess a shell built out of identical tiles) or even "envelope viruses" (that have a membrane made of lipids and carbohydrates borrowed from host cells) (Dimmock *et al.* 2009). But there is also considerable diversity even within morphotypes: for instance, a survey of bacteriophage diversity in a European lake reported thirty-nine morphologically distinct types of head-and-tail viruses (Demuth *et al.* 1993).

Another strong reason to consider viruses – as well as all other types of non-cellular micro-entities – as a key component of microbial diversity is their extremely tight functional integration into larger ecosystems. Because viruses critically depend on their host organisms to reproduce (hence the genetic and morphological diversity we just saw above), they end up being involved in the ecosystems in which their hosts partake, sometimes playing critical eco-systemic roles. It has been shown, for instance, that viruses can modify the spatial distribution of certain plants (Wren *et al.* 2006). But viruses can also have much larger geochemical effects. An example is the role viruses have in increasing the mortality of marine microorganisms, which, in turn, results in an increase of dissolved carbon in the oceans, hence also affecting the levels of atmospheric carbon (Suttle 2005). Another telling example is the case of those marine viruses that infect specific microalgae and induce them to produce dimethylsulphide, the latter having an impact on cloud formation, and hence on global climate (Ayers and Cainey 2007).

Another reason to extend microbial diversity so as to include not only cellular microorganisms but also non-cellular ones as well such as viruses, is that all such non-cellular entities use the same biomolecules – mostly nucleic and amino acids – as cellular organisms. An additional reason is the fact that these non-cellular entities have been shown to be capable of Darwinian evolution through natural selection (e.g. Li *et al.* 2010 in the case of prions). However, such an extension of microbial diversity raises an intriguing question as it forays into a domain of entities whose status as "living" – and hence whose justification to be part of the "bio" of *bio*diversity – is no longer clear. Viruses for instance do display some, though not all, of the characteristics associated with living organisms, and their status as living entities is debated (Forterre 2010, Moreira and López-García 2009). Some recently discovered endosymbionts are classified as bacteria – hence as living unicellular entities – despite lacking many of the critical life-sustaining genes (Nakabachi *et al.* 2006). The question therefore is whether microbial diversity – and by the same token *bio*diversity – should extend beyond the domain of clearly living entities. And if yes, whether one can draw a non-arbitrary line at some point beyond which it would no longer make sense to count entities as units of biodiversity (Malaterre 2013). These questions, of course, depend on the reasons why we are interested in biodiversity in the first place, and whether these reasons also extend to the microscopic world.

Microorganisms and conservation

Interest in biodiversity often goes hand-in-hand with interest in conservation, and it is frequently the case that conservation measures are taken when biodiversity is threatened. Hence there is an interest in biodiversity as a means of identifying what is at risk and what needs to be conserved (e.g. Sarkar 2005). Microbial diversity, however, is rarely studied with an immediate view toward conservation. Interest in microbial diversity generally stems from a broad set of applied objectives and theoretical goals, and concern about preserving microorganisms is rarely voiced. Besides the fact that microorganisms are not easily spotted with the naked eye, two major reasons are often offered why conservation efforts should *not* target microorganisms. First, microorganisms are extremely abundant and ubiquitous. Such characteristics, it is said, clearly reduce the apparent need to conserve them. Indeed, why would one care about conserving something that comes in billions of copies per cubic centimeter of soil or milliliter of ocean water? Second, microorganisms are known to be capable of evolving quite rapidly, thereby adapting easily to changes in the environment. For this reason, the conservation of microorganisms would not only be a never-ending task, it would also be totally futile: microorganisms do extremely well by themselves when it comes to surviving, even billions of years of evolution, and it is not at all obvious that they would need us at all in this respect.

On the other hand, one could still be interested in the conservation of *some* microorganisms, while not necessarily caring about *all* microorganisms and their total diversity (Colwell 1997, Cockell and Jones 2009). One point that may put some sense of urgency onto microbial conservation measures is the fact that extinctions of microorganisms have already occurred, and are still occurring as we speak (Weinbauer and Rassoulzadegan 2007). This is typically the case of microorganisms whose exclusive hosts have been driven to extinction (though such microorganisms could possibly still be around in a dormant state). Targeting hosts that are at risk could therefore prevent the extinction of their respective microbial communities, as could be the case, for instance, with the microbial diversity of marine sponges.

One of the most obvious reasons to conserve at least *some* microorganisms is their *direct-use value*. Numerous microorganisms are pathogens, and it is useful to have them, so to speak, in store so as to be able to develop and market appropriate diagnostic tools, drugs, and antibiotics. Microorganisms are also used in a plethora of agro-industrial processes in the food industry, hence additional reasons to conserve them, not to mention the utility of microorganisms in a broad range of other applications, from energy production to environmental remediation, as seen above (see section "Why study microbial diversity?"). So, like larger plants and animals, microorganisms provide marketable commodities in a broad range of ways, from food and medicine, to biological control and industrial raw materials. Note, however, that the rationale behind conservation measures is articulated differently. For macroorganisms, the question is whether current patterns of exploitation are sustainable or not, a perspective that often leads to species extinction issues and the idea of an irreversible biodiversity crisis (e.g. Wilson 1988). For microorganisms, on the other hand, sustainable exploitation is not a central issue at all as they are taken to be extremely abundant, ubiquitous, and evolvable. Rather, the issue is one of convenience and efficiency: the objective is to make sure microorganisms are available where and when we need them to be.

Another reason to put microorganisms on the conservation agenda is their *indirect-use value*. As seen above, microorganisms provide a broad range of ecosystemic services through biogeo-chemical cycling: microorganisms digest dead plants, animals, and microbes, and contribute to the cycling of organic compounds. They also carry out key geochemical processes in the cycling of gases and other compounds. Without microorganisms, it is fair to say that macrobial life would be virtually impossible.

The overall impact of microbial diversity in nutrient cycling is complex, due to the high number of metabolic pathways involved, their redundancies, and their intricate interrelatedness. Furthermore, the scales at which this microbial diversity works are planetary scales. For these reasons, conserving microorganisms with a view to preserving the balance of current planetary environmental parameters, such as CO_2 atmospheric content, is not something as easy to envision as might be, for instance, the conservation of endangered mammals in specific local areas. Nevertheless, the importance of microbial diversity with regards to their geochemical roles stresses the need to further investigate the question of microbial conservation (e.g. Cockell and Jones 2009, Bodelier 2011).

The conservation of microbial diversity also matters when one takes into account the fact that many microorganisms could prove useful in the future. Similarly to larger plants and animals, microorganisms can be said to have an *option value*: they are potential sources for yet-to-be-discovered medicines, industrial applications, and other marketable commodities. Not only is there an unknown future value of known species, but also an unknown value of unknown species (or other units of diversity) (Takacs 1996). Note that such an argument thereby encompasses all microorganisms, thereby giving microbial conservation a much larger scope.

One further question that concerns microbial conservation is whether microorganisms have an *intrinsic value* similar to the intrinsic value that is sometimes argued for in the case of larger plants and animals (Taylor 1989, Cockell 2008). Such an intrinsic value could reflect our ethical feelings; alternately, it could reflect our duties towards microorganisms, as if they had rights to exist comparable to those of macroorganisms. One way to approach this issue is to ask ourselves whether it is acceptable, from an ethical point of view, to intentionally drive microorganisms to extinction, including a pathogenetic bacteria or even a virus. Another way is to find out whether it is sensible to promote conservation measures that are solely motivated by microbial diversity, as has been suggested, for instance, for microbial rock-dwelling communities found on the cliffs of the Niagara Escarpment, Canada (Gerrath *et al.* 2000) or for the microbial communities that inhabit pristine ice-sealed lakes such as Lake Vida, east Antarctica (Doran *et al.* 2008).

Conservation of microorganisms can be implemented in two ways: either *in situ* or *ex situ*. As is the case for macroorganisms, *in situ* conservation typically focuses on the corresponding habitats of these organisms, thereby targeting whole ecosystems or communities. The above case of the microbial communities of the cliffs of the Niagara Escarpment is an example of microorganism conservation through the conservation of their habitat. Another example is the conservation of the hyperthermophilic microorganisms at Yellowstone, USA through the conservation of the hot springs they inhabit. One intriguing question surfaces with the case of microorganisms that dwell in polluted or toxic – from an anthropocentric point of view – environments: should these habitats be conserved as well? It is well known, for instance, that some sites polluted by heavy metals host specific tolerant microorganisms (Stierle *et al.* 2007). Should such habitats be on our conservation agenda for the sake of conserving the microorganisms they host? One way to answer this question could be to turn to *ex situ* conservation measures. *Ex situ* conservation typically consists of isolating particular species – usually under the form of pure strains – and maintaining them alive by culturing them with the appropriate nutrients and physico-chemical conditions. A fair number of microbial resource centers (MRCs), or microbial depositories, have been created worldwide, with the specific aim of developing and maintaining *ex situ* microbial collections. However, despite the development of novel techniques for culturing previously unculturable microorganisms (e.g. Alain and Querellou 2009), the problem remains that a very large number of taxa still cannot be cultured, raising a critical issue for *ex situ* microbial conservation.

Contrary to what is usually possible for larger plants and animals, microorganisms can also be conserved *ex situ* thanks to a broad range of methods that have nothing in common with the natural habitats of these microorganisms. Such conservation methods include cryopreservation, whereby microorganisms are conserved through deep freezing at -196°C, and lyophilization, a method that consists of vacuum desiccation (Prakash *et al.* 2013). Another method that also works by temporarily suspending life-sustaining processes is to induce a state of cell dormancy, a mechanism of cell survival in response to starvation and environmental stress. The conservation of microorganisms can therefore be implemented through a much broader range of methods than is usually available for the conservation of larger plants and animals – at least for now. In addition, the conservation of smaller non-cellular organisms is also possible not just through DNA banks, but also through DNA databases. Such possibilities open up conservation means that are entirely virtual. An interesting question is thereby whether or not such dematerialized conservation changes the nature of our duties towards those organisms that can be conserved in this way. Another question is whether or not – if one believed that all living things, including microorganisms, have intrinsic value and should have the ability to survive and flourish in their natural habitats – such procedures would be ethically acceptable.

Of course, prior to implementing conservation measures, decisions must be taken as to what should be conserved. This is a real challenge for microbial conservation, as it is often impractical to target individual species for conservation in their natural environments (Cockell 2008). The small size of microorganisms, their abundance and diversity at very small spatial scales make it impossible, in practice, to ensure the conservation of any given species individually *in situ*. *In situ* conservation will therefore often target communities of microorganisms and consist of the conservation of their joint habitat. On the other hand, what is often assessed to be of most value in microorganisms is the function they perform, not the fact of belonging to such and such taxa. Therefore, microbial conservation efforts may target the conservation of specific functions that are performed across several taxa, either by aiming at the conservation of communities of microorganisms that all perform the same function, or by specifically focusing on one of these taxa as a representative member of the functional trait. In that case, conservation can be done *ex situ*, for instance by maintaining cultures, or by cryopreservation or lyophilization. Because such functions are encoded in genetic material – including mobile genetic material such as plasmids – conservation can also focus exclusively on the genetic sequences of interest. In any case, the abundance of microorganisms makes it virtually impossible to preserve all of the species of any given genus, let alone all of the strains of any given species. So prioritization is a key element that needs to be further investigated in light of the reasons that justify microbial conservation, as well as of the *in situ* and *ex situ* conservation techniques that are available, including dematerialized digital conservation tools.

Conclusion

Microbial diversity is an area of intense research work at the junction of microbiology and ecology. Fascinating living entities in their own rights, microorganisms raise not only challenging taxonomic issues, but puzzling ecological questions as well. Microbial diversity also raises serious questions of philosophical import, ranging from a renewed debate on the units of biodiversity to an extension of the valuation problem to unicellular and possibly even non-cellular entities at the edge of life. Not only is the study of microbial diversity necessary to understand specific ecological phenomena – as is the case when microbial functional diversity explains how specific ecosystem services take place – it also plays a central theoretical role in our modeling of biodiversity in ways that complement those produced by the study of macroorganism diversity, thereby challenging our very intuitions about biodiversity.

Acknowledgments

I am indebted to Steven Kembel, Purificación López-García, and Maureen O'Malley for very helpful comments on earlier versions of this chapter. I also wish to thank Justin Garson, Anya Plutynski, and Sahotra Sarkar for inviting me to contribute to this collective volume and for providing insightful comments on the manuscript. This chapter benefited from collaborations stimulated by European COST Action TD 1308. Financial support from UQAM Research Chair in Philosophy of Science, from the Canada Research Chairs program and from Canada SSHRC grant 435-2014-0943 is gratefully acknowledged.

References

Alain, K., and J. Querellou. 2009. "Cultivating the Uncultured: Limits, Advances and Future Challenges." *Extremophiles* 13(4): 583–594.

Ayers, G. P., and J. M. Cainey. 2007. "The CLAW Hypothesis: A Review of the Major Developments." *Environmental Chemistry* 4(6): 366–374.

Bapteste, E., and J. Dupré. 2013. "Towards a Processual Microbial Ontology." *Biology & Philosophy* 28(2): 379–404.

Breitbart, M., P. Salamon, B. Andresen, J. M. Mahaffy, A. M. Segall, D. Mead, F. Azam, and F. Rohwer. 2002. "Genomic Analysis of Uncultured Marine Viral Communities." *Proceedings of the National Academy of Sciences* 99(22): 14250–14255.

Catling, D. C., K. J. Zahnle, and C. P. McKay. 2001. "Biogenic Methane, Hydrogen Escape, and the Irreversible Oxidation of Early Earth." *Science* 293(5531): 839–843.

Cho, I., S. Yamanishi, L. Cox, B. A. Methé, J. Zavadil, K. Li, Z. Gao, D. Mahana, K. Raju, I. Teitler, H. Li, A. V. Alekseyenko, and M. J. Blaser. 2012. "Antibiotics in Early Life Alter the Murine Colonic Microbiome and Adiposity." *Nature* 488(7413): 621–626.

Cockell, C. S. 2008. "Environmental Ethics and Size." *Ethics & the Environment* 13(1): 23–39.

Cockell, C. S., and H. L. Jones. 2009. "Advancing the Case for Microbial Conservation." *Oryx* 43(4): 520–526.

Colwell, R. R. 1997. "Microbial Diversity: The Importance of Exploration and Conservation." *Journal of Industrial Microbiology & Biotechnology* 18(5): 302–307.

Demuth, J., H. Neve, and K.-P. Witzel. 1993. "Direct Electron Microscopy Study on the Morphological Diversity of Bacteriophage Populations in Lake Plußsee." *Applied and Environmental Microbiology* 59(10): 3378–3384.

Dimmock, N., A. Easton, and K. Leppard. 2009. *Introduction to Modern Virology*. New York: John Wiley & Sons.

Doran, P. T., C. H. Fritsen, A. E. Murray, F. Kenig, C. P. McKay, and J. D. Kyne. 2008. "Entry Approach into Pristine Ice-sealed Lakes – Lake Vida, East Antarctica, a Model Ecosystem." *Limnology and Oceanography: Methods* 6(10): 542–547.

Dykhuizen, D. E. 1998. "Santa Rosalia Revisited: Why Are There So Many Species of Bacteria?" *Antonie van Leeuwenhoek* 73(1): 25–33.

Ereshefsky, M. 2010. "Microbiology and the Species Problem." *Biology & Philosophy* 25(4): 553–568.

Falkowski, P. G., T. Fenchel, and E. F. Delong. 2008. "The Microbial Engines that Drive Earth's Biogeochemical Cycles." *Science* 320(5879): 1034–1039.

Forterre, P. 2010. "Defining Life: The Virus Viewpoint." *Origins of Life and Evolution of Biospheres* 40(2): 151–160.

Franklin-Hall, L. R. 2007. "Bacteria, Sex, and Systematics." *Philosophy of Science* 74(1): 69–95.

Gerrath, J. F., J. A. Gerrath, U. Matthes, and D. W. Larson. 2000. "Endolithic Algae and Cyanobacteria from Cliffs of the Niagara Escarpment, Ontario, Canada." *Canadian Journal of Botany* 78(6): 807–815.

Gevers, D., F. M. Cohan, J. G. Lawrence, B. G. Spratt, T. Coenye, E. J. Feil, E. Stackebrandt, Y. Van de Peer, P. Vandamme, F. L. Thompson, *et al.* 2005. 'Re-evaluating Prokaryotic Species.' *Nature Reviews Microbiology* 3(9): 733–739.

Gogarten, J. P., W. F. Doolittle, and J. G. Lawrence. 2002. "Prokaryotic Evolution in Light of Gene Transfer." *Molecular Biology and Evolution* 19(12): 2226–2238.

Green, J. L., B. J. M. Bohannan, and R. J. Whitaker. 2008. "Microbial Biogeography: From Taxonomy to Traits." *Science* 320(5879): 1039–1043.

Horner-Devine, M. C., K. M. Carney, and B. J. M. Bohannan. 2004. "An Ecological Perspective on Bacterial Biodiversity." *Proceedings of the Royal Society B: Biological Sciences* 271(1535): 113–122.

International Committee on Taxonomy of Viruses. 2015. Available online at www.ictvonline.org/virusTaxInfo.asp (Accessed 9 June 2016).

Javaux, E. J., C. P. Marshall, and A. Bekker. 2010. "Organic-walled Microfossils in 3.2-billion-year-old Shallow-marine Siliciclastic Deposits." *Nature* 463(7283): 934–938.

Kallmeyer, J., R. Pockalny, R. R. Adhikari, D. C. Smith, and S. D'Hondt. 2012. "Global Distribution of Microbial Abundance and Biomass in Subseafloor Sediment." *Proceedings of the National Academy of Sciences* 109(40): 16213–16216.

Kembel, S. W., E. Jones, J. Kline, D. Northcutt, J. Stenson, A. M. Womack, B. J. Bohannan, G. Z. Brown, and J. L. Green. 2012. "Architectural Design Influences the Diversity and Structure of the Built Environment Microbiome." *The ISME Journal* 6(8): 1469–1479.

Kembel, S. W., T. K. O'Connor, H. K. Arnold, S. P. Hubbell, S. J. Wright, and J. L. Green. 2014. "Relationships between Phyllosphere Bacterial Communities and Plant Functional Traits in a Neotropical Forest." *Proceedings of the National Academy of Sciences* 111(38): 13715–13720.

La Scola, B., S. Audic, C. Robert, L. Jungang, X. de Lamballerie, M. Drancourt, R. Birtles, J.-M. Claverie, and D. Raoult. 2003. "A Giant Virus in Amoebae." *Science* 299(5615): 2033–2033.

Li, J., S. Browning, S. P. Mahal, A. M. Oelschlegel, and C. Weissmann. 2010. "Darwinian Evolution of Prions in Cell Culture." *Science* 327(5967): 869–872.

López-García, P. 2005. "Extremophiles." In *Lectures in Astrobiology*, ed. M. Gargaud, B. Barbier, and J. Reisse, 657–679. Heidelberg: Springer.

Maclaurin, J., and K. Sterelny. 2008. *What is Biodiversity?* Chicago, IL: University of Chicago Press.

Magurran, A. E., and B. J. McGill, eds. 2011. *Biological Diversity: Frontiers in Measurement and Assessment.* Oxford: Oxford University Press.

Malaterre, C. 2013. "Microbial Diversity and the 'Lower-limit' Problem of Biodiversity." *Biology & Philosophy* 28(2): 219–239.

Martiny, J. B. H., B. J. M. Bohannan, J. H. Brown, R. K. Colwell, J. A. Fuhrman, J. L. Green, M. C. Horner-Devine, M. Kane, J. A. Krumins, C. R. Kuske, P. J. Morin, S. Naeem, L. Øvreås, A.-L. Reysenbach, V. H. Smith, and J. T. Staley. 2006. "Microbial Biogeography: Putting Microorganisms on the Map." *Nature Reviews Microbiology* 4(2): 102–112.

Moreira, D., and P. López-García. 2002. "The Molecular Ecology of Microbial Eukaryotes Unveils a Hidden World." *Trends in Microbiology* 10(1): 31–38.

Moreira, D., and P. López-García. 2009. "Ten Reasons to Exclude Viruses from the Tree of Life." *Nature Reviews Microbiology* 7(4): 306–311.

Morgan, G. J. 2010. "Evaluating Maclaurin and Sterelny's Conception of Biodiversity in Cases of Frequent, Promiscuous Lateral Gene Transfer." *Biology & Philosophy* 25(4): 603–621.

Nakabachi, A., A. Yamashita, H. Toh, H. Ishikawa, H. E. Dunbar, N. A. Moran, and M. Hattori. 2006. "The 160-Kilobase Genome of the Bacterial Endosymbiont *Carsonella.*" *Science* 314(5797): 267–267.

OECD. 2007. *OECD Best Practice Guidelines for Biological Resource Centres.* Paris: Organization for Economic Co-operation and Development. Available online at www.oecd.org/science/biotech/38777417.pdf (accessed 2 April 2014).

O'Malley, M. A., and J. Dupré. 2007. "Size Doesn't Matter: Towards a More Inclusive Philosophy of Biology." *Biology & Philosophy* 22(2): 155–191.

Øvreås, L., and T. Curtis. 2011. "Microbial Diversity and Ecology." In *Biological Diversity: Frontiers in Measurement and Assessment,* ed. A. Maguran and B. McGill, 221–236. Oxford: Oxford University Press.

Pace, N. R. 1997. "A Molecular View of Microbial Diversity and the Biosphere." *Science* 276(5313): 734–740.

Postgate, J. 1998. *Nitrogen Fixation.* Cambridge: Cambridge University Press.

Prakash, O., Y. Nimonkar, and Y. S. Shouche. 2013. "Practice and Prospects of Microbial Preservation." *FEMS Microbiology Letters* 339(1): 1–9.

Prosser, J. I., B. J. Bohannan, T. P. Curtis, R. J. Ellis, M. K. Firestone, R. P. Freckleton, J. L. Green, L. E. Green, K. Killham, J. J. Lennon, *et al.* 2007. "The Role of Ecological Theory in Microbial Ecology." *Nature Reviews Microbiology* 5(5): 384–392.

Rohwer, F. 2003. "Global Phage Diversity." *Cell* 113(2): 141.

Sarkar, S. 2005. *Biodiversity and Environmental Philosophy: An Introduction.* Cambridge: Cambridge University Press.

Sforna, M. C., M. A. van Zuilen, and P. Philippot. 2014. "Structural Characterization by Raman Hyperspectral Mapping of Organic Carbon in the 3.46 Billion-year-old Apex Chert, Western Australia." *Geochimica et Cosmochimica Acta* 124: 18–33.

Staley, J. T., and J. J. Gosink. 1999. "Poles Apart: Biodiversity and Biogeography of Sea Ice Bacteria." *Annual Reviews in Microbiology* 53(1): 189–215.

Stierle, D. B., A. A. Stierle, and B. Patacini. 2007. "The Berkeleyacetals, Three Meroterpenes From a Deep Water Acid Mine Waste Penicillium." *Journal of Natural Products* 70(11): 1820–1823.

Suttle, C. A. 2005. "Viruses in the Sea." *Nature* 437(7057): 356–361.

Takacs, D. 1996. "The Idea of Biodiversity: Philosophies of Paradise." Available online at http://philpapers. org/rec/TAKTIO (accessed 23 January 2015).

Tannock, G. W. 1990. "The Microecology of Lactobacilli Inhabiting the Gastrointestinal Tract." In *Advances in Microbial Ecology,* ed. K. C. Marshall, 147–171. New York: Plenum.

Taylor, P. W. 1989. *Respect for Nature: A Theory of Environmental Ethics.* Princeton, NJ: Princeton University Press.

Torsvik, V., L. Øvreås, and T. F. Thingstad. 2002. "Prokaryotic Diversity – Magnitude, Dynamics, and Controlling Factors." *Science* 296(5570): 1064–1066.

Treseder, K., T. Balser, M. Bradford, E. Brodie, E. Dubinsky, V. Eviner, K. Hofmockel, J. Lennon, U. Levine, B. MacGregor, J. Pett-Ridge, and M. Waldrop. 2012. "Integrating Microbial Ecology into Ecosystem Models: Challenges and Priorities." *Biogeochemistry* 109(1–3): 7–18.

Waldrop, M. P., T. C. Balser, and M. K. Firestone. 2000. "Linking Microbial Community Composition to Function in a Tropical Soil." *Soil Biology and Biochemistry* 32(13): 1837–1846.

Wang, X., C. Zhou, K. Tang, J. Lan, Y. Zhou, and Z. Li. 2008. "Preliminary Studies on Species and Distribution of Citrus Viroids in China." *Agricultural Sciences in China* 7(9): 1097–1103.

Weinbauer, M. G., and F. Rassoulzadegan. 2007. "Extinction of Microbes: Evidence and Potential Consequences." *Endangered Species Research* 3(2): 205–215.

Whitman, W. B., D. C. Coleman, and W. J. Wiebe. 1998. "Prokaryotes: The Unseen Majority." *Proceedings of the National Academy of Sciences* 95(12): 6578–6583.

Wilson, E. O., ed. 1988. *Biodiversity*. Washington, DC: National Academy Press.

Wren, J. D., M. J. Roossinck, R. S. Nelson, K. Scheets, M. W. Palmer, and U. Melcher. 2006. "Plant Virus Biodiversity and Ecology." *PLoS Biology* 4(3): e80.

Further reading

Barton, L., and Northrup, D. E. 2011. *Microbial Ecology*. Hoboken, NJ: Wiley-Blackwell. (A comprehensive introduction to microorganisms, how they interact with micro- and macroorganisms, and contribute to ecosystem services.)

Bodelier, P. L. E. 2011. "Toward Understanding, Managing, and Protecting Microbial Ecosystems." *Frontiers in Microbiology* 2(80). (A perspective paper on the status of microbial diversity and conservation from an ecosystemic point of view.)

Madigan, M. T., J. M. Martinko, K. S. Bender, D. H. Buckley, D. A. Stahl, and T. Brock. 2014. *Brock Biology of Microorganisms*, 14th edn. Boston, MA: Benjamin Cummings. (Very comprehensive biology textbook – 1,032 pages on microorganisms, with specific chapters on diversity and ecology. Everything you ever dreamed of knowing about microorganisms.)

O'Malley, M. 2014. *Philosophy of Microbiology*. Cambridge: Cambridge University Press. (Extremely well-documented volume on the philosophy of microbiology. See in particular chapter 5 on microbial ecology.)

Ogunseitan, O. 2005. *Microbial Diversity: Form and Function in Prokaryotes*. Malden, MA: Blackwell. (An introduction to microbial diversity, assessment, and applications.)

PART III

Why protect biodiversity?

11

IS BIODIVERSITY INTRINSICALLY VALUABLE? (AND WHAT MIGHT THAT MEAN?)

Katie McShane

Introduction

Michael Soulé famously claimed that the intrinsic value of biodiversity is one of the central normative postulates defining the field of conservation biology (Soulé 1985: 731). Other writers, while they haven't gone quite that far, have nonetheless argued that intrinsic value claims are part of the philosophy behind the very idea of biodiversity conservation (Reyers *et al.* 2012), and that conservationists should insist on voicing their support for the intrinsic value of biodiversity (McCauley 2006). These claims have been quite controversial. Other writers have argued that intrinsic value claims are not useful – and are even counterproductive – in conservation (Maguire and Justus 2008).

While the discussion of intrinsic value has a very long history in environmental ethics, the debates about it and about its application to biodiversity have been rekindled recently by the increase in the popularity of the "ecosystem services" approach to assessing the value of the natural world. The ecosystem services approach assesses the value of some part of the natural world – an ecosystem, a species, etc. – by determining the economic value of the services it provides to humans – filtering water, preventing erosion, etc. Many criticisms have been raised of this approach (see e.g. McCauley 2006, Redford and Adams 2009, Hansen 2011, Conniff 2012), and some critics have pointed to claims about the intrinsic value of biodiversity as an alternative – and perhaps preferable – position.

However, as veterans of the intrinsic value debates in environmental ethics will attest, people often mean very different things when they use the term "intrinsic value," and debates in which the participants assume different meanings of key terms typically don't go well (O'Neill 1992, McShane 2007, Jamieson 2008). The recent literature in conservation biology is no exception. As will become evident below, there are very different understandings of what intrinsic value *is* that underlie the different positions in this debate. The first task, then, in trying to answer the question "does biodiversity have intrinsic value?" is getting clear about what one means by "intrinsic value." In the next section, I will give a rough overview of the most common meanings that people seem to presuppose when they claim that biodiversity does or doesn't have intrinsic value. For each meaning, I will briefly describe what one would be committed to in claiming that biodiversity is intrinsically valuable in that sense. In the final section, I will consider which meanings render the claim that biodiversity has intrinsic value most plausible, and describe other claims that might better capture theorists' worries about the ecosystem services approach.

Meanings of "intrinsic value"

To ask whether biodiversity has intrinsic value is to ask an unclear question, for there are many things that people use both the terms "biodiversity" and "intrinsic value" to mean. As earlier chapters in this book have made clear, "biodiversity" can refer to variety that exists along a number of different dimensions. First, there are many different things we might want diversity *of*: phenotypes, genotypes, ecosystemic niches, species, populations, and so on. Second, there are many different things we might mean by the "diversity" of such things: the sheer number of different kinds represented; how far such kinds depart from the norm; how many departures from the norm are represented, and so on (Sarkar 2005, Maclaurin and Sterelny 2008).[1] Together, these are what Maclaurin and Sterelny (2008) refer to as the "units-and-differences problem," and as they rightly point out, a solution to it is required before it will be clear what assessments of biodiversity are assessments *of*.

Let us for the moment, however, set aside difficulties with the multiple meanings of "biodiversity." Pick a meaning, and take that to be what is at issue in the value claims that follow. We will return to this issue in the last section of the chapter, but first, further clarifications are needed. Most immediately, in order to determine what it would mean for biodiversity to have intrinsic value, we need to see what is involved in ascribing any kind of value (intrinsic or otherwise) to biodiversity. This requires us to get clearer about what kind of a thing, conceptually speaking, biodiversity is. In ordinary talk, we might say that biodiversity isn't really a *thing* (as, say, ecosystems are) so much as a *way that things can be*.[2] Biodiversity might be described as a property (or perhaps a cluster of related properties) that things (communities, regions, ecosystems, etc.) can instantiate, perhaps to varying degrees. Technically speaking, then, we might understand our ascription of value to biodiversity in at least two different ways.

One claim we might be making when we say that biodiversity is valuable is that the instantiation of states of affairs wherein objects have this property is valuable. The more things in the world that are biodiverse – and biodiverse to a higher degree – the better the world is. There is nothing peculiar about this sort of claim (contra Maier 2012: 22). It precisely parallels what hedonistic utilitarians say about happiness. The more happiness there is in the world, the better. Similarly, one might be claiming, the more biodiversity there is in the world, the better.

The other claim we might be making when we say that biodiversity is valuable is that the possession of this property by an object makes a positive contribution to the value of the object (e.g. the place, the ecosystem, the population, etc.). To conceive of the value of biodiversity this way is to think of it as what philosophers call a "value-adding property" of objects rather than a valuable aspect of states of affairs. On this view, the claim isn't that the property of biodiversity itself is valuable, or that states of affairs are more valuable the more biodiversity they involve, but rather that being biodiverse makes particular concrete things (places, ecosystems, etc.) more valuable.[3] This is also a familiar kind of claim within ethics.[4] For example, some ethicists claim that the possession of virtue adds to the value of a person – to be a virtuous person is to be a better person.[5]

There is nothing about the structure of ethical theories or about the kind of thing biodiversity is that would rule out the attribution of value to biodiversity. How to understand what exactly it means to say that biodiversity is valuable might be different for different theories, but it would be wrong to think that because biodiversity is a property, it is not the kind of thing that can be a bearer of value.

But what about the possession not just of value, but of *intrinsic* value? This is the claim we're trying to evaluate, after all. What might we mean by making that claim, and could it be true?

Moral standing

There is one thing that some people mean by "intrinsic value" (sometimes called "inherent value" or "inherent worth") that we can rule out quickly. For something to have intrinsic value in this sense is for it to have moral standing or moral considerability, for it to be the kind of thing whose interests ought to matter to moral agents in their decisions about what to do. This is an attribution that one sees often in the literature on animal ethics (Regan 1983) and in some arguments for individualist biocentrism (Taylor 1980). It is most common in deontological ethical theories.[6] Yet it is fairly clear that "biodiversity" doesn't itself have interests. It isn't a thing, after all, but rather an aspect of things. It might figure in our assessments of what counts as a harm or benefit to things that do have interests, but if having intrinsic value requires its bearers to have interests of their own, then biodiversity cannot have intrinsic value of this kind. So if this is what one means by "intrinsic value," then biodiversity cannot have intrinsic value. However, this is far from the only thing that people mean by "intrinsic value."

Objective value

Another meaning that people sometimes seem to have in mind when they talk of intrinsic value is "objective value," that is to say, value that a thing has independently of whether it is in fact valued (O'Neill 1992, Maclaurin and Sterelny 2008).[7] This idea of the objectivity value is often a point of misunderstanding between scientists and ethicists. Claims about the objectivity of value are quite common in ethics, but sometimes baffling to those who work in the sciences. When ethicists say that a value is objective they typically mean that it's something a person could be mistaken about, and there are many different accounts of what mistakenness might consist of within ethical theory. But roughly, the idea is this: I could believe that it's perfectly fine to murder innocent people, for example, and yet be wrong. My belief that it's ok to murder innocent people doesn't make it ok (not even just for me). In fact, we might all believe that it's perfectly fine to murder innocent people, and we all might be wrong. Even if every single living person believed that it was ok, that wouldn't be enough to make it ok. Most ethicists think that ethics needs to allow for the possibility of widespread, perhaps even universal, error in our ethical beliefs. There is much more to say about this issue, and the field of metaethics is full of sophisticated treatments of the matter. For our purposes, it is simply important to note that ethical theories today typically accept this very minimal kind of objectivity for at least some values, though which values and exactly how to understand this objectivity differ from theory to theory.[8]

I think that this fact is often puzzling to scientists for at least two reasons. First, the sciences take as their object of study the valu*ed*, not the valu*able*. That is to say, the sciences ask what people value (and how they value it, how much they value it, etc.), not what they ought to value. The social sciences examine preferences, and preferences that are especially deeply held and central to a person's identity are deemed "values" (e.g., Malle 2004: 259, n.2). Thus when a sociologist studies values, she is asking what people prefer, not whether they are right to do so. Ethics, however, asks not about the existence of preferences but about their justification. When an ethicist studies values, she is usually asking which things are worth preferring, not which things are in fact preferred. So scientists and ethicists often have in mind very different things when they talk about "values."[9] Second, scientists often have a different sense of what objectivity consists of than ethicists do. From the perspective of the sciences, it is easy to think that for some matter to be objective isn't just for it to be the sort of thing that one could be wrong about, but also for its truth to be built into the world and discernable – at least in principle – through certain kinds of investigation, such as observation and experimentation. Some version of the scientific method, then, is the way that one gets at objective truths about the world. From this point of view, there is a close connection between a particular methodology (the scientific method) and objectivity itself. Ethics uses a different methodology: that of philosophy rather than that of the sciences. Philosophers are less inclined to believe that the scientific method is the only way to discern objective truths about the world, and this makes room for a broader understanding of what objectivity might involve.[10]

There is, of course, much more to say about all of this. The relation between philosophy, the sciences, and other fields of inquiry, not to mention the nature of objectivity itself, are interesting and complex issues. However, with all of this said, I think that the objectivity of value is a red herring in the discussions about the intrinsic value of biodiversity. It might be true that my inner ethicist would be inclined to say that some values are objective and that your inner scientist would be inclined to disagree. But in discussions of biodiversity, the issue is really whether biodiversity is valuable in a different way – and perhaps a more important way – than other things are.[11] This isn't a debate about whether any values are objective; it's a debate about how the value of biodiversity compares to the value of other things. Is the value of biodiversity just a matter of aesthetic taste (e.g. I like variety) or the career interests of scientists (e.g. you need a good topic for your next grant proposal)? Or is it about something that matters in a deeper, more important, or simply different way? This is a question that can't be answered by metaethics; it requires a substantive investigation into what (if anything) it is about biodiversity that makes it good.

Non-instrumental value

Another thing we might mean by saying that something has intrinsic value is that it has value that goes beyond just serving our interests.[12] Critics of the "ecosystem services" approach often seem to have this kind of value in mind when they argue for intrinsic value claims as an alternative to the ecosystem services approach. Their criticism is that the ecosystem services approach reduces the value of the natural world to what it can do for us, and that this view is too narrow – surely the value of the natural world isn't exhausted by the ways in which it serves human ends (Redford and Adams 2009, Schröter *et al.* 2014)?

There are a couple of aspects of this discussion worth noticing. First, technically, for a thing to be non-instrumentally valuable is for it to be valuable independently of the ways in which it serves the interests of others – that is, the interests of *any* others, not just human beings. If biodiversity's value is a matter of how it serves the interests of all living things, then its value is merely instrumental. For this reason, the denial of anthropocentrism (the view that the non-human natural world matters only insofar as it serves *human* interests) doesn't imply the denial of instrumental value (the claim that a thing has value only insofar as it serves the interests of *any* other thing). To show that biodiversity has value beyond its usefulness to humans isn't to show that it has value beyond its usefulness to all other things.

Second, "ecosystem services" are not the only way that the natural world might be thought to serve human interests. While economists and policy-makers can sometimes leave us thinking that our only interests are financial, environmental philosophers have long argued that human interests are broad and varied: we also have scientific interests, aesthetic interests, religious/spiritual interests, and so on. Many philosophers have argued that these values aren't well captured either by actual ecosystem services assessment tools or by the economic modes of valuation meant to account for them – existence value, option value, and the like (Davidson 2013, Jax *et al.* 2013). Taking these broader interests seriously would still provide grounds for criticizing the ecosystems services approach (at least in its current forms) as too narrow without abandoning the claim that ultimately the value of the natural world is a matter of its serving human interests.

So concerns about the ecosystem services approach don't lead quite so directly to claims about intrinsic value, even in the non-instrumental sense. But even if these claims can't be derived from the inadequacies of the ecosystem services approach to valuation, might it nevertheless be true that biodiversity has non-instrumental value? To think about this, imagine that biodiversity was of no use whatsoever to anything. Imagine that increasing or decreasing the biodiversity of the world and its parts would have no effect on the interests of anything that has interests. Would biodiversity still be worth caring about? And if so, why? It is a difficult thought experiment to run because so much of our talk about biodiversity is about its importance for the robustness of populations, the survival of species, the resilience of ecosystems, etc. In the actual world, biodiversity seems to be deeply connected to the interests of many things, including ourselves. But perhaps that should tell us something. Perhaps that is in fact why we value biodiversity – because of its importance to the interests of so many things. If so, then its value might be instrumental after all.

Non-extrinsic value

Another meaning of "intrinsic value" is slightly broader than non-instrumental value: it amounts to value that a thing possesses independently of *any* connection to *any* valuable thing.[13] Notice that non-instrumental value is the value that a thing has independently of a certain kind of relationship to another valuable thing: serving that thing's interests. But there might be ways that a thing can get its value from another thing besides simply serving its interests. Value theorists have proposed many other kinds of relationship that might confer value on an otherwise valueless thing. Some commonly discussed relationships are:

- **Constitutive value**: value that a thing gets by being a part of something else that is valuable (e.g. a single brushstroke on the "Mona Lisa").
- **Symbolic value**: value that a thing gets by representing something else that is valuable (e.g. a memento).
- **Evidential value**: value that a thing gets by being evidence of something else that is valuable (e.g. the lab results showing you are cancer-free).
- **Historical value**: value that a thing gets by being part of the history of something else that is valuable (e.g. the pen that James Madison used to sign the US Constitution).

Non-extrinsic value would be the value a thing has independently of any of these kinds of relationships. However, if it seemed implausible above to think that biodiversity could be valuable if it didn't serve anything's interests, it will likely seem even more implausible to think that biodiversity is valuable in the absence of any value-conferring relationship with other things. That is what one would be claiming in claiming that biodiversity has non-extrinsic value.

Final value

Whether a thing has final value or not is a matter of how one ought to care about it. Something has final value if it is the sort of thing that one ought to care about for its own sake, rather than for the sake of some other good.[14] Unlike non-extrinsic value, which is a matter of where a thing gets its value from, final value is a matter of what kind of attitude we should take toward it. Final value is sometimes called "end value," reflecting the idea that one ought to care about such goods as ends rather than merely as means. Many ethical norms are about how one ought to care about things, not just about how much one ought to care about them. For example, it is a common view that one ought to care about one's friends as ends, not merely as means. To care about a person simply as a means to your other ends is to use her rather than to befriend her. Likewise, it is a common view that one ought to care about money only as a means and not as an end. Money is good because of the other goods in life that it can provide for one; to want lots of money simply for the sake of having it is to care about money in the wrong way.

The difference between caring about something as an end and caring about it as a mere means is in the first instance a difference in one's psychological orientation toward it. Yet the difference is not merely psychological – these different orientations are connected to different social practices, supported by different kinds of reasons, and so on. In Western cultures at the very least, there is an important difference between the way that one feels about and acts toward things that one regards as merely useful and the way that one feels about and acts toward things that one regards as worthy of respect or love.[15]

My own view is that these psychological differences are in the background of many discussions of what has intrinsic value and what doesn't. What many people find objectionable about the view that nature has value only insofar as it serves human interests is that it seems to relegate nature to the category of "things that are useful" rather than "things that deserve our respect." If we were to agree with Kant that everything has either a dignity or a price, then putting a price on nature can feel like a failure to accord it the dignity – and thus respect – that it deserves.[16]

So what would it mean to say that biodiversity has final value? It would mean that biodiversity is something we should care about for its own sake. Some people make this kind of claim about knowledge (also: virtue, kindness, integrity). The idea is that we should value these things for their own sake, not for the sake of other goods they might be a means to. But should we care about biodiversity for its own sake? There are reasons one might give for answering this question affirmatively. Perhaps a biodiverse world is simply a better world. That could be because diversity in general is a good thing or because diversity specifically in the biological world is a good thing (Miller 1982, Mikkelson 2014). Either of these claims, of course, would stand in need of justification. Why would the world be better for having more kinds of things – or perhaps more kinds of biological things – in it? Is this just at bottom an aesthetic preference that creatures like us happen to have (we get bored easily) or is there something about variety that is inherently linked to goodness?

Alternatively, one might try to give a more pragmatic justification for this claim. Perhaps creatures like us are more careful about values that we regard as ends rather than as means; perhaps regarding something as a mere means makes us more likely to underestimate its importance. Humans have all sorts of cognitive biases, moral blind spots, and motivational failures: there might be certain vices that we are more likely to fall into when we value certain kinds of things as mere means. If this were true – if we tend to undervalue or mistreat things we regard as means – then one could argue that in circumstances such as the ones we are in, where undervaluing biodiversity would be especially disastrous, we ought to value biodiversity as an end.

Pragmatic justifications such as these are very controversial within ethics. On the one hand, there are problems with recommending to people that they value something in a way that it does not actually merit. If we were to hide from people the fact that biodiversity does not actually deserve to be valued for its own sake, we would be acting disingenuously and manipulatively. If we were not to hide that fact from people, we would be asking them to adopt attitudes that they know to be mistaken. Neither of these allows for an open, honest, straightforward ethic.

On the other hand, insofar as ethics makes recommendations to us humans about what to do, it must take account of the psychology that we actually have. Final value is explicitly formulated to do that: what defines it and distinguishes it from value as a mere means is a matter of the psychological orientation of valuers: it amounts to caring about something one way rather than another way. If final value is meant to address the question "how should we care about that thing?," then the ways of caring of which humans are capable, and the implications of caring about things in those ways, is clearly relevant. From this point of view, the pragmatic justification is not so much disingenuous as context-sensitive. Imagine: if humans could only care about things in four ways, and if three of them tended to go terribly wrong in cases such as these, then recommending the fourth is precisely what we *should* be doing. It would be a recommendation based on facts about us as well as facts about what we are valuing, but that wouldn't undermine its legitimacy. On this view, ethical recommendations should always take account of both the nature of the objects we are responding to and what is involved in responding to it one way rather than another.

We cannot resolve the debate between these positions here; doing so would take us deep into metaethics and moral psychology. Suffice it to say that should these issues be successfully resolved, this might be a way to argue that biodiversity has – or should be treated as having – final value.

Unconditional value

Another thing one might mean in claiming that biodiversity has intrinsic value is that it has value regardless of the context it is in (Korsgaard 1983, Rolston 1988). Some writers have claimed that this is what the intrinsicness of value must amount to, since to be intrinsic to a thing is to be internal to it. Hence, they have argued, things that have intrinsic value must carry their value with them rather than acquiring (or losing) it though relations with other things (Rolston 1988, Moore 1993). On this view, then, intrinsic value is value that a thing has in its own right – no matter what context it finds itself in.[17] This is different from non-extrinsic value: since intrinsic properties might nonetheless be accidental properties, they needn't be possessed in every context.

Though not impossible, it would be surprising if biodiversity had value of this kind. One could imagine, after all, a world with too much biodiversity: a world in which there are so many different kinds of things so different from one another that there is hardly any unity, cohesion, or integrity possible within the natural world. It is difficult to see why biodiversity would still be valuable in such a world. On the face of it, our valuing of biodiversity seems to be a response to the threats we face in the present moment. Our behaviors are dramatically reducing the levels of biodiversity, largely by killing off so many things. The result is a world with fewer things, and fewer kinds of things, than it once had. This is a real loss. But the fact that it is a loss doesn't mean more diversity is always better. To give an analogy, if the iron levels in my blood drop dramatically, that might reduce my health greatly. But I shouldn't conclude from this that increasing the amount of iron in a person's blood, regardless of the initial level, will always improve her health. Likewise with biodiversity: depending on the initial conditions in a region and the kind of diversity being added, an increase in biodiversity could threaten its ecological stability rather than improve it. In that case, we might think that more biodiversity is a bad thing.

Overriding value

Another meaning that people sometimes seem inclined to give to intrinsic value is a kind of value that overrides other values.[18] It is true that some ethicists, particularly deontologists, think that certain values override others. The value of the life of the drowning child I save, for example, overrides the value of my shoes, which will be ruined by rescuing her. The claim here is not that certain values merely outweigh others – all ethicists allow that this can happen. The value of $6 outweighs the value of $5. Rather, the claim is that some kinds of value take priority over others in virtue of their kind, not just in virtue of their amount. The value of the child's life outweighs the value of my shoes because *it is the value of a person's life*. Lives trump shoes. It doesn't matter how expensive the shoes are or how many of them will be ruined by my efforts; the child's life will always be more valuable. So one thing we might mean by intrinsic value is this kind of status: a value that wins out over others in cases of conflict simply in virtue of the kind of value that it is.

If this is what we mean by intrinsic value, does biodiversity have it? One might be tempted to think so when thinking about some of the values that often conflict with biodiversity. If I want to build a parking lot over an area with a particularly high degree of biodiversity, an environmentalist might describe this as a case where profit and biodiversity are in conflict with one another and argue that the value of biodiversity should override the value of generating profits. Surely we shouldn't let a species go extinct or destroy critical habitat just to improve this year's bottom line. Trade-offs such as these often seem unacceptable, and one way to explain their unacceptability would be to say that it comes from the kinds of goods involved. It doesn't matter how much profit I can make, one might think, destroying biodiversity for any amount of profit is unacceptable.

Aside from the obvious fact that one would need to show *why* such trade-offs are impermissible, another concern about these kinds of claims about overriding value is that they are absolute in a way that can be problematic. The above claim implies that no amount of biodiversity can be destroyed for any amount of profit. But what if the reduction in biodiversity would be so small as to be both unnoticeable and not have any effect on the functioning of the biological systems involved? And what if the profit were to go toward alleviating a considerable amount of human or animal suffering? What if we lived in a world where biodiversity was abundant and profits were both rare and crucial for survival? Overridingness claims are very strong ones. They can make arguments in favor of conservation easy to formulate, but that ease comes at a cost: they impose a fairly rigid set of constraints on our ethical reasoning, and we might have good reasons for wanting an ethics that is more sensitive to changes in context.

Non-anthropocentric value

The last meaning that is sometimes attached to "intrinsic value" in claims about the intrinsic value of biodiversity is in some ways not really about the intrinsicness of value at all. This is the claim that for a thing to be intrinsically valuable is for it to be valuable independently of its benefit to humans. We saw above that non-instrumental value and non-extrinsic value both concern whether a thing's value depends on its usefulness or other relationship to anything outside of itself, not just to humans. But particularly in the literature on ecosystem services, the question is often whether all value depends on benefits to humans. This isn't usually considered a question about the intrinsicness of value, either in ethical theory or in environmental ethics. There it is considered a question about whether all value is anthropocentric value. However, I include it here because of its centrality in the literature on ecosystem services.

It is worth noticing that the ecosystem services approach does focus on benefits to humans rather than on benefits to members of other species or to wholes such as ecosystems, entire species, etc. And yet, there is no ethical reason why benefits to humans should be all that count toward a thing's value. From an ethical perspective, it is difficult to see why providing clean water to humans has worth but providing clean water to chimpanzees does not. We all need clean water; we all suffer if we do not get it. Of course, chimps are not particularly inclined to pay for those ecosystem services that they cannot get for free, and they are terrible at filling out contingent valuation surveys, so assessing the value of the benefit to them would be difficult. But it is important to distinguish between things that have no worth and things with a worth that is difficult to measure. If the benefit of an ecosystem's services to non-humans is an important part of its value, and if our tools for pricing ecosystem services fail to take account of that value, then we should regard those tools as seriously deficient.

So what reasons are there for thinking that the value of biodiversity isn't simply a matter of its benefits to us? One reason might be that we think it has value in its own right, in the sense of having non-instrumental value, non-extrinsic value, and/or unconditional value. If it had one of these kinds of value, then its value wouldn't just be a matter of its benefit to us, or to anything else. Perhaps, as mentioned above, a biodiverse world is simply a better world. If that were true, then biodiversity wouldn't need to be beneficial to anyone or anything in order to have value. But another reason for thinking that the value of biodiversity isn't just a matter of its benefits to us might be that, while we accept that biodiversity is only valuable insofar as it is beneficial to something, the "something" involved needn't be a human. Here, I think, is where many authors' concerns really lie and where I think there is the strongest case to be made. To treat the natural world as though it has value only to the extent that it confers benefits on us is precisely *not* to view ourselves, as Leopold urged, as "plain members and citizens of the biotic community" (Leopold 1970). It is to view ourselves as owners, consumers, and/or masters of the natural world, whose role is simply to be serviced by our ecosystems. As Sian Sullivan (2009) aptly points out, there is nothing reciprocal about this relationship: the ecosystem provides services to us; we do not provide services to it – and even if we did, they would have no value.

Whether biodiversity does confer benefits on whichever parts of the natural world we think it is good to benefit is an empirical question. How to answer it turns on what kind of biodiversity we have in mind, and which things we think have a welfare that matters morally. However, it is worth noticing that the facts we find are likely to be complicated. We know that too little diversity of certain kinds can be a problem and that too much can also be a problem. Increasing the species richness of an area sometimes increases the stability or resilience of the ecosystem and sometimes reduces it. Increasing the genetic diversity of a population sometimes leads to greater rates of survival over the long term and sometimes leads to lower ones – much depends on what the relevant genes do for and in their host organisms. However, even if the facts do turn out to be complicated, we shouldn't conclude from this that the value of biodiversity is equivocal. We do, after all, live in a time when we are losing a great deal of it, and that fact is causing problems for many of Earth's inhabitants, ourselves included. Whether we should think of our environmental problems as primarily about the extinction of particular species, or the increase in ecosystemic instability, or the reduction in biodiversity, etc., is an open question – there are lots of things to worry about. But even if biodiversity isn't everything that matters, we shouldn't conclude that we don't need to worry about it.

Conclusion

As the preceding analysis shows, there is much more that would need to be said to show that biodiversity is not just valuable, but intrinsically valuable. One might try to show that a more bio-diverse world is simply a better world – that biodiversity is itself a value-adding property. That is a very difficult kind of argument to construct, and past attempts have not gone so well, but perhaps there is a better argument still to be made.[19] Or one might try to show that in virtue of the kind of value it is, biodiversity can never be traded off against certain other kinds of values. Again, an argument would need to be given for this claim and the relevant kinds made more explicit – this would be no small task. Somewhat more promisingly, one might try to show that biodiversity is the sort of thing we should care about for its own sake, due to some combination of facts about biodiversity and facts about human psychology. This strategy would require a more detailed explanation of what it is to care about something for its own sake and what it is about biodiver-sity that makes it the fitting object of such an attitude (McShane 2007). Finally, one might focus less on the intrinsicness of biodiversity's value and more on the non-anthropocentricness of its value. This strategy would probably be the least difficult of the four, as it could rely on arguments for non-anthropocentrism already extant within the environmental ethics literature. This kind of claim would still allow for robust criticism of the ecosystem services approach without the added burden of arguing for some of the more difficult positions described here.

There are also alternative claims that might better capture what opponents of the ecosystem services approach are after, for example, claims that particular organisms, species, ecosystems, etc., have intrinsic value. Showing that we should care about blue whales or rainforests for their own sakes is quite different from showing that we should care about biodiversity for its own sake. While there might be arguments that could succeed in establishing the final and/or non-anthropocentric value of biodiversity, my own view is that intrinsic value claims about particular parts of the world tend to do a better job of capturing concerns that many people express in terms of biodiversity, which are often concerns about particular things in the world that are threatened. It's an open question whether my view is correct; there is no way to evaluate the ultimate adequacy of any of these positions without first solving the units-and-differences problem – that is, defining the kind of biodiversity that we are claiming has value.

Debates about the intrinsic value of biodiversity reveal quite a bit about the way that values are handled in conservation and conservation biology, and I would argue that there is room for improvement in this regard. One of the most frequently cited problems with intrinsic values is that they can't be measured. If they can't be measured, the argument goes, they can't be quanti-fied, and if they can't be quantified, we can't make rational choices about trade-offs. If we can't make rational choices about trade-offs, then we can't choose the best (or the least bad) policies to adopt (Maguire and Justus 2008, Justus *et al.* 2009, see also Maclaurin and Sterelny 2008). All of this is complete nonsense. Measurement is not the only path to quantification, and in any case, many philosophers have been perfectly happy to quantify intrinsic values (Bentham 1789). But quantification isn't the only way to make decisions about trade-offs. Rules of lexical priority, for example, can achieve the same result without quantifying anything (Taylor 1980, Regan 1983). Finally, policy choices needn't be the product of a mechanical decision procedure (numerical or otherwise) to count as rational. They need to be justified against the alternatives, but neither the economist's math nor the lawyer's system of lexical priorities is necessary in order for a choice to be justified (Rawls 1971, Anderson 1993, Scanlon 1998, Parfit 2011). If we really want to assess the value of different components or qualities of the natural world, we would do well to give up the assumptions that underlie these criticisms and develop a broader and deeper understanding of the variety of values that characterize our world.

Notes

1 To take an example, consider the difficulty of answering the following question. Which group has the greatest height diversity: (a) one person who is 4'6", thirteen people who are 5'9", one person who is 9'; (b) three people who are 5'8", three people who are 5'9", three people who are 5'10", three people who are 5'11", three people who are 6'; or (c) seven people who are 4'6", one person who is 5'9", one person who are 9'?

2 Here I depart from the "collection of collections" model used by Maier (2012).

3 Oddie (2005) argues that properties are the primary bearers of value. On this view, the property of biodiversity would itself be valuable.

4 For a discussion of the differences between attributing intrinsic value to states of affairs and attributing it to concrete individuals within environmental ethics, see McShane (2014). Most ethicists deny that one is reducible to the other. See, for example, the discussion of this matter in Parfit (2011) and Anderson (1993). But see Sarkar (2005: 53).

5 I'm setting aside some complications here for Aristotelian forms of virtue ethics, wherein "a better person" is to be understood as "better at being a person." The relationship between the Ancient Greek understanding and our modern conception of value is too complicated to take up here.

6 Though there are utilitarian versions of it. Singer's early arguments for animal welfare seem to be arguments in favor of the moral considerability of sentient organisms (Singer 1990). Some have criticized Singer, however, claiming that it is really happiness and suffering that have value/disvalue on this view, not the creatures that experience the happiness and suffering. See Regan (1983).

7 This seems to be the meaning that Maclaurin and Sterelny (2008) have in mind when criticizing intrinsic value claims. Strangely, they present subjectivist accounts of value as the accepted view within philosophy and they criticize intrinsic (objective) values for violating it. ("Something is valuable because agents value it. Theories of intrinsic value seem to cut this link. To say that biodiversity is intrinsically valuable is to say that it would be valuable even if nobody were to actually value it," p. 150.) Many ethicists would find these claims puzzling.

8 My discussion glosses over a very important issue within ethics, namely what deserves to be called objectivity (as opposed to subjectivity, intersubjectivity, etc.) and why. I am deliberately ignoring this issue, which is a matter of genuine controversy within philosophy, for reasons of space. But it is worth pointing out that some accounts of value which authors describe as subjectivist rather than objectivist (e.g. Callicott 1986, 1992) nonetheless meet the criterion for objectivity that I describe here.

9 For a helpful discussion of the different uses of the term "values" in ethics, economics, sociology, psychology, and political science, see Dietz *et al.* (2005).

10 For an excellent discussion of this issue in the context of conservation biology, see Odenbaugh (2003).

11 See O'Neill (1992) and Jamieson (2002) for other arguments for the irrelevance of the question of objectivity to the question of whether nature has intrinsic value.

12 Most proponents of the intrinsic value of biodiversity seem to be focused on the claim that biodiversity has non-instrumental value in this sense, which is why I focus on this particular construal of the instrumental/non-instrumental distinction. However, in the conservation literature, particularly that on ecosystem services, the distinction between instrumental and non-instrumental value is sometimes at the same time treated as equivalent to the economic distinction between use value and non-use value. I believe this is a mistake. If what economists call "existence value" and "option value" (both non-use values) served human interests, they would be considered instrumental values on this construal.

13 This seems to be the understanding of Oksanen (1997).

14 This is the meaning articulated by Reyers *et al.* (2012). See also Sullivan (2009) for a discussion of the way that viewing nature as a service-provider is connected to other social norms and values.

15 I am unsure how far this claim generalizes outside of Western cultures. Often our modes of caring and the concepts that we use to classify them are bound up with cultural categories, and not all such categories are universal. I thus take it to be an empirical question how widely the means–end distinction, so central to much of Western thinking, is shared outside of Western cultural contexts.

16 "In the kingdom of ends everything has either a price or a dignity. What has a price can be replaced by something else as its equivalent; what on the other hand is raised above all price and therefore admits of no equivalent has a dignity" (Kant 1996).

17 There is a slip here, I think, between intrinsic properties and essential properties. Formulations such as "value in itself" are ambiguous between these two possibilities, and writers do not always take care to distinguish them. Accidental intrinsic properties will not be possessed in every context; likewise,

some essential properties (possessed in every context) might be extrinsic rather than intrinsic. See Rabinowicz and Rønnow-Rasmussen (2000). That said, I include the line of reasoning above because it is so prevalent in the literature.

18 This seems to be the view of McCauley (2006), who describes the claim that biodiversity is intrinsically valuable as the view that it is "priceless" and that its value is "infinite."

19 For a discussion of Leibniz's version of this kind of argument in the context of biodiversity, see Maier (2012: 64–66).

References

Anderson, E. 1993. *Value in Ethics and Economics*. Cambridge, MA: Harvard University Press.

Bentham, J. 1789. *The Principles of Morals and Legislation*. London.

Callicott, J. B. 1986. "On the Intrinsic Value of Nonhuman Species." In *The Preservation of Species: The Value of Biological Diversity*, ed. B. G. Norton. Princeton, NJ: Princeton University Press.

Callicott, J. B. 1992. "Can a Theory of Moral Sentiments Support a Genuinely Normative Environmental Ethic?" *Inquiry* 35: 183–198.

Conniff, R. 2012. "What's Wrong with Putting a Price on Nature?" *Environment360*. Available online at http://e360.yale.edu/feature/ecosystem_services_whats_wrong_with_putting_a_price_on_nature/2583/.

Davidson, M. D. 2013. "On the Relation between Ecosystem Services, Intrinsic Value, Existence Value and Economic Valuation." *Ecological Economics* 95: 171–177.

Dietz, T., A. Fitzgerald, and R. Shwom. 2005. "Environmental Values." *Annual Review of Environmental Resources* 30: 335–372.

Hansen, J. 2011. "Ecosystem Services: Critics and Defenders Debate." *PLoS*, available online at http://blogs.plos.org/blog/2011/04/20/ecosystem-services-critics-and-defenders-debate/.

Jamieson, D. 2002. "Values in Nature." In *Morality's Progress*. Oxford: Oxford University Press.

Jamieson, D. 2008. *Ethics and the Environment: An Introduction*. Cambridge: Cambridge University Press.

Jax, K., D. N. Barton, K. M. A. Chan, R. de Groot, U. Doyle, U. Eser, C. Görg, E. Gómez-Baggenthun, Y. Griewald, W. Haber, R. Haines-Young, U. Heink, T. Jahn, H. Joosten, L. Kerschbaumer, H. Korn, G. W. Luck, B. Matzdorf, B. Muraca, C. Neßhöver, B. Norton, K. Ott, M. Potschin, F. Rauschmayer, C. von Haaren, and S. Wichmann. 2013. "Ecosystem Services and Ethics." *Ecological Economics* 93: 260–268.

Justus, J., M. Colyvan, H. Regan, and L. Maguire. 2009. "Buying into Conservation: Intrinsic Versus Instrumental Value." *Trends in Ecology and Evolution* 24: 187–191.

Kant, I. 1996. "Groundwork of the Metaphysics of Morals." In *Practical Philosophy*, ed. M. J. Gregor. Cambridge: Cambridge University Press.

Korsgaard, C. M. 1983. "Two Distinctions in Goodness." *Philosophical Review* 32: 169–195.

Leopold, A. 1970. *A Sand County Almanac with Essays on Conservation from Round River*. New York: Ballantine Books.

Maclaurin, J., and K. Sterelny. 2008. *What Is Biodiversity?* Chicago, IL: University of Chicago Press.

Maguire, L. A., and J. Justus. 2008. "Why Intrinsic Value Is a Poor Basis for Conservation Decisions." *BioScience* 58: 910–911.

Maier, D. S. 2012. *What's So Good About Biodiversity? A Call for Better Reasoning About Nature's Value*. Dordrecht: Springer.

Malle, B. F. 2004. *How the Mind Explains Behavior: Folk Explanations, Meaning, and Social Interaction*. Cambridge, MA: MIT Press.

McCauley, D. J. 2006. "Selling Out on Nature." *Nature* 443: 27–28.

McShane, K. 2007. "Why Environmental Ethics Shouldn't Give Up on Intrinsic Value." *Environmental Ethics* 29: 43–61.

McShane, K. 2014. "The Bearers of Value in Environmental Ethics." In *Consequentialism and Environmental Ethics*, ed. A. Hiller, R. Ilea, and L. Kahn. London: Routledge.

Mikkelson, G. M. 2014. "Richness Theory: From Value to Action." *The Ethics Forum* 9: 99–109.

Miller, P. 1982. "Value as Richness: Toward a Value Theory for an Expanded Naturalism in Environmental Ethics." *Environmental Ethics* 4: 101–114.

Moore, G. E. 1993. *Principia Ethica*. Cambridge: Cambridge University Press.

O'Neill, J. 1992. "The Varieties of Intrinsic Value." *The Monist* 75: 119–137.

Oddie, G. 2005. *Value, Reality, and Desire*. Oxford: Clarendon.

Odenbaugh, J. 2003. "Values, Advocacy and Conservation Biology." *Environmental Values* 12: 55–69.

Oksanen, M. 1997. "The Moral Value of Biodiversity." *Ambio* 27: 541–545.

Parfit, D. 2011. *On What Matters*. Oxford: Oxford University Press.

Rabinowicz, W., and T. Rønnow-Rasmussen. 2000. "A Distinction in Value: Intrinsic and for Its Own Sake." *Proceedings of the Aristotelian Society* 100: 33–51.

Rawls, J. 1971. *A Theory of Justice*. Cambridge, MA: Harvard University Press.

Redford, K. H., and W. M. Adams. 2009. "Payment for Ecosystem Services and the Challenge of Saving Nature." *Conservation Biology* 23: 785–787.

Regan, T. 1983. *The Case for Animal Rights*. Berkeley: University of California Press.

Reyers, B., S. Polasky, H. Tallis, H. A. Mooney, and A. Larigauderie. 2012. "Finding Common Ground for Biodiversity and Ecosystem Services." *BioScience* 62: 503–507.

Rolston III, H. 1988. *Environmental Ethics*. Philadelphia, PA: Temple University Press.

Sarkar, S. 2005. *Biodiversity and Environmental Philosophy: An Introduction*. Cambridge: Cambridge University Press.

Scanlon, T. M. 1998. *What We Owe to Each Other*. Cambridge, MA: The Belknap Press.

Schröter, M., E. H. van der Zanden, A. P. E. van Oudenhoven, R. P. Remme, H. M. Serna-Chavez, R. S. de Groot, and P. Opdam. 2014. "Ecosystem Services as a Contested Concept: A Synthesis of Critique and Counter-Arguments." *Conservation Letters* 7: 514–523.

Singer, P. 1990. *Animal Liberation*. New York: Avon Books.

Soulé, M. E. 1985. "What is Conservation Biology?" *BioScience* 35: 727–734.

Sullivan, S. 2009. "Green Capitalism, and the Cultural Poverty of Constructing Nature as Service Provider." *Radical Anthropology* 3: 18–27.

Taylor, P. 1980. *Respect for Nature: A Theory of Environmental Ethics*. Princeton, NJ: Princeton University Press.

12

WHAT GOOD IS IT, ANYWAY?

J. Baird Callicott

A phenomenological exploration of intrinsic value

Many of my colleagues in ecology and conservation biology – and, indeed, many of my col-leagues in environmental philosophy – think that intrinsic value is an exotic concept: new, mysterious, abstract, elusive (Shrader-Frechette 1981, Collar 2003). Nothing could be further from the truth. While the *name* may be new or at least unfamiliar and its referent may be unclear, the thing itself is familiar to everyone and something that, under the right circumstances, rushes palpably to the forefront of consciousness. Edwin P. ("Phil") Pister – until his retirement, an Associate Fishery Biologist in the California Department of Fish and Game – tells a story that very nicely makes my point.

Pister's territory was the arid Owens River Valley of eastern California in the rain shadow of the Sierra Nevada mountain range to the west and flanked on the east by the White and Inyo mountains. Fed by Sierra rain and snow, the Owens River terminated in salty Owens Lake at the southern end of the valley. By the 1960s, water from the Owens River had been diverted to Los Angeles for half a century, in ever-increasing quantities, desiccating Owens Lake and reducing the lower river to a bare trickle. The once-plentiful Owens pupfish (*Cyprinodon radiosis*), a fish about the size of your thumb, was down to its last refuge – a few puddles of water contained the entire extant population – and those puddles were about to dry up. Pister (1992) rescued the species in two buckets of water.

Soon thereafter, Pister formed the Desert Fishes Council in 1969 to save these and similar little fishes living in scattered pools of water throughout the arid country of the American Southwest – an archipelago of aquatic islands, so to speak, dotting a sea of sands and alkali flats. Having evolved in isolation, these species of "desert fishes" are endemics. Subsequently, the Devils Hole pupfish (*Cyprinodon diabolis*) living in a pool of water at the upward-facing mouth (which is about as big around as a hot tub) of a deep flooded cavern in Nevada's Amargosa Desert found itself threatened by a precipitous drop in the groundwater table due to crop irrigation. The water is deep, but the pupfish spawn on a shallow limestone ledge, which was about to be left high and dry. Nowhere to spawn; no more Devils Hole pupfish. Under the provisions of the newly enacted US Endangered Species Act (ESA: *7 U.S.C. § 136, 16 U.S.C. § 1531 et seq.*), Pister took its case (*Cappaert v. United States*) all the way to the US Supreme Court … and won.

The state agency for which Pister worked is largely supported by hunting and fishing license sales and most of his colleagues were devoted to managing the fish and wildlife of California for recreation and consumption by sportspersons. The concern and care lavished by Pister on these tiny non-game species of fish baffled his colleagues (who called them "dicky fish"). Of each such species rising to the attention of a judge, instead of a fly, they would ask him, what good is it, anyway? For years Pister struggled to answer that question. For example, some of these fish thrived in salt-saturated brine; so maybe research on their remarkable kidneys could provide information applicable in medicine. But would such speculative option value – to put the issue in economistic terms – outweigh the value of drinking water for thirsty LA and agricultural, commercial, and residential development in western Nevada? Hardly. His quest for an effective answer to the what-good-is-it-anyway question led Pister to *Environmental Ethics* (the journal). And there, in the concept of intrinsic value, he found the answer that had eluded him.

That answer – species of desert fish have intrinsic value – certainly satisfied Phil Pister, who now had a term and a body of academic literature to justify his own intuitive application of the concept to endangered species. But it didn't satisfy his associates, one of whom said, "Phil, when you start talking about morality and ethics, you lose me" (Pister 1987: 228). Pister finally found a rejoinder that has provided us environmental philosophers with as much insight and rhetorical leverage as we ever provided him. He answered the question, what good is it, anyway?, with a question of his own: what good are you?

When confronted with that question, one abruptly comes face to face with the distinction between instrumental and intrinsic value. Most of us (I hope) want to be instrumentally valuable – useful to family, to friends, to our employers, to society, perhaps even to humanity in the abstract. In addition to our instrumental value, we human beings claim to have intrinsic value. We do not subject human beings to a benefit–cost analysis, such that the social benefits must exceed the costs to justify their continued existence and prevent their summary disposal. Indeed, only in cases of the most extreme social costs imposed by wanton criminals does their instrumental disvalue outweigh their intrinsic value and warrants the termination of their existence (execution) or some lesser abrogation of their intrinsic value (incarceration). Our claim to intrinsic value is nothing less than the foundation of our human rights (Regan 1983). Thus intrinsic value is utterly familiar and will be viscerally experienced when one is unexpectedly asked the question, what good are you?

Let the foregoing anecdote be regarded as a phenomenological exploration of intrinsic value; and to the extent that phenomenology is a kind of philosophy, we have thus far advanced a philosophical understanding of intrinsic value. We have introspectively discovered what it is or at least what it feels like to claim it for ourselves.

What's new, mysterious, and elusive is not the concept of intrinsic value, but the application of it beyond the human sphere. In addition to human beings, what other kinds of things might have intrinsic value? Is biodiversity, more particularly, included in the class of things that have it?

Many other, more basic philosophical questions surrounding intrinsic value, however, also remain first to be answered. What is the ontological status of intrinsic value? What practical difference does having intrinsic value make? Does intrinsic value come in degrees, such that some kinds of things have more of it than do others? If so, how might intrinsic value be quantified? How do intrinsic value and instrumental value interact in the real world of trade-offs?

The ontological status of intrinsic value: objective

The term "intrinsic value" and the less used alternative term "inherent worth," mean, lexically speaking, pretty much the same thing. Merriam Webster's Collegiate Dictionary, tenth edition, defines "intrinsic" thus: "belonging to the essential nature or constitution of a thing." And it defines "inherent" thus: "involved in the constitution or essential character of something … intrinsic." The English word "value" comes from the Latin word "*valere*" – "to be worth, to be strong"; and "worth" comes from the Old English word "weorth," meaning "worthy, of value." Lexically speaking, thus, to claim that the value (or worth) of something is intrinsic (or inherent) is to claim that its value (or worth) belongs to its essential nature or constitution. Therefore, on the face of it, the lexical meaning of "intrinsic" would suggest that intrinsic value is an objective property of the things that have it, like mass is an objective property of material objects or like sentience is an objective property of vertebrate animals.

Mass, however, can be measured and sentience can be experimentally assessed, but value itself seems to be altogether different. If value is an objective property, it is no more empirically accessible than the property of being created in the image of God. True, we can vividly experience our own intrinsic value introspectively. We can also vividly experience jealousy and love introspectively. But jealousy and love are not properties; they are intentional affects – subjective feelings directed toward objective entities. Value is not an emotion like jealousy or love, but, like jealousy or love, it is an intentional state of consciousness. The objective intentions, the targets, of conscious valuing are valuable because they are valued, just as the objective intentions of conscious desiring are desirable because they are desired. To think of value – any kind of value – as itself an objective property is a "category mistake." Value is no more an objective property of things that "have" value, than a name is an objective property of things that "have" names.

Honoring what we might call lexical fidelity to the meaning of "intrinsic," some philosophers regard it, in effect, as a supervenient property – that is, a property piggybacking, as it were, on a genuinely objective property – although "supervenient property" is a term that rarely appears in the literature of animal and environmental ethics. To *identify* intrinsic value with a genuinely objective (that is, "natural") property, such as rationality or sentience, exposes those philosophers who do so to the allegation that they are committing the naturalistic fallacy à la G. E. Moore (1903). The notion of a supervenient property (*sensu* Hare 1952, 1984) comes to their rescue. Philosophers have, in effect, identified several "base properties" on which intrinsic value might supervene.

The classic base property on which intrinsic value might supervene is reason or rationality, most strongly defended by Immanuel Kant (1785/1959). Making reason or rationality the base property would limit the having of intrinsic value to rational beings. (I know, that's a tautology, but stick with me; it's a tautology worth stating.) A philosophical tradition going back to Aristotle posits reason to be the defining characteristic of human beings. Humans, Aristotle believed, were animals, but, uniquely among all other animals, he believed that humans are *rational* animals. Kant seems to have assumed the cogency of this ancient tradition and accepted it implicitly. He vaguely suggests that there may be other-than-human rational beings, but they are not to be found among other earthly animals – perhaps he had in mind the heavenly host or extraterrestrials; it's hard to say. Thus, for all practical purposes, Kant's ethics is thoroughly anthropocentric.

If reason or rationality, however, is construed to be a property that can be measurably assessed, in fact all human beings are not rational – not pre-rational infants; not sub-rational and post-rational adults. This class of non-rational humans – insensitively called the "marginal cases" by Regan (1979) – would lack intrinsic value, if intrinsic value supervenes on reason or rationality. Members of the class of marginal cases therefore could, within the bounds of morality and propriety, be used as we presently use some other animals. But the prospect of performing painful medical experiments on infants; hunting the mentally disabled; making dog food out of the demented – is thoroughly repugnant (I also hope) to everyone's moral sensibilities. If all human beings are to have intrinsic value – the foundation of human rights – the base property on which intrinsic value supervenes must be differently specified (Regan 1979).

Regan's "Argument From Marginal Cases" was formulated in service of his "case for animal rights." Regan (1983) rather modestly selected being the "subject of a life" that could go better or worse from the subject's point of view as the base property on which intrinsic value supervenes. But once the door had been opened by his Argument from Marginal Cases, other philosophers rushed in to defend other base-property choices. Sentience, the capacity to experience pleasure and pain, found a staunch advocate in Peter Singer (1975). It is also a classic base-property choice on which intrinsic value might supervene, as it had been nominated for that role (or something like it) by Kant's contemporary, Jeremy Bentham. Quoted ad nauseam by partisans of animal liberation, about sentient animals Bentham (1789: ch. xvii, sec. 1, n. 121) wrote, "the question is not, Can they *reason*? Nor Can they *talk*? But Can they *suffer*?"

While the ethical enfranchisement of one or another set of animals is the ultimate goal of animal ethicists – iconically represented by Tom Regan (1983) and Peter Singer (1975) – their end point was the starting point for environmental ethicists attempting to enlarge the scope of ethics even further. Kenneth E. Goodpaster (1978), building on animal liberation, argued that being alive or having life should be the base property on which intrinsic value supervenes – because, from an evolutionary point of view, sentience evolved in animals as a means to the end of staying alive. Joel Feinberg (1974), building on animal rights, identified conativity – the capacity to strive toward an end and thus to have interests – as the base property on which intrinsic value and ultimately moral rights supervene. Paul Taylor (1981) made it being a "teleological center of life."

In most cases, especially that of Taylor, this way of conceiving intrinsic value limited the having of it to individual organisms – including plants as well as animals. The one exception is Goodpaster (1978), who thought that the "biosystem" (aka the biosphere) might also be alive and thus have intrinsic value, citing James Lovelock's Gaia hypothesis in support of this possibility. Lawrence E. Johnson (1991) to the contrary notwithstanding, the objects of genuine conservation concern – species, biotic communities, ecosystems, biodiversity – are not subjects of a life or sentient or alive or conative or teleological centers of a life. So the attempt to provide an *objective ontology* for intrinsic value – however soundly or unsoundly argued – leads to a dead end, from a conservation point of view, as Norton (1987) observes.

The ontological status of intrinsic value: subjective

The alternative, which Norton (1987) did not consider, is a subjective ontology. "Value" is a verb and "intrinsic" is an adverb. We humans value (verb transitive) things in two primary ways: instrumentally and intrinsically. And valuing things in both these ways is not arbitrary; we have reasons, and often good reasons, for valuing things.

We instrumentally value our houses and cars, obviously because the former shelter us and the latter transport us. Some object may be lying around one's house or yard that one does not value instrumentally – say a broken dinner plate or an unwanted toy. Someone else might convince one that the toy, at least, might be instrumentally valued by an underprivileged child and that it would be better to donate it to Good Will than to toss it in the trash. We intrinsically value ourselves, as we discovered in the first section. But we also intrinsically value other people. Make a phenomenological try of it. Think of a close relative or friend and feel the way you value them. Indeed, one mark of false friendship is to value a putative friend only instrumentally; and when he or she is no longer useful to oneself to cut them out of one's life. Less viscerally and more ideally, we ascribe intrinsic value to all human beings.

Racists – a term of reproach – do not value all human beings intrinsically. But many good reasons can be offered to convince racists that all human beings should be valued intrinsically, if only in the abstract. And all of the arguments generated over the past forty years for the objective intrinsic value of individual animals and plants can be converted into reasons for intrinsically valuing them. If "value" is a verb and "intrinsic" is an adverb, then reasons can also be offered for intrinsically valuing the objects of conservation concern – species, biotic communities, ecosystems, biodiversity. Whether or not such reasons are convincing may depend on one's wider worldview. For those who believe in God and that God created the world in the way described in Genesis, humans should intrinsically value the creation and its biodiversity because it is the handiwork of God (Cone 2012). For those who embrace a scientific worldview, other species are, in the words of Aldo Leopold (1949: 109), "fellow voyagers … in the odyssey of evolution." From an evolutionary point of view, other species are our phylogenetic kin and thus we might value each intrinsically in a way similar to the way that we intrinsically value our familial kin. To the scientifically literate, the astonishing magnitude and exuberance of biodiversity is a spellbinding wonder and certainly something to be valued for its own sake, something to be valued intrinsically.

How ontologically subjective values are epistemologically objectified

A close examination of Kant's understanding of intrinsic value reveals that he soberly regards values to be ontologically subjective, but he also thinks that they may somehow become objective:

> [R]ational nature exists as an end in itself [which is one definition of intrinsic value: the value of something as an end in itself]. Man necessarily thinks of his own existence this way [here recall Phil Pister's question, what good are you?]; thus far it is a *subjective principle* of human actions. Likewise does every other rational being think of his own existence based on the same rational ground which holds also for myself; thus it is at the same time an *objective principle* from which, as a supreme practical ground, it must be possible to derive all laws of the will. The practical imperative, therefore, is the following: act so that you treat humanity, whether in your own person or in that of another, always as an end withal and never as a means only.
>
> (Kant 1785/1959: 47)

So how does subjectively valuing oneself as an end in oneself magically become objective by the mere consideration that other beings of various kinds also value themselves as ends in themselves? (Let's be up with the Jacob-von-Uexküll and Imanishi-Kinji times and allow that many kinds of beings are self-valuing, not just rational human beings.) Intrinsic value becomes objective not in the ontological sense of the word, but in the epistemological sense. That is, in the sense that an umpire or referee is "objective" if unbiased in enforcing the rules of a game or in the sense that a scientist's research is "objective" if it is unbiased by a pet theory (like cold fusion) or a funding source (like the American Petroleum Institute). At the heart of ethics, Kant is saying, if I ask others to acknowledge and respect my self-ascribed intrinsic value and act accordingly, I am obliged to acknowledge and respect their self-ascribed intrinsic value and act accordingly – to act in such a way as to treat others as well as myself always as an end withal and never as a means only.

Holmes Rolston III (1994) makes a slightly different move than Kant in transforming subjective into objective intrinsic value, exploiting another ambiguity in the word "objective." Going beyond Uexküll, Imanishi, and their contemporary exponents, Rolston (1994) points out that other organisms value themselves intrinsically, although they may not be conscious of doing so or may not even be conscious at all. Primates (Rolston's example, more particularly, is lemurs) do consciously value themselves intrinsically, although they may not be able to express it in words. Rolston descends the slippery slope he sets up, going from primates on to vertebrates (Rolston's example, more particularly, is warblers), who may not consciously value themselves as ends in themselves, but birds and other vertebrates manifest their self-valuing by striving to live and to reproduce themselves. At the bottom of Rolston's slippery slope are plants that are by no means passive organisms, but which manifest their self-valuing by means of allelopathic chemical excretions and other self-promoting "behaviors." Therefore, Rolston argues, there is intrinsic value out there, objectively, in nature. Of course, he has a point; but for Rolston as for me, "value" is a verb first and "intrinsic" an adverb. There is a lot of self-valuing going on out there objectively in nature. And the way all these self-valuing organisms value themselves is intrinsically.

Kant's move from ontologically subjective to epistemologically objective intrinsic value is thus also open to Rolston. So far, Rolston has established by argument what requires no argument for Kant. Kant takes it for granted that "every other *rational being* thinks of his own existence by means of the same rational ground which holds also for myself." It is not so obvious that every other *living being* values its own existence in ways similar to the way we human beings value our own existence – that is, intrinsically. Thus Rolston must effectively argue that they do. Once that is convincingly established, then the way is clear for him to echo Kant and say, so far it is an ontologically *subjective principle* of the actions of living beings. It becomes an epistemologically *objective principle* of the actions of *moral beings* (moral agents) if we moral beings regard their self-valuing and our own self-valuing in an unbiased or impartial manner. (While Rolston's field of moral patients is as extensive as Paul Taylor's, his field of moral agents is populated only by human beings.)

How ontologically subjective values are socially objectified

There is yet another sense in which subjective values become objective – the social sense.

Look around you and you will everywhere see the objectification of socially aggregated subjective *instrumental* values: the McDonalds, the McMansions, the SUVs, the iPhones, the strip malls, the convenience stores, the aisles and aisles of junk food in grocery stores – examples enough of what is meant.

How are our subjective individual *intrinsic values* socially objectified? I may value endangered species and biodiversity intrinsically and you may not. You value whatever you value intrinsically and I value whatever I value intrinsically, just as you have your favorite color and I have mine. Are we stalemated? Are values surds and as beyond the reach of rational argument and persuasion as two children arguing over the prettiest color or the tastiest candy? Surely, one does not just up and prefer emancipation to slavery or the right of women to vote. Values are not immune to rational argument and those for which there is the strongest rational support are objectified – eventually anyway – through political processes (Sagoff 1988). They are reified in our national constitutions, in legislation, in jurisprudence. And they evolve.

For example, the US Constitution of 1789 did not recognize the intrinsic value of enslaved persons of African descent. That changed in 1868 when the Fourteenth Amendment was ratified. Why did it change? Because sound reasons were offered in favor of equal treatment under the law for all Americans. Had the South won the Civil War, almost certainly the constitution of the Confederate States of America would also have eventually been amended to socially objectify the intrinsic value of persons of African descent, because the reasons for doing so became ever more indisputable and compelling.

Animal and environmental ethics have shifted the ways that animals and some other natural entities are valued from the instrumental end of the socially objectified values spectrum toward the intrinsic end. The use of animals in agriculture, sport, and medical research has increasingly become regulated by law, in the USA beginning in 1966 with the Animal Welfare Act (7 *U.S.C § 2131 et seq.*); and such use has become ever more restrictive by subsequent amendments, the most recent enacted in 2008. I have argued that, while not mentioning intrinsic value by name, the US Endangered Species Act of 1973 implicitly ascribes intrinsic value to listed species (Callicott 2006).

Examples of the social objectification of intrinsic value abound. In his field-defining paper, Michael Soulé (1985: 731, emphasis in original) wrote, "*biotic diversity has intrinsic value*, irrespective of its instrumental or utilitarian value." That "there is intrinsic value in the natural diversity of organisms, the complexity of ecological systems, and the resilience created by evolutionary processes" is the first among five "organizational values" listed in the Strategic Plan of the Society for Conservation Biology (2011), which Soulé helped to found. With exquisite sensitivity to the distinction between intrinsic and instrumental values, the preamble of the UN Convention on Biodiversity (1992) begins with the former kind of value and then lists several categories of the latter kind: "Conscious of the intrinsic value of biological diversity and of the ecological, genetic, social, economic, scientific, educational, cultural, recreational, and aesthetic values of biological diversity and its components, …" The first of four principles of the UN Earth Charter (2004) is a statement of one definition of intrinsic value: "Recognize that all beings are interdependent and every form of life has value regardless of its worth to human beings."

Transformative value as an alternative to intrinsic value

Bryan G. Norton (1987: 185) would countenance only those values that can "establish viable bases for [environmental] policy formation." While many well-established anthropocentric instrumental values (of clean air and water, for example) certainly establish viable bases for environmental policy, not all (perhaps not even most) of the objects of conservation concern have well-established, widely recognized anthropocentric instrumental value – the pupfish with which we began being cases in point. For biodiversity and the other holistic objects of conservation concern to be intrinsically valued widely "would require," according to Norton (1987: 185), "an assault on a widely accepted metaphysical dichotomy" between humans and everything else – an assault on anthropocentrism, in a word. Norton (1987: 187) acknowledges that anthropocentrism is vulnerable to assault on behalf of individual non-human beings, but "locating intrinsic value in species themselves or in ecosystems … go[es] beyond the simple annexation of standard ethical rules and principles … ." Holistic non-anthropocentrists "must define a nonindividualistic conception of intrinsic value and then state some positive characteristic standing as the mark of such value. Only then can they begin to argue that species and ecosystems have the relative characteristic and to derive policies from those values."

As we see, Norton restricts a cogent theory of intrinsic value for species, ecosystems, and biodiversity to an objective ontology of intrinsic value outlined in the third section. But if we assume a simple and phenomenologically corroborated subjective ontology of intrinsic value, we may value (verb transitive) anything at all intrinsically. And while some people intrinsically value some things in irrational and idiosyncratic ways, most of the things that most people value intrinsically are so valued for good reasons. The evolving social objectification of subjectively conferred intrinsic value via policy and legislation – as in the case of the Fourteenth Amendment to the US Constitution and the UN Convention on Biodiversity – is based on rational considerations.

The assault on anthropocentrism has been relentlessly prosecuted during the thirty years since Norton believed it to be a quixotic errand. And the examples of policy and law noted in the previous section, which either implicitly or explicitly invoke the concept, indicate that the intrinsic value of endangered species and biodiversity does in fact establish a viable basis for conservation policy formation. As Norton suggests, whether one grants or does not grant that the class of intrinsically valuable entities is more extensive than the class of humans depends on one's deeply rooted metaphysics. For metaphysically resolute anthropocentrists, Norton (1987: 188) formulates the concept of transformative value as an alternative to intrinsic value, distinguishing between "felt preferences" and "considered preferences." One might, for example, feel a strong preference for smoking cigarettes and then consider that tobacco use may cause one to suffer heart disease and lung cancer. Thus one may ultimately come to loathe smoking. Knowledge of tobacco's harmful effects on human health thus transforms one's values. Similarly, according to Norton (1987: 189), "wild species and undisturbed ecosystems [can be] occasions to alter … felt preferences … Experience of nature can promote questioning and rejection of overly materialistic and consumptive felt preferences."

Norton may put too much faith in the experience per se of nature unaccompanied by an appropriate cognitive framing of that experience. For example, consider the difference in how one might experience it when the Congo basin is cognitively framed as a "jungle" and the "heart of darkness" versus when cognitively framed as a "rain forest" and "biodiversity hot spot." As Sahotra Sarkar (2012) suggests, a field course in conservation biology, including both experience of nature and an appropriate cognitive framing of that experience, may indeed lead to reconsidered preferences. Biodiversity conservation might well come to be preferred over, say, rhino-horn aphrodisiacs, ivory piano keys, and mahogany furniture.

Norton's stated rationale for positing the transformative value of biodiversity is to establish a viable basis for conservation policy formation – on the assumption that the intrinsic value of biodiversity is not a viable basis. The intrinsic value of biodiversity has, however, not only been cogently theorized (the subjective-ontology theory), it has actually been invoked in policy formation at the highest institutional level (for example, the UN Convention on Biodiversity). That biodiversity has transformative value has not been invoked in any conservation policy statement or piece of conservation legislation. Indeed, the intrinsic value of biodiversity has been such a popular basis for conservation policy formation in the conservation community that the suggestion by Karieva *et al.* (2012) that conservation policy be based exclusively on anthropocentric instrumental values in the form of ecosystem services has provoked outrage (Soulé 2013, Vucetich *et al.* 2015). The instrumental-value-versus-the-intrinsic-value-of-biodiversity kerfuffle became so acrimonious that it attracted attention in the popular media (Max 2014) and prompted 238 members of the conservation community to co-sign a plea for peace and reconciliation between the adversaries (Tallis and Lubchenco 2014). The concept of transformative value has aroused no such impassioned conflict in the conservation community.

Finally, an empirical study of European arguments for conservation found that:

> Non-instrumental arguments are among the oldest and most widespread arguments for a value of nature. They contributed largely to a policy shift in the 1970s and 1980s, which brought environmental problems to the forefront of public awareness (Naess 1973, Callicott 1989[1]) … and they have influenced European environmental

governance and policy since. In our study we found that non-instrumental arguments were quite frequently used – both in the general and the in-depth assessments … It seemed that non-instrumental arguments were a widely accepted paradigm.

(Mueller and Maes 2015: 19)

Transformative value was not discussed by Mueller and Maes (2015); it has had no documentable bearing on conservation policy formation; and it has aroused no heated debate in the conservation community. That all may be due to the maladroit categorization of the psychological impact on humans of cognitively well-framed experience of other forms of life as a species of *value*. Sarkar (2012: 56) more precisely denominates the potential of other species, ecosystems, and biodiversity to transform our felt preferences into considered preferences as their "transformative *power*." Apart from the question of which is the more viable basis for conservation policy formation, the intrinsic value of biodiversity and the transformative value (power) of biodiversity differ not only metaphysically (non-anthropocentric vs anthropocentric, respectively), but also axiologically. The upshot of transformative value (or power) is a (re)considered human *preference*; but intrinsic value is a way of valuing different from having preferences. Again, do we merely prefer emancipation to slavery? Or do we have a moral obligation, irrespective of our preferences, whether felt or considered, to respect human dignity and rights? Do we merely prefer, after due consideration, a biodiverse world? Or do we have a moral obligation to preserve biodiversity whether we prefer it or not? We objectify our preferences in the marketplace; we objectify our values in policy and legislation (Sagoff 1988). The two concepts differ also in a third way – pragmatically, as detailed in the next section.

The pragmatic meaning of intrinsic value

Just as examining intrinsic value phenomenologically leads to a revelatory encounter with one's own intrinsic value and that of at least some other people, a pragmatic analysis of intrinsic value is equally illuminating. So let's ask, what does having intrinsic value do for the things that have had theirs socially objectified in policy and law? What is the operational or pragmatic meaning of "having intrinsic value"?

The most obvious and immediate effect of something's intrinsic value being socially objectified is to make traffic in it illegal. Clearly, this is the case in regard to what we call human trafficking – buying and selling human beings in labor and sex markets. That is also the effect of socially objectifying the intrinsic value of listed endangered species, as the Convention on International Trade in Endangered Species (CITES) clearly illustrates. Ancillary to this is the moral discomfort we feel when something that has intrinsic value is priced. Pricing – even shadow pricing – seems to be a dangerous step toward commodification and trafficking. Pricing implicitly moves something that has socially objectified intrinsic value from the political realm to the economic, from the parliament to the market. Kant himself gets at this in his own stiff style:

> In the realm of ends everything has either a *price* or a *dignity*. Whatever has a price can be replaced by something else as its equivalent; on the other hand, whatever is above all price, and therefore admits of no equivalent, has a dignity.
>
> That which is related to general human inclinations [i.e. wants] and needs has a *market price*. That which, without presupposing any need, accords with a certain taste, i.e., with pleasure in the mere purposeless play of our faculties [e.g. scenic beauty], has an *affective price* [i.e. a shadow price]. But that which constitutes the condition under

176

which alone something can be an end in itself does not have a mere relative value, i.e., a price, but an intrinsic value, i.e., a *dignity*.

(Kant 1785/1959: 53)

It can be said without cynicism – just a sad fact – that one effect of socially objectifying something's intrinsic value is to create a black market in it. There is, unfortunately, a flourishing trade in enslaved human beings. And the black markets in ivory and horn threaten the extinction of two genera of elephant (*Loxodonta* spp. and *Elephas maximus*) and five species of rhinoceros (*Ceratotherium simum*, *Dicerorhinus sumatrensis*, *Diceros bicornis*, *Rhinoceros unicornis*, and *R. sondaicus*). The only good thing to be said about a black market in a prohibited "commodity" – be it ivory, horn, or enslaved human beings – is that it increases the price of that "commodity" because of the risk of prosecution and imprisonment or victimization (by other outlaws) for those who traffic it. That might at least slow down the trade in the things that should not have a price because they are the sorts of things that we have collectively decided should have a dignity.

Another effect of socially objectifying something's intrinsic value is to shift the burden of proof from those who would protect it to those who would harm it (Fox 1993). If only the instrumental value of something is recognized in policy and law, and one wants to protect it from being destroyed or consumed, one has – usually by means of a benefit–cost analysis (BCA) – to show that its conservation value exceeds its exploitation value. For example, if one wants to protect coastal mangroves from being replaced by shrimp farms, one must show that the instrumental value of the ecosystem services rendered by the mangroves, expressed in a monetary metric, exceeds the projected value of the proposed shrimp farms, expressed in the same metric.

Consider another example. Suppose Phil Pister followed Norton (1987) as emended by Sarkar (2012), not Leopold (1949) as emended by Callicott (2006), and attributed transformational value (power), not intrinsic value, to desert fishes. Further suppose that Pister eloquently described how his experience of these little fishes transformed his initial felt preference for promoting game fish into a considered preference for the preservation of the Cyprinodontidae. His personal considered preference and that of those whom he might convince to share it would have to compete on a level playing field with the felt preferences of millions of people in Los Angeles and the powerful agro-industrialists in Nevada for a plentiful water supply. In other words, the benefits of the preservation of the Cyprinodontidae, calculated in terms of the willingness to pay of Pister and his ilk for knowing that these species continue to exist, would be compared to the higher costs incurred by water consumers in southern California and Nevada for a resource made scarcer by its diversion to biodiversity preservation. Preferences are preferences, whether felt or considered, and which preferences should be satisfied is indicated by BCAs and ultimately by market forces. On the other hand, if the intrinsic value of something is socially objectified in policy and law, any proposal to destroy or consume it must show that the lost instrumental value of not doing so is intolerably great.

Alan Randall suggests how the Safe Minimum Standard (SMS) alternative to BCA in environmental economics is pragmatically equivalent to the intrinsic-value approach to conservation in respect to shifting the burden of proof:

> Whereas the … BCA approach starts each case with a clean slate and painstakingly builds from the ground up a body of evidence about the benefits and costs of preservation, the SMS approach starts with the assumption that the maintenance of the SMS for any species is a positive good. The empirical economic question is, "Can

we afford it?" Or, more technically, "How high are the opportunity costs of satisfying the SMS." The SMS decision rule is to maintain the SMS unless the opportunity costs of doing so are intolerably high. In other words, the SMS approach asks, how much will we lose in other domains of concern by achieving a safe minimum standard of biodiversity? *The burden of proof is assigned to the case against maintaining the SMS.*

(Randall 1988: 221, italics added)

While Randall (1988: 221) clearly states that the SMS approach to biodiversity preservation, no less than the BCA approach, is "utilitarian, anthropocentric, and instrumentalist in the way it treats biodiversity," the SMS approach, in the economic sphere, has the same pragmatic effect as the intrinsic-value approach in the political sphere. In this respect, the SMS approach to biodiversity conservation is more nearly the pragmatic equivalent of the intrinsic-value approach than is the transformative value (power) approach, which has no documentable burden-of-proof effect. The burden-of-proof effect of socially objectified intrinsic value is most familiarly manifest in the jurisprudential principle of "innocent until proven guilty." If we propose to deprive an intrinsically valuable human being of property (in the form of a seizure or fine), liberty (in the form incarceration), or life (in the form of execution), the instrumental disvalue to society of not imposing some such punishment must be intolerable – the disintegration of public safety and the annihilation of private property. Shifting the burden of proof is most dramatically documented in the history of the Endangered Species Committee (aka "the God Squad") of the US ESA. It was created by an amendment to the ESA in 1978 to allow projects of great instrumental value to trump the intrinsic value of one or more listed species. This seven-member, cabinet-level committee must approve an exemption to the Act by a five to two vote. It has convened only a handful of times and granted such an exemption in only a few instances (Plater *et al.* 2004).

Vucetich *et al.* (2015) have identified a third pragmatic effect of socially objectifying the intrinsic value of biodiversity: doing so ensures the objectivity (in the epistemic sense) of conservation biologists. If the only way Phil Pister could answer the question about the Devils Hole pupfish (and other endemic desert fishes) – what good is it, anyway? – was in terms of instrumental value, there is a temptation to skew data to support the attribution of instrumental value. Pister himself suggested that research on the physiology of these fishes – which tolerate high salinity, high temperatures, and low oxygen in the waters they inhabit – might lead to breakthroughs in medicine. But then again, such research may not. In the absence of the ESA socially objectifying their intrinsic value, a physiologist sharing Pister's moral commitment to saving these species might be tempted to put his thumb on the data scale. This pragmatic effect was actually anticipated by Aldo Leopold:

> When one of these non-economic categories is threatened and we happen to love it, we invent subterfuges to give it economic importance. At the beginning of the century songbirds were supposed to be disappearing. *Ornithologists jumped to the rescue with some distinctly shaky evidence to the effect that insects would eat us up if birds failed to control them.* The evidence had to be economic to be valid.
>
> It is painful to read these circumlocutions today. We have no land ethic yet, but we have at least drawn nearer the point of admitting that birds should continue as a matter of biotic right, regardless of the presence or absence of economic advantage to us.

(Leopold 1949: 211)

Compatibility of instrumental and intrinsic value

Are instrumental and intrinsic value mutually exclusive, such that things that have the one kind cannot also have the other? For all he has to say about some things having a price and others having a dignity, one might think that, for Kant, instrumental and intrinsic value are indeed mutually exclusive. But note how Kant (1785/1959) carefully phrases the second formulation of his categorical imperative: "Act so that you treat humanity, whether in your own person or in that of another, always as an end withal and never as a means *only*." It's okay to treat other human beings (who, Kant thinks, are the only intrinsically valuable entities) as means to one's own ends as long as you do not treat them *only* as means. You must also treat them as ends in themselves.

The mutual compatibility of instrumental and intrinsic value is also utterly familiar. In the context of small businesses, where employers and their employees work closely together, employees are valued primarily as means to the employers' ends, but in many cases such employers also palpably value their employees as ends in themselves. And they may express how they intrinsically value their employees in many ways – for example, by not laying them off when business slows down, despite the economic loss suffered by the employers. That is the price of the dignity of their employees – if putting it so is not an affront to Kant. In the context of very large businesses, employers may not palpably value their employees intrinsically, but coldly value them as means only. To protect employees of such impersonal corporations, we have socially objectified their intrinsic value in a suite of labor laws – laws concerning occupational health and safety, minimum wage, overtime pay, health benefits, pension plans, etc. – insuring that the dignity of employees is respected and that they are not treated as means only.

Even in those human and cross-species relationships in which intrinsically valuing others is the primary way we value them, we also value them instrumentally. Spouses perform many and various services for one another. Children are assigned chores. Their parents use the children's grandparents as babysitters. The family dog is relied on to alarm the family of possible intruders (although usually it's a false alarm, you never know). Pleas to end the bitter wrangling in the conservation community between proponents of the intrinsic value and those of the instrumental value of biodiversity have been based, in part, on the fact that the two kinds of value are not mutually exclusive (Peterson *et al.* 2013, Tallis and Lubchenco 2014).

Quantifying intrinsic value

Again, Kant's talk of price and dignity might suggest that he himself thought that if something has intrinsic value, that its value is absolute, inviolable, even infinite. And if so great a philosopher as Kant – the very font and source of all subsequent philosophical reflection on intrinsic value – thought so, we had better give his thought about the matter careful consideration. So, aren't intrinsically valuable things priceless, according to Kant, and doesn't "priceless" mean of infinite value? Not necessarily; indeed, not at all. What Kant actually means is that instrumental value and intrinsic value are expressed in different conceptual domains – the one in the domain of the market, the other in the domain of the law. For Kant, the latter domain is the *moral* law, but subsequent philosophical thought about intrinsic value should not be bound by the letter of Kantian law, if I may put it so. In the context of the social objectification of intrinsic value, the kind of law that we are considering is not *jus* but *lex* – "the formal product of a legislative or judicial body" – cognate with the Greek *lexis*, speech.

In any case, the foregoing considerations indicate that the question – does intrinsic value come in degrees, such that some kinds of things have more of it than do others? – has a decisive *empirical* answer. Judging from the way in which intrinsic value is socially objectified, as a matter of fact it does come in degrees. For example, consider the difference in the policies and laws protecting other animals from harm and those protecting humans. Laws protect domestic animals from cruel and abusive treatment from their human masters. But such animals can be legally trafficked. And they can be euthanized – note *eu*thanized, given a good death – when they come to the end of their usefulness.

If something varies in degree, then it is something that can be quantified. By its very nature, intrinsic value cannot be quantified in a *monetary metric* – because one essential function of socially objectifying the intrinsic value of something is to remove it from the realm of price and legal market trade. The proper realm of intrinsic value is that of law and policy, the political realm (Sagoff 1988). And that consideration points to quantifying how much intrinsic value something has in terms of the penalty for violating the laws that objectify its intrinsic value and protect its dignity. Intrinsic value may, in short, be roughly quantified in a *penalty metric*. Penalties for murder vary by jurisdiction, but are generally quite severe. As to human trafficking in illegal sex markets, the US Mann Act of 1910 (*18 U.S.C. § 2421–2424*), amended in 1978 and 1986, prohibits the transport of minors and the coercion of adults to cross state lines for the purpose of engaging in commercial sex. Both crimes are punishable by up to twenty years in federal prison.

The UN Convention on Biodiversity is a so-called "soft law" and specifies no penalties for reducing or eroding biodiversity. The US ESA and CITES, protecting listed endangered species, however, may serve as surrogates for laws that objectify and protect the intrinsic value of biodiversity. CITES relegates enacting laws to protect listed endangered species and associated penalties for violating those laws to the signatory Parties. The US ESA is one such law and provides up to one year in prison and/or up to a $50,000 fine for the "taking" (which is elaborately specified in the law) of a specimen of a listed species. In comparison with the shadow pricing of such things as ocean views and national parks, the quantification of the intrinsic value of listed endangered species and biodiversity as a whole in a non-monetary metric is a completely unexplored and undeveloped domain of quantitative axiology. But hopefully it will become a future subspeciality of political science, just as environmental and ecological subspecialties have developed in economics.

Summary and conclusion

At the beginning of this chapter a number of philosophical questions were posed and then systematically answered. Here in sum are those questions and their answers.

Is biodiversity the sort of thing that can have intrinsic value? The answer to that question depends on another: what is the ontological (and grammatical) status of intrinsic value? It is not an objective property of things that have it, although it has been plausibly analyzed as a supervenient property referenced to a genuinely objective base property – such as rationality and sentience – of the things that might have it. On such an analysis, however, biodiversity cannot have intrinsic value because it has no suitable base property on which intrinsic value might supervene. By way of an alternative, simpler, and therefore a more elegant analysis, all value is conferred on things by intentional acts of valuing subjects. Value is a verb first and a noun second. Things have value if and only if they are valued. And things may be valued in two basic ways: instrumentally and intrinsically. On this analysis, biodiversity can have intrinsic value if it is intrinsically valued; and it is in fact intrinsically valued by many reputable valuers. Valuing things, both instrumentally and intrinsically, is not arbitrary. We value things for various reasons, and often they are good reasons.

Subjectively conferred intrinsic value can, however, be objectified in two ways: epistemologically and socially. The first trades on two distinct meanings of "objective": (i) existing independently of consciousness in the real world, (ii) unbiased judgment. Although based on an equivocation, epistemologically objectifying subjectively conferred intrinsic value can play an important role in ethics, as it does in Kant's ethics. According to Kant and other ethicists, we should be unbiased – impartial – in consideration of the intrinsic value of others as well as our own. Subjectively conferred intrinsic value is socially objectified in law and policy – and law and policy exist independently of consciousness in the real world. Law and policy evolve in response to rational argument and persuasion.

Having intrinsic value makes a big practical difference. It takes the things that have it out of legal markets and, unfortunately, also creates black markets. Things that have intrinsic value may be traded off for instrumentally valued things – say, an endangered fish species for drinking water or hydroelectric power. But having intrinsic value shifts the burden of proof onto those who might want to trade off the things that have it for things of mere instrumental value. That's how intrinsic value and instrumental value interact in the real world.

As objectified in law and policy, intrinsic value comes in degrees. Intrinsic value can be roughly quantified – roughly at least at this initial stage of quantitative axiology – by reference to the penalties specified for violating the laws that objectify their intrinsic value and protect the things that have it. Under the US ESA – which may be treated as a benchmark for quantifying the socially objectified intrinsic value of biodiversity – the maximum penalty for taking a specimen of a listed endangered species is one-twentieth the degree of severity for taking a minor into the commercial sex market. That indicates that the amount of intrinsic value had by a specimen of a listed endangered species is at most one-twentieth the amount of intrinsic value had by a human being.

Acknowledgments

Research for this chapter was supported in part by the National Socio-Environmental Synthesis Center (SESYNC) (NSF award # DBI-1052875); I benefited from discussions about intrinsic value with members of the SESYNC working group on Ecological Restoration and Ecosystem Services, especially Alan Randall.

Note

1 The citations are to "Naess A. 1973. The shallow and the deep, long-range ecology movements: a summary. *Inquiry* 16: 95–100" and "Callicott JB. 1989. *In Defense of the Land Ethic: Essays in Environmental Philosophy*. State University of New York Press, Albany, 325pp."

References

Bentham, J. 1789. *Introduction to the Principles of Morals and Legislation*. Oxford: The Clarendon Press.

Callicott, J. 2006. "Explicit and Implicit Values." In *The Endangered Species Act at Thirty: Conserving Biodiversity in Human-Dominated Landscapes*, Vol. 2, ed. J. Michael Scott, Dale D. Goble, and Frank W. Davis, 36–48. Washington, DC: Island Press.

Collar, M. J. 2003. "Beyond Value: Biodiversity and the Freedom of the Mind." *Global Ecology and Biogeography* 12: 265–269.

Cone, C. 2012. *Redacted Dominionism: A Biblical Approach to Grounding Environmental Responsibility*. Eugene, OR: Wipf & Stock.

Convention on Biological Diversity. 1992. Available online at www.cbd.int/convention/articles/default.shtml?a=cbd-00 (accessed 13 January 2015).

Earth Charter. 2004. Available online at www.earthcharterinaction.org/content/pages/Read-the-Charter.html (accessed 13 January 2015).

Feinberg, J. 1974. "The Rights of Animals and Unborn Generations." In *Philosophy and Environmental Crisis*, ed. W. Blackstone, 43–68. Athens: University of Georgia Press.

Fox, W. 1993. "What Does the Recognition of Intrinsic Value Entail?" *Trumpeter* 10: 101.

Goodpaster, K. E. 1978. "On Being Morally Considerable." *Journal of Philosophy* 75: 308–325.

Hare, R. M. 1952. *The Language of Morals*. Oxford: Oxford University Press.

Hare, R. M. 1984. "Supervenience." *Aristotelian Society Supplementary* 58: 1–16.

Johnson, L. E. 1991. *A Morally Deep World: An Essay on Moral Significance and Environmental Ethics*. Cambridge: Cambridge University Press.

Kant, I. 1785/1959. *Grundlegung zur Meataphisic der Sitten*. Berlin: L. Heinmann; *Foundations of the Metaphysics of Morals*, trans. L. W. Beck. New York: Bobbs-Merrill.

Karieva, P., M. Marvier, and R. Lalasz. 2012. "Conservation in the Anthropocene: Beyond Solitude and Frugality." Available online at http://thebreakthrough.org/index.php/journal/past-issues/issue-2/conservation-in-the-anthropocene (accessed 29 August 2015).

Leopold, A. 1949. *A Sand County Almanac and Sketches Here and There*. New York: Oxford University Press.

Max, D. T. 2014. "Green is Good." *The New Yorker*, May 12, 54–63.

Moore, G. E. 1903. *Principia Ethica*. Cambridge: Cambridge University Press.

Mueller, A., and J. Maes. 2015. "Arguments for Biodiversity Conservation in Natural 2000 Sites; An Analysis Based on LIFE Projects." *Nature Conservation* 12: 1–26.

Norton, Bryan G. 1987. *Why Preserve Natural Variety?* Princeton, NJ: Princeton University Press.

Peterson, M. N., M. J. Peterson, T. R. Peterson, and K. Leong. 2013. "Why Transforming Biodiversity Conflict is Essential and How to Begin." *Pacific Conservation Biology* 19: 94–103.

Pister, E. P. 1987. "A Pilgrim's Progress from Group A to Group B." In *Companion to A Sand County Almanac: Interpretive and Critical Essays*, ed. J. B. Callicott, 221–232. Madison: University of Wisconsin Press.

Pister, E. P. 1992. "Ethical Considerations in Conservation of Biodiversity." *Transactions of the 57th North American Wildlife and Natural Resources Conference*, 355–364.

Plater, Z. J. B., R. H. Abrams, W. Goldfarb, and R. L. Graham, L. Heinzerling, and D. A. Wirth. 2004. *Environmental Law and Policy: Nature, Law, and Society*, 3rd edn. New York: Aspen Publishers.

Randall, A. 1988. "What Mainstream Economists Have to Say about the Value of Biodiversity." In *Biodiversity*, ed. E. O. Wilson and F. M. Peter, 221–227. Washington, DC: National Academy Press.

Regan, T. 1979. "An Examination and Defense of One Argument Concerning Animal Rights." *Inquiry* 22: 189–219.

Regan, T. 1983. *The Case for Animal Rights*. Berkeley: University of California Press.

Rolston III, H. 1994. *Conserving Natural Value*. New York: Columbia University Press.

Sagoff, M. 1988. *The Economy of the Earth: Philosophy, Law, and the Environment*. New York: Cambridge University Press.

Sarkar, S. 2012. *Environmental Philosophy: From Theory to Practice*. Malden, MA: Wiley-Blackwell.

Shrader-Frechette, K. S. 1981. *Environmental Ethics*. Pacific Grove, CA: Boxwood Press.

Singer, P. 1975. *Animal Liberation: A New Ethics for Our Treatment of Animals*. New York: New York Review.

SCB Society for Conservation Biology. 2011. "2011–2015 SCB Strategic Plan: Enhancing the Impact of Conservation Science." Available online at www.conbio.org/images/content_about_scb/2011SCBStrategicPlan_Branded_edite d.pdf (accessed 13 January 2015).

Soulé, M. 1985. "What is Conservation Biology?" *BioScience* 35: 727–734.

Soulé, M. 2013. "The 'New Conservation'." *Conservation Biology* 27: 895–897.

Tallis, H. and J. Lubchenco. 2014. "A Call for Inclusive Conservation." *Nature* 515: 27–28.

Taylor, P. 1981. "The Ethics of Respect for Nature." *Environmental Ethics* 3: 197–218.

Vucetich, J., J. T. Bruscotter, and M. P. Nelson. 2015. "Evaluating Whether Nature's Intrinsic Value is an Axiom of or Anathema to Conservation Biology." *Conservation Biology* 29: 321–322.

13

ECONOMIZING ON NATURE'S BOUNTY

Lisa Heinzerling

Should we protect biodiversity? Why? To what extent?

One way to try to answer these questions is to determine how much a given quantum of biodiversity is worth in monetary terms and to compare that value to the cost of protecting it. A standard framework for making this comparison is cost–benefit analysis. In fact, under executive orders issued by United States presidents over the last several decades, cost–benefit analysis has become the executive branch's dominant evaluative framework for federal agencies' rules, including rules aimed at protecting biodiversity (Heinzerling 2014). It is thus worth paying attention to cost–benefit analysis and the signals it sends about the reasons for protecting biodiversity and the proper extent of our commitment to it.

Cost–benefit analysis in the United States takes place under a highly systematized, rather rigid set of rules established by the Office of Regulatory Affairs (OIRA) within the Office of Management and Budget (OMB) (OMB 2003), two offices within the Executive Office of the President. The evaluative framework that emerges is one that systematically but subtly undermines the case for protecting biodiversity. Indeed, this framework either dismisses or undercuts some of the most basic reasons why we might think it important to preserve biodiversity.

In this chapter, I first describe the essential mechanics of cost–benefit analysis, with special reference to its application to policies that protect biodiversity. Then I will turn to an explanation of the ways in which the central elements of cost–benefit analysis are in tension with a thoroughgoing commitment to the protection of biodiversity.

The mechanics of cost–benefit analysis

Cost–benefit analysis has three basic steps: quantification, monetization, and discounting (Ackerman and Heinzerling 2004). These steps are performed for both the costs and the benefits of a regulatory policy, although they tend to be more complicated in the case of benefits. Many of the costs of environmental regulation – such as installation of pollution control technology – are, in principle, straightforward to quantify and monetize, as they come in discrete units and their price can be determined through market exchanges. Many of the benefits, however, are quite difficult to quantify, let alone monetize. I will focus here on the benefits side of the equation, as this is the aspect of cost–benefit analysis that offers the greatest challenge to efforts to protect biodiversity.

The first step, quantifying the benefits of efforts to protect biodiversity, might seem relatively straightforward, but in fact it is quite complex. In order to quantify a consequence of a regulatory policy, one must have in mind what the desired consequence is, with some degree of specificity. "Increase in biodiversity" or "averted decrease in biodiversity" is hardly precise enough to permit quantification. Here, ongoing debates about what exactly biodiversity is complicate the quantitative task (Adams *et al.* 2000); in order to quantify an effect on biodiversity, one needs first to have an understanding of what biodiversity is. As earlier chapters in this volume make clear, we do not yet have such a settled understanding.

It seems possible, however, to stipulate that at the very least the protection of species and of ecosystems is part of the project of protecting biodiversity. Quantifying the effect of a regulatory policy on a species or an ecosystem is certainly more manageable than quantifying a vague "increase in biodiversity." Thus, in conducting the first step of cost–benefit analysis in this domain, one might aim to specify, for example, the size and biological content of a specific geographical area protected, or the portion of a particular species that will be saved if one protects the entire species from extinction. The more particularly one describes the goal of the regulatory intervention, the more tractable the quantitative analysis will be. But of course aiming for greater particularity has the drawback of pulling an ecosystem or a species out of its natural context and analyzing it in an atomistic fashion, an analytic method that may not capture the nature and value of biodiversity.

Even if one is able to specify the regulatory goal without counterproductive reductionism, scientific uncertainty will often make it difficult to quantify the effects of a regulatory policy with any assuredness. One may know, for example, that one wants to protect the polar bear from becoming extinct due to climate change, but understanding the effect of any particular policy intervention on this goal may be challenging, if not impossible. Cost–benefit analysis asks for a number or a range of numbers for the boxes marked "benefits of regulation," yet those boxes are hard to fill.

Once one has a passable quantitative estimate for the benefits of a regulatory policy, the next step in cost–benefit analysis is to translate the quantified estimate into monetary terms. If one has identified a particular ecosystem, for example, as the object of a regulatory intervention, then one must determine how much protection of that ecosystem – or the parts of it that will be affected by the intervention – is worth, in dollars. Likewise, if one has identified a particular species or a set of species as the target of the intervention, one must figure out how much protecting the species is worth.

Needless to say, this can be tricky. Many ecosystems and species have some commercial value; indeed, many have a great deal of commercial value. To the extent they do, then one might use the value of the natural resource as traded in the commercial marketplace as an estimate of the worth of the resource in cost–benefit analysis. If one is protecting a fishery, for example, one might consider the value of the fish caught and sold as at least a lower bound on the value of protecting the fishery. Or if one is protecting a pristine wilderness, one might consider how much individuals are willing to pay to visit the wilderness in trying to understand how much the wilderness is worth. In addition, many natural resources play a role in ecosystems that, if the resources were damaged or depleted, would have to be performed (if this were even possible) by a manmade piece of equipment or other technology that itself has a market price. Here, one might take the market price of the substitute equipment or technology as a sign of the economic value of the resource; this is, in simplified form, the economic component of the notion of "ecosystem services." In each of the cases just described, one would, in the parlance of cost–benefit analysis, be aiming to understand the "use" value of the resources to be protected.

In many instances, however, one wants to protect elements of biodiversity – such as species and ecosystems – not only because humans directly use them, but also because humans believe these resources have value even when not in use by humans, that is, they have "non-use" value. One might, for example, care very much about the northern right whale even if one never expects to use its blubber for oil, or to use its baleen to make tools or clothing, or even to see one. One might derive satisfaction from knowing that the northern right whale exists ("exist-ence" value), or from knowing that one could see one in the wild if one wanted to ("option" value), or from knowing that one's descendants will have this opportunity ("bequest" value) (Walsh *et al.* 1984). The jargon is wooden, but the sentiment is not; it reflects humans' belief that natural resources have value apart from our direct use and experience of them.

In many cases, these non-use values will be high, and will even dwarf the use values associ-ated with a natural resource. To take one famous example, the commercial fisheries damaged by the *Exxon Valdez* oil spill in 1989 recovered approximately US$300 million in court-ordered damages (Ackerman and Heinzerling 2004: 154). This amount reflected the use value of the resources damaged by the spill. But economists charged with trying to understand the non-use value of the damaged resources came up with a far larger figure for this aspect of the damage, on the order of at least US$9 billion in value lost as a consequence of the spill (Ackerman and Heinzerling 2004: 156). Although the technical methods used to determine non-use values – predominantly a method called "contingent valuation" – have been in use for decades, they remain contentious.

The last step in estimating the benefits of regulatory policy aimed at protecting biodiversity is to take the quantified, monetized benefits one has calculated and to discount them from the date when the benefits are likely to be realized back to the date on which costs to realize them are incurred. There are two basic theories underlying discounting in the regulatory context (Heinzerling 1998). One is that money in the future is worth less than money today because one could, if one had the money today, invest it to earn even more money in the future. The other is that money in the future is worth less than money today because people are simply impatient; they would prefer to have a good thing now than to wait for the good thing to come to them in the future. Either way, the notion is that future benefits must be reduced – "discounted" – by some amount in order to reflect their lesser worth compared to current benefits.

Discounting has an enormous effect on estimates of the benefits of regulatory policy. Current OMB guidance on cost–benefit analysis directs agencies to discount future benefits (and costs) at rates of 3 percent and 7 percent (OMB 2003). Although OMB does allow agencies to use discount rates of 1–3 percent in cases where intergenerational issues loom large, these lower rates may be used only in sensitivity analyses and not in the primary estimates. Moreover, even in a context where intergenerational issues predominate – climate change – OMB, in concert with an interagency working group, has stipulated that discount rates from 2.5 to 5 percent are appropriate (Interagency Working Group on Social Cost of Carbon 2013) – thus, strangely, opt-ing for a higher discount rate for the specific context of climate change than for the sensitivity analysis accepted for other areas in which intergenerational equities figure prominently. When discounted at the rates reflected in OMB guidance, even very large future benefits can be ren-dered very small. As Derek Parfit observed years ago, use of a 5 percent discount rate implies that the death of a billion people in 500 years from now is less serious than the death of one person next year (Parfit 1984: 357). The choice of a discount rate, and indeed the decision whether to discount future benefits at all, can make all the difference in determining whether a given policy passes or fails the cost–benefit test.

There are of course many complexities and wrinkles in the method of cost–benefit analysis. I have tried to draw a simple but not overly simplified picture of the basic methodology. As I have explained, the cost–benefit analyst has three main tasks to perform in attempting to estimate the benefits of protecting biodiversity: quantification, monetization, and discounting. In describing these tasks, I have hinted at some of the problems posed by cost–benefit analysis in the context of protecting biodiversity. Now I turn to more detailed examination of these problems.

Quantification

Quantification is the most basic step in cost–benefit analysis. If one does not or cannot quantify the benefits of a regulatory policy, then one will have a hard time moving on to the other aspects of cost–benefit analysis, monetization and discounting – and, perhaps more fundamentally, one will have a hard time defending the regulatory policy against critics who can often point to well-defined regulatory costs. Even the most ardent fans of cost–benefit analysis recognize that an inability to quantify the benefits of a regulatory policy may require adoption of decision-making frameworks other than cost–benefit analysis (Sunstein 2014).

In the context of regulatory policies protecting biodiversity, quantification can be challenging even when the aim of a regulatory policy is seemingly quite precise. Good illustrations of this point come from the context of endangered species. The Endangered Species Act is perhaps the most prominent of the US laws aimed at protecting biodiversity. Its central constraint is a prohibition of federal agency programs and projects that jeopardize the continued existence of species deemed by the federal government to be endangered or threatened (*16 U.S.C. § 1536*). Regulatory policies undertaken pursuant to this legal framework thus aim to prevent extinction of covered species.

The benefits of even this apparently precise regulatory goal, however, frequently resist the kind of quantification that is essential to economic analysis. It will often be inappropriate simply to say that the answer to quantification in the context of a specific policy aimed at protecting a species from extinction is the number "1": the benefit of the regulatory policy is that one species will be protected. It is possible, for example, that other regulatory activities, undertaken concurrently, also will protect the species, and thus the regulatory policy being evaluated may not be responsible for the total protective impact on the species. Moreover, species protection often occurs in a setting of great scientific uncertainty, in which policy-makers undertake actions – protecting the critical habitat of endangered or threatened species, say – that they hope will forestall extinction, but they do not really know what proportion of the species will remain once the actions have taken hold. In these circumstances, the economic analyst cannot simply rest with the assumption that the benefit of a regulatory measure aimed at preventing extinction is the protection of the entire population of the species in question.

Recent economic analyses of decisions of the US Fish and Wildlife Service (FWS), designating critical habitats for species covered by the Endangered Species Act, bear out these observations. In assessing the benefits of designating the critical habitat of the jaguar, for example, the FWS was able to quantify the total number of acres that would be affected by the designation but was not able to quantify the actual impacts on jaguar populations. The agency reported that 858,137 acres would be included in the designation (US FWS 2014: 1–3) – but its quantitative precision ended there. The agency could not say what percentage of jaguars would be preserved by its action, nor even assay a numerical range. The agency candidly summarized its inability to quantify the benefits of jaguar habitat protection, for jaguar populations, as follows:

> The primary goal of critical habitat designation for the jaguar is to support its long-term conservation .... The extent to which critical habitat designation for the jaguar may improve the species' population is unknown. That is, information is not available on the potential percent increase in jaguar populations, or the incremental change in the probability of recovery, generated by the incremental conservation efforts described in this analysis. Absent information on the incremental change in jaguar populations or recovery potential associated with this rulemaking, we are unable to monetize associated incremental use and non-use ... economic benefits.
>
> (US FWS 2014: 11–1)

Although the FWS supplemented its benefit assessment with the observation that protecting the critical habitat of the jaguar would also benefit biodiversity more broadly insofar as other species living in the relevant areas would also be advantaged by protection of their habitat, here, too, the agency's analysis was entirely qualitative (US FWS 2014: 11–6–11–7). The economic analysis accompanying other recent decisions on critical habitat designation, such as a decision involving a set of six amphibians in Texas, exhibits a similar incapacity to attach numbers to the benefits of the designation (US FWS 2013).

Things are, if anything, even more challenging when the policy aimed at protecting bio-diversity is highly diffuse. In that case, it may not even be clear exactly which ecosystems and species may be affected by the policy. Nationally applicable rules limiting air or water pollution, for example, will often affect biodiversity in some fashion, but the precise effects may be hard to determine. For example, the Environmental Protection Agency (EPA) has long been bedeviled by the challenge of quantitative estimation of the impacts of its ambient air quality standards on flora and fauna. In the economic analysis of its recent proposal to strengthen the national air quality standard for ozone, the EPA described a long list of potential effects of tropospheric ozone on vegetation and wildlife, including everything from damaged plant function to changes in the composition of plant communities to decline of habitats that support endangered species to decline of aquatic species – and beyond (US EPA 2014: 6–1–6–17). The EPA encapsulated its analysis thus: "[o]zone can affect ecological systems, leading to changes in the ecological community and influencing the diversity, health, and vigor of individual species" (US EPA 2014: 6-3). The agency even observed that, with respect to ecosystem diversity, "ozone may be the pollutant with the greatest potential for region-scale forest impacts" (US EPA 2014: 6-3). Despite the large body of scientific evidence the agency marshaled in describing the potential adverse effects of ozone on biodiversity, however, it was able to quantify only what it called a "small portion" of the ecological effects of possible changes to the ozone standard – namely, the effects on consumers and producers of forest and agricultural products (US EPA 2014: 6-6). It is worth noting that the EPA's economic analyses of proposed changes to national air quality standards are among the most sophisticated regulatory impact analyses undertaken within the US government today. Yet even here, the challenge of quantifying the benefits of protecting biodiversity has severely limited the picture drawn by economic analysis.

The upshot of these observations is that if we value biodiversity by applying cost–benefit analysis, biodiversity probably will not fare very well. Benefits that cannot be quantified often do not count for much in cost–benefit analysis. Frequently, unquantified benefits are treated as if the number that should be attached to them is zero. Even when such benefits are described qualitatively in some detail, they are often no match for costs that are both quantified and monetized. This is especially so if cost–benefit analysis is a legally required predicate for a regulatory policy; in that case, a lopsided relationship between costs and benefits can spell doom for the policy. In invalidating an EPA rule banning asbestos, for example, a federal appeals court chided the EPA for relying in substantial part on unquantified benefits to justify the costs of its rule, even suggesting that the agency might have "arbitrarily" limited its calculations on purpose in order to "preserve a large unquantified portion" (*Corrosion Proof Fittings v. Environmental Protection Agency* 1991: 1219).

However, attempting to correct for the problem of quantification by fixating on whichever aspect of a policy relating to biodiversity can be quantified is in some tension with rationales for protecting biodiversity itself. Trying to quantify the benefits of designating critical habitat for the jaguar, for instance, by counting the number of individual animals that will survive if the habitat is protected treats the jaguar as a discrete, independent entity within the environment. It abstracts from the place of the jaguar in the larger ecosystem. This kind of reductionism may sometimes prove helpful in economic analysis, as it may make quantification possible. But it is not a helpful or even realistic way to think about biodiversity. Aldo Leopold, as usual, put it best: "Everybody knows … that the autumn landscape in the north woods is the land, plus a red maple, plus a ruffed grouse. In terms of conventional physics, the grouse represents only a millionth of either the mass or the energy of an acre. Yet subtract the grouse and the whole thing is dead" (Leopold 1966: 146).

Monetization

Even if one could overcome the challenge of reliably quantifying the benefits of protecting biodiversity, another, equally daunting, problem looms: that of translating the quantified benefits into monetary terms. Just as unquantified benefits may be poor contenders when compared to quantified costs, so, too, non-monetized benefits may appear insignificant when weighed against monetized costs. Thus, the economic analyst who wishes to give biodiversity a fighting chance must press on from quantification to monetization.

To the extent humans actually use an ecosystem, say, or a species, monetization may be relatively straightforward. In that case, the economic analyst will ask how much we are willing to pay for that use – how much, in other words, we are willing to pay to obtain the lumber for our houses or for the game meat for our tables or for the thrill of seeing a humpback whale. However, the magnitude of these "use" values often pales in comparison to the magnitude of values not associated with actually using the relevant resources. As noted earlier, the economic accounting of the consequences of the *Exxon Valdez* oil spill found non-use values greatly in excess of the commercial losses associated with the spill. Similarly, the "use" of whales through commercial whale watching in the United States brings in hundreds of millions of dollars a year, while the "non-use" value reflected in contingent valuation surveys comes to many billions of dollars (Ackerman and Heinzerling 2004: 161). If only use values are included in economic analysis of policies relating to biodiversity, the analysis will likely slight the actual value of biodiversity.

Getting a handle on non-use value is, however, dicey. Contingent valuation tries to ascertain what individuals are willing to pay to protect natural resources by asking them to complete detailed questionnaires. The idea, in simplified form, is to elicit the non-use value of resources by asking direct questions rather than by observing the operation of markets. Worries include the idea that individuals who are being asked what they are willing to pay to protect a resource do not actually have to pay the money they say they are willing to pay, and thus the survey may overstate the true value they place on that resource. Another concern is that responses to these surveys do not always seem to distinguish precisely between loss of, say, one or a small subset of members of a species, and loss of the entire species. For reasons like these, the use of contingent valuation to determine non-use values remains contentious.

Often, moreover, estimates of non-use value are simply not available for the aspect of biodiversity under discussion. When the FWS analyzed the economic impacts of designating critical habitat for the jaguar, for example, it found no valuation studies that matched the circumstances of its policy choice. The quantitative studies the FWS found did not involve the jaguar at all, and the agency concluded that studies involving wolves or sheep were not good indicators of the value of the jaguar (US FWS 2014: 11–6). Yet if we do not have economic studies of the non-use value of the jaguar – a charismatic, keystone species familiar to many – then one can only shrink from the prospect of developing non-use values for the multitude of lesser known and less appealing species and ecosystems.

These practical difficulties are matched by issues of principle. First, the challenges of monetization press the economic analyst toward reductionism, away from holism, in the same way the challenges of quantification do.

Second, the individualized focus of monetization ignores the nature of biodiversity as a public good. The process of monetization entails asking people, individually, what they are willing to pay to protect some element of biodiversity. Natural resources are, however, classic examples of public goods; one cannot buy individual portions of them. The rational actor assumed in cost–benefit analysis will pay nothing individually to protect a common resource, since she will not expect her individual payment to do any good. This does not mean, however, that she will pay nothing toward a collective effort to protect the common resource.

Recognizing this dilemma, economic analysts have tried to place questions about resource valuation in a collective context, such as by asking how much study participants would be willing to pay in additional taxes to protect a resource. There is reason to worry, however, that simply reframing the question does not convince participants in valuation studies that they are in the midst of a collective enterprise. Despite such reframing, participants in a survey assessing the value of a marsh in England, for example, doubted that their own individual monetary contributions would protect the marsh. "[W]hat good would it be," asked one participant, "if I had said 'oh yes, I'd give a thousand pounds?' I mean, in isolation that is absolutely no good anyway, is it?" (Clark *et al.* 2000: 50).

Participants in valuation studies, in other words, have been astute enough to question whether public goods can be valued by individuals acting in isolation from one another. Yet the closer contingent valuation comes to an exercise in collective deliberation, the further it moves from the central economic premise on which it was founded, namely, the idea that individual willingness to pay is a reliable measure of value. More profoundly, to the extent that cost–benefit analysis is used to evaluate and, often, undermine regulatory policies implementing laws that protect biodiversity, one must seriously question which collective enterprise – economic analysis, or legislative action – should be given pride of place. Even if one could, in principle, refine the process of eliciting values so that all involved in contingent valuation studies knew they were engaged in a collective process, aimed at valuing collective goods, the question would remain: why should *this* collective process trump the collective process that led to legislative action?

Last, the project of valuing biodiversity according to individual willingness to pay largely assumes away the ethical content of choices concerning biodiversity. It assumes that natural resources are ordinary economic goods, just like chairs or shoes or cutlery. Indeed, reacting to the fact that participants in valuation studies often express moral reservations about "buying" or "selling" natural resources (Stevens *et al.* 1991), some scholars have argued that the introduction of moral impulses into valuation studies renders the whole enterprise economi-cally suspect. They contend that respondents giving voice to such impulses are not valuing public goods but moral satisfaction, and that "[t]he amount that individuals are willing to pay to acquire moral satisfaction should not be mistaken for a measure of the economic value of public goods" (Kahneman and Knetsch 1992: 69). From this perspective, economic analysis of natural resources deliberately excludes the ethical impulses that might undergird the very value the analysis is trying to measure.

Discounting

Suppose we can, despite the obstacles discussed above, develop reliable quantitative and mon-etized estimates of the benefits of protecting some aspect of biodiversity relevant to a policy choice. The sad fact is that we will have come this far only, in all likelihood, to have those esti-mates rendered minuscule by the process of discounting.

Recall that cost–benefit analyses of US regulatory policies use discount rates from 2.5 percent (the low-end discount rate used in calculating the social cost of carbon) to 7 percent (the high-end discount rate for other contexts). Given the long timeframes encountered in many policies related to biodiversity, these discount rates will steeply reduce the perceived benefits of such policies.

The Endangered Species Act again provides a useful illustration. The US Fish and Wildlife Service has listed the polar bear as a threatened species under this statute (US FWS 2008). Although the agency found that polar bear populations are even now under stress due to cli-mate change, the agency also concluded that the species is not now in danger of extinction (the predicate for finding a species "endangered" rather than threatened under the statute). Instead, the FWS concluded that the polar bear is likely to be in danger of extinction "within the fore-seeable future" (US FWS 2008: 28212). The FWS defined the "foreseeable future" for the polar bear as approximately forty-five years – three polar bear generations (US FWS 2008: 28239–40). "Not all populations will be affected evenly in the level, rate, and timing of effects," the agency reported, "but we have determined that, within the foreseeable future, all polar bear populations will be negatively affected" (US FWS 2008: 28275).

The OMB-generated guidelines for federal agencies' economic analysis not only specify discount rates as high as 7 percent, but they also instruct agencies to discount regulatory benefits over the interval between now and the time when those benefits would occur (OMB 2003). In the case of the polar bear and its threatened extinction, therefore, an economic analysis per-formed according to these guidelines would need to discount the benefits of saving the polar bear over a forty-five-year period.

Discounting benefits at 7 percent over forty-five years shrinks those benefits considerably. Suppose that the economic value of preserving the polar bear were calculated to be US$10 billion. Discounted at OMB's preferred 7 percent rate over forty-five years, this value would shrink to less than half a billion dollars in present-value terms. Because the polar bear is not expected to become extinct for decades, therefore, the economic analysis currently practiced by the federal agencies would, via discounting, take a huge slice out of whatever monetary value could be attached to the protection of this species.

As with quantification and monetization, therefore, discounting introduces a systematic bias against robust protections for biodiversity, if those protections are evaluated by using cost–benefit analysis. Indeed, the very notion of discounting, premised as it is on the relative unim-portance of the future, is at odds with our central legal frameworks for addressing conflicts over natural resources. The laws protecting endangered species, national parks, national monuments, national forests, wilderness areas, and other public lands all aspire to long-term protection of natural resources. Not one of these laws has a "best by" date – a date on which resource protection becomes unimportant, or even trivial. Discounting disserves these laws – and, more fundamentally, the perspectives on biodiversity that animate them – by systematically, even inherently, privileging short-term gains over long-term protection.

Conclusion

Cost–benefit analysis is, at best, a severely incomplete way of representing the value of biodiversity. At each step along the way toward a formal cost–benefit analysis – quantification, monetization, and discounting – the benefits of protecting biodiversity appear to get smaller and smaller. Many do not even survive the first cut, when efforts are made just to quantify the benefits of an action aimed at protecting biodiversity. Biodiversity benefits that make it past this first stage then fall off the radar if they do not involve direct use by humans or if no one has spent the time and resources to complete a study of non-use values. Discounting finishes the job by exponentially shrinking any benefits that remain after running the gauntlet of quantification and monetization. In the realm of biodiversity, it is not terribly off to say that cost–benefit analysis is a very complicated way of getting to zero.

More work, however, such as redoubled efforts to quantify and monetize the benefits of biodiversity, will not make cost–benefit analysis an appropriate framework for valuing biodiversity. The framework is itself in at odds with the holistic, public, future-oriented nature of biodiversity. Even if perfectly executed, the cost–benefit technique will miss too much.

Acknowledgment

I am grateful to Matthew Johnson for excellent research assistance.

References

Ackerman, F., and L. Heinzerling. 2004. *Priceless: On Knowing the Price of Everything and the Value of Nothing.* New York: The New Press.
Adams, J. S., B. A. Stein, and L. Kutner. 2000. "Biodiversity: Our Precious Heritage." In *Precious Heritage: The Status of Biodiversity in the United States*, ed. B. A. Stein *et al.* Oxford: Oxford University Press.
Clark, J., J. Burgess, and C. M. Harrison. 2000. "'I Struggled with this Money Business': Respondents' Perspectives on Contingent Valuation." *Ecological Economics* 33: 45.
Corrosion Proof Fittings v. Environmental Protection Agency. 1991. 947 F.2d 1201 (5th Circuit).
Heinzerling, L. 1998. "Regulatory Costs of Mythic Proportions." *Yale Law Journal* 107: 1981.
Heinzerling, L. 2014. "Inside EPA: A Former Insider's Reflections on the Relationship Between the Obama EPA and the Obama White House." *Pace Environmental Law Review* 31: 337–381.
Interagency Working Group on Social Cost of Carbon, US Government. 2013. "Technical Support Document: Technical Update of the Social Cost of Carbon for Regulatory Impact Analysis Under Executive Order 12866," available online at www.whitehouse.gov/sites/default/files/omb/inforeg/social_cost_of_carbon_for_ria_2013_update.pdf.
Kahneman, D., and J. L. Knetsch. 1992. "Valuing Public Goods: The Purchase of Moral Satisfaction." *Journal of Environmental Economics and Management* 22: 57.
Leopold, A. 1966. *A Sand County Almanac.* New York: Oxford University Press.
Office of Management and Budget (OMB). 2003. "Circular A-4: Regulatory Analysis (2003)", available online at www.whitehouse.gov/omb/circulars_a004_a-4/.
Parfit, D. 1984. *Reasons and Persons.* Oxford: Oxford University Press.
Stevens, T. H., J. Echeverria, R. J. Glass, T. Hager, and T. A. More. 1991. "Measuring the Existence Value of Wildlife: What Do CVM Estimates Really Show?" *Land Economics* 67: 390–400.

Sunstein, C. R. 2014. "The Limits of Quantification." *California Law Review* 102: 1369–1422.

US Environmental Protection Agency (EPA). 2014. "Regulatory Impact Analysis of the Proposed Revisions to the National Ambient Air Quality Stndards for Ground-Level Ozone," available online at www.epa.gov/ttnecas1/regdata/RIAs/20141125ria.pdf.

US Fish and Wildlife Service (FWS). 2008. Final Rule, "Endangered and Threatened Wildlife and Plants; Determination of Threatened Status for the Polar Bear (*Ursus maritimus*) Throughout Its Range." *Federal Register* 73: 28212–28303.

US FWS. 2013. "Final Economic Analysis of Critical Habitat Designation for the Phantom Cave Snail, Phantom Springsnail, Diminutive Amphipod, Diamond Y Spring Snail, Gonzales Springsnail, and Pecos Amphipod," available online at www.fws.gov/southwest/es/Documents/R2ES/West_TX_Inverts_fLfCH_FEA_20130507.pdf.

US FWS. 2014. "Final Economic Analysis of Critical Habitat Designation for the Jaguar," available online at www.fws.gov/southwest/es/arizona/Documents/SpeciesDocs/Jaguar/20140115_Jaguar_fCH_FEconAnalysis2_.pdf.

Walsh, R. G., J. B. Loomis, and R. A. Gillman. 1984. "Valuing Option, Existence, and Bequest Demands for Wilderness." *Land Economics* 60: 14–29.

14

PROTECTING BIODIVERSITY AND MORAL PSYCHOLOGY; OR WHY PHILOSOPHERS ARE ASKING THE WRONG QUESTIONS

Jay Odenbaugh

Introduction

In this essay, I contend philosophers have been asking the wrong questions about the value of biodiversity. By and large, they have concerned themselves with the intrinsic value of non-human organisms, species, ecosystems, and biodiversity more generally. First, I consider the most important argument, the "Argument from Teleology," for claiming they have intrinsic value. I argue that it fails. Second, I argue that *even if* it succeeds, psychological studies suggest that claims of intrinsic value are comparatively poor motivators of pro-environmental behavior. Third, I claim that focusing on ecosystem services offers a better chance of protecting biodiversity.

Environmental philosophy and the Argument from Teleology

For the purposes of this essay, think of biodiversity as genes, species, and ecosystems along with their variation. This definition is too inclusive and imprecise. But nothing here hangs on getting such an account exactly correct. Protecting the biodiversity of some such group consists in protecting the group and its variation. For example, the biodiversity of a group of species would be the species and the variation in their properties. But, *why* should we protect species and their intraspecific and interspecific variation?

The most straightforward answer is this. Biodiversity is necessary for present and future humans' well-being. We should conserve whatever is necessary for present and future humans' well-being. Therefore, we should conserve biodiversity (Norton 1987).[1] Additionally, we should not cause unnecessary suffering to sentient, non-human animals (Regan 1985, Singer 1995). If this is correct, and reducing biodiversity will cause unnecessary suffering to them, then we have an additional reason beyond human well-being for conserving biodiversity (Jamieson 1998).

In general, these arguments have left environmental philosophers unsatisfied. First, many philosophers have claimed that non-sentient organisms deserve direct moral consideration.

Second, several philosophers have claimed that collectives such as populations, metapopulations, species, communities, and ecosystems deserve it as well (Callicott 1987, Leopold 1989, Johnson 1993, Rolston 2012).[2] Thus, environmental philosophy has concerned whether and to what extent this motley crew, and by extension, biodiversity, has *intrinsic value*. Until recently, this has been the *sine qua non* of environmental philosophy.

"Intrinsic value" is a notoriously unclear term especially as used by environmentalists. We can distinguish between several senses (O'Neill 1992). First, something has intrinsic value just in case its value depends solely on its intrinsic properties. An intrinsic property is one that an object has independently of context. Suppose that the rarity of a species with *n* members is characterized by $1/n$. If species have intrinsic value in virtue of their intrinsic properties, then rarity is such a property (Sarkar 2005, 57). Second, something has intrinsic value just in case it is valuable for its own sake. As Aristotle pointed out, if everything is only instrumentally valuable, then we have either an infinite regress or a tight circle.

Given the unacceptability of either, something is valuable for its own sake. Maybe organisms, species, etc., are valuable as ends and not merely as means. Third, something has intrinsic value just in case its value does not depend on human valuers. Suppose California coastal redwood trees (*Sequoia sempervirens*) have value even if there were no humans. Then this value would be mind-independent or "inherent."[3]

Though there is a plethora of intrinsic value concepts, there are fewer types of arguments for intrinsic value. Aristotle's argument at best shows that there is something intrinsically valuable, but is silent as to what it is. One of the first arguments that some such non-human, non-sentient entities have intrinsic value is owed to Richard Routley (later Sylvan) (1973). Routley argued Western ethics was beholden to a principle of "basic human chauvinism"; namely, something is permissible insofar as it harms no one irreparably (Routley 1973: 207). He then has us consider the *last person example*:

> The last man (or person) surviving the collapse of the world system lays about him, eliminating, as far as he can, every living thing, animal or plant (but painlessly if you like, as at the best abattoirs). What he does is quite permissible according to basic chauvinism, but on environmental grounds what he does is wrong.
>
> (Routley 1973: 207)

Routley's argument is this. If the principle of basic human chauvinism is correct, then a human is morally permitted to act how they choose provided he does not harm any human irreparably. However, what the last person does is not morally permissible though it harms no human irreparably. Therefore, the principle of basic human chauvinism is false.

The argument is contentious. First, not everyone shares Routley's biocentric convictions. If you do not, then you will reject his second premise. Second, even if you do, it says nothing about *why* it would be morally wrong to do this. Third, one might prefer that the non-human world continue to exist, but deny it is a moral matter. For example, one might prefer the non-human world to continue for aesthetic reasons (Sagoff 1974, Elliott 1997).

Routley's argument was supposed to show the limitations of traditional normative ethics and thereby herald a distinctive *environmental* ethic. Some date environmental ethic's beginning to Kenneth Goodpaster's (1978) essay on moral considerability. In effect, he is answering the question of why we should accept Routley's second premise. He argues that living things matter because they have *interests*. That is, we can act in such a way as to harm or benefit them. But, if we can harm or benefit something, it is because they have interests and whatever has interests has intrinsic value. Therefore, non-human living things have intrinsic value. For reasons that will become clear, I will call this the *Argument from Teleology.*

1. Living things have interests.
2. If something has interests, then it has intrinsic value.
3. Living things have intrinsic value.

Of course, if something has intrinsic value, we must determine what the appropriate response to such value is. For example, should it be promoted, honored, respected, etc. (Anderson 1995)? The Master Argument has been articulated in varying terms by many different philosophers including Nicholas Agar (2001), Lawrence Johnson (1993), Holmes Rolston III (2012), Paul Taylor (2011), and Gary Varner (1998). In what follows I consider the argument and why it fails.

Goodpaster articulated this position with regard to living individuals and not non-living collectives, as have other philosophers. For example, Paul Taylor in his *Respect for Nature* argues along very similar lines. According to Taylor, something has a good just in case it can be benefited or harmed without reference to another entity.[4] For Taylor, something having a good is grounded in their being a "teleological center of life." He writes,

> To say it is a teleological center of life is to say that its internal functioning as well as its external activities are all goal-oriented, having the constant tendency to maintain the organism's existence through time and to enable it successfully to perform those biological operations whereby it reproduces its kind and continually adapts to changing environmental events and conditions. It is the coherence and unity of these functions of an organism, all directed toward the realization of its good, that make it one teleological center of activity.
>
> (Taylor 2011: 121–122)

Taylor's theory is an elegant and powerful.[5] The philosophers mentioned above have added added details to the biocentrism staked out by Goodpaster and Taylor. Still, the core Argument from Teleology has remained the same.

I now want to consider two challenges to the Argument from Teleology. The first objection is that it is incorrect to move from teleology to intrinsic value. The second objection is that even if the first inference was sound, this approach cannot be extended to biodiversity. I will take these objections in turn.

Interests and intrinsic value

Philosophers of science find teleological claims about biological systems perplexing. If biological systems are not designed, then what makes functional claims true? In the twentieth century, the most important attempt to understand these claims is Larry Wright (1973, 1976). On his account, the function of x is to z means x is there because it zs, and z is a consequence of x's being there. For example, the function of the human heart is to circulate blood means the heart is there because it circulates blood, and circulating blood is a consequence of human hearts being there. This account is subject to counterexamples such as the following (Boorse 1976). Suppose there is a gas leak in a scientist's lab, which renders the scientist unconscious. It is there because it renders them unconscious and their being unconscious is a consequence of the leak. But, no one says a function of the leak is to render the scientist unconscious.

Many philosophers think we may avoid these counterexamples by invoking evolution by natural selection. Suppose the function of a trait is that for which the trait evolved by natural selection in the recent past. A trait evolves by natural selection just in case that trait is heritable, the trait exhibits variation, and trait-bearers have greater reproductive success compared to alternatives. Thus, the human heart has the function of circulating blood just in case having said heart is heritable, there is variation among hearts, and humans with such a heart have greater expected reproductive success. We can say that a trait evolves by natural selection just in case it exhibits heritable variation in fitness. This is the selected effects account of function (Millikan 1984, Neander 1991, Godfrey-Smith 1994).

On this account of function, if a trait has the function to *F*, then it is supposed to *F*. It is malfunctioning if it does not *F*. Some have claimed the selected approach allows us to reduce norms to facts. If you think that normative claims are teleological in nature and teleology just is what evolution by natural selection designed traits to do, then normative claims are species of factual ones. Moreover, if ethical claims are a type of normative claim, then ethical claims too are just factual ones (Foot 2001, Casebeer 2003, Post 2006). We have solved the "is–ought" problem.

I reject this reduction (Odenbaugh 2015). Here is an example to make this point. Consider lions (*Panthera leo*) that live in the Serengeti National Park in Tanzania. Prides contain between three and twelve adult females and one to six adult males. Females are all related to one another and they reproduce between four and eighteen years old. Males leave their pride around age three, and eventually attempt to take over the prides of others. Even if successful, they themselves will be removed in a few years. Bertram (1975) noticed two interesting phenomena. First, females have a synchronized estrus cycle. Second, when a new male takes over a pride, they often kill the cubs. The proposed explanation of this is that it increases the expected reproductive success of males since it returns females to estrus more quickly. If the cubs are not killed, it can take twenty-five months for females to come into estrus. By killing the cubs, females shorten that period to nine months. However, even if the behavior of killing cubs when entering a new pride has this selected effects function, we do not think this as of moral value. Even if young males are "supposed" to do this, the Serengeti is not better morally for it. We cannot reduce moral norms to selected functions.

Let me describe my argument more carefully. Premise (2) of the Argument from Teleology claims that if something has interests, it has intrinsic value. Suppose from the claim that male lions are "teleological centers of life," they have interests. For example, they have interests in killing cubs as they enter a new pride; frustrating those interests lowers their fitness. Premise (2) implies that since male lions have interests then they have intrinsic value. Surely insofar as something has intrinsic value we should promote that value. But next to none of us think we have an obligation or duty of helping males kill cubs. Thus, since we do not have this obligation and we would if they have intrinsic value, premise (2) of the Argument from Teleology should be rejected.

One response to this counterargument is that promotion is not the only fitting attitude to take towards intrinsic value; for example, sometimes it is honor or respect. So, by creating a national park in Tanzania for lions and other living things, we are expressing our respect for them. But, nature is red in tooth and claw, and it is questionable whether this carnage deserves honor or respect morally construed.[6] If there is no non-ad-hoc way to settle this question, then we are back where we started.

Here is another objection to premise (2). Consider artifacts such as automobiles, cell phones, and thermostats. These objects are designed to perform certain tasks. As such, they have interests. Moreover, if whatever has interests has intrinsic value, then they have intrinsic value. But there is nothing wrong with me destroying my iPhone per se. Thus, we have a counterexample to (2) as well. Hence, their interests do not "create" intrinsic value.

You might be thinking artifacts only have interests insofar as we design them. As Taylor puts it, their interests are realized insofar as they contribute to the interests of something besides themselves. However, on the selected effects account of functions, the function of some entity is always in reference to a reproductive community and an environment. Consider domesticated animals and plants. We have designed them at least in part to have features we desire. It would also follow that our pets for example would have no more value than the pieces of technology we use. Insofar as there is nothing wrong with destroying my iPhone, similarly I could treat my dog Charlie as I wish. But this is unacceptable too. Thus, premise (2) should be rejected. The Argument from Teleology fails.

Interests and biodiversity

Even if the Argument from Teleology was sound, there would still be a deep problem. The Argument from Teleology does not apply to collectives, which is required if we are to claim that biodiversity has intrinsic value. First, populations, species, and ecosystems are not alive. But, some environmentalists believe that they have intrinsic value. Thus, they have revised (1) as follows,

(1★) Anything that exhibits teleological behavior or functions has interests.[7]

I will argue that our (1★) does not apply to species or ecosystems. As such, the Master Argument cannot be used to undergird our obligations to biodiversity.

Population biologists have long recognized that birds often have fewer viable offspring than they can. Wouldn't evolution by natural selection select for the greatest number of offspring? One hypothesis is that birds forgo having more offspring for the good of the group to avoid overshooting the carrying capacity of the environment (Wynne-Edwards 1962). Thus, individual sacrifice was in the interest of the species. Clutch size evolved by group, or species, selection.

Since 1947, the great tit (*Parus major*) has been studied in Wytham Woods around Oxford, UK, initially by David Lack (1954). Most of the breeding pairs have eight to nine offspring. However, if more eggs are added they can incubate them with success. Still, as the number of hatchlings in the brood increases, then average weight decreases. This is due to the hatchlings receiving less and lower-quality food (e.g. caterpillars). Heavier chicks have a greater probability of survival and reproduction. In experiments, it has been demonstrated that the optimal clutch size is approximately eight to nine eggs.[8] Lack and others argued individual selection explained clutch size; group selection is not required.

Most evolutionary biologists think group selection can occur under certain restrictive circumstances, and has occurred in the history of life occasionally. However, it is a general consensus that it occurs rarely (though see Sober and Wilson 1999). If this is correct, then species rarely exhibit teleological behavior of their own. At best, any teleological behavior exhibited is a by-product of that of its constituent organisms. As evolutionary biologist George Williams (2008) pointed out, there is a big difference between a *fleet herd* of deer and a herd of *fleet deer*. The upshot then is that species do not have interests and thus they do not have intrinsic value. Thus, the Master Argument simply does not apply to an ethic of conserving biodiversity.[9]

Let me now extend this argument to ecosystems. Philosopher Lawrence Johnson writes,

> Just as we may think of an individual organism as an ongoing life process, manifested in a continually changing combination of material elements, and a species as an ongoing process progressively embodied in different individuals, so may we think of an ecosystem as an ongoing process taking place through a complex system of interrelationships between organisms, and between organisms and their nonliving

environment … Normally, an ecosystem maintains its stability through an intricately complex feedback system … However, an ecosystem can suffer stress and be impaired … In short, an ecosystem has well-being interests and therefore has moral significance.

(Johnson 1993: 217)

Johnson is correct that ecosystem ecologists make functional claims. Nitrogen, and nitrogen-containing compounds, move through our biosphere. In our atmosphere, we have a reserve of nitrogen in gaseous form (N_2). N_2 is converted to ammonia or nitrate through nitrogen fixation. One type of nitrogen fixation occurs through organisms like *Rhizobium* bacteria in the root structures of plants, and from there nitrogen can be assimilated into the plant. Ecosystem ecologists claim that Rhizobia function to fix nitrogen contributing to the nitrogen cycle. If Rhizobia have the function of fixing nitrogen, then (a) fixing nitrogen is a heritable trait among Rhizobia, (b) fixing nitrogen contributed to the reproductive success of Rhizobia relative to alternative traits in the recent past, and (c) there was variation in fixing nitrogen among Rhizobia. *Maybe* Rhizobia has these selected effects functions. However, 10 percent of nitrogen fixation occurs through abiotic components. Lightning and volcanoes can fix nitrogen too. Thus, an ecosystem ecologist might claim that volcanoes have the function of fixing nitrogen thereby contributing to the nitrogen cycle. But, lightning and volcanoes do not exhibit heritable variation in fitness. Thus, ecosystems do not have selected effects functions and hence do not have interests (though see Swenson *et al.* 2000). As such, they do not have intrinsic value. If we think we ought to conserve ecosystems and their variation, the Argument from Teleology does not help.

The critic might retort that other accounts of functions exist. Specifically, Robert Cummins (1975) offered the systemic capacity account of functions. Suppose we have a system with some disposition. Moreover, suppose a part of that system has a disposition, which contributes to the disposition of the system. Using the example above, Rhizobia have a disposition to fix nitrogen, which contributes to the ecosystem's contribution to the nitrogen cycle. Cummins claims that when the sub-disposition of the part contributes to the disposition of the whole, then the part has a function of contributing to the whole. Thus, Rhizobia have a function of fixing nitrogen because it contributes to the ecosystem's contribution to the nitrogen cycle. And, more significantly, we can make the same claim of lightning and volcanoes. Cummins's account applies to functional claims made on behalf of ecosystems.

There is a rub here for moving from Cummins's functions to normative claims. On the selected effects account, one can argue that the functional claims are normative, the reason being that selected effects functions provide us with "norms of performance." If the human heart has the function of circulating blood since past ones that did this had greater reproductive success, then ones that do not are malfunctioning. They are not doing what they are "supposed to." However, on the systemic capacity account, there is no normativity (Davies 2001). A part simply no longer has or executes a sub-disposition. Thus, even if ecosystems exhibit teleology, they do not have interests and thus do not have intrinsic value. The Argument from Teleology simply does not apply to biodiversity.

Thus, the Argument from Teleology is unconvincing. Moreover, I do not know of any other plausible argument for the intrinsic value of living things, species, and ecosystems; that is, biodiversity.[10] *However*, suppose that the above arguments were sound. Even still, these arguments ignore a crucial issue. We can see this by outlining the following questions.

1. What are the values of organisms and collectives of them?
2. What reasons are there for accepting they have these values?
3. What reasons will motivate people to protect them?

Even if organisms and collectives of them have intrinsic value and there are warranted reasons for accepting this, evidence suggests most people's motivations lie elsewhere. Environmental philosophers have been largely concerned with (1) and (2). If environmental ethics is to be a *practical* discipline – an area of *applied* ethics – then (3) matters. I have argued the Argument from Teleology fails and have suggested there are no other good arguments for the intrinsic value of biodiversity. What then motivates people and gives them reasons to care about biodiversity? It is to these issues that I now turn.

Moral psychology, sentimentalism, and cognitive biases

Humans care about the natural world for a variety of reasons. Stephen Kellert (1997) provided a classification of the types of values that Americans associate with the natural world. Table 14.1 describes these types of value, their definition, and the roles they play in our lives.

Kellert also documented the overall ranking of values and their importance to Americans as represented in Fig. 14.1.

From his work, we recognize that Americans first have a strong emotional attachment to the natural world. In this, they find companionship and cooperation. We can even speak of their "love" of a place and the natural flora and fauna therein. Second, the natural world has spiritual associations and generates ethical concerns. It provides meaning and offers opportunities to be better people (Norton 1987). This category is not restricted to those who care about the natural world for its own sake. Third, the natural world provides occasions of "fear and trembling"; it is something from which we need security and protection.

In 2002, the public opinion firm Belden Russonello and Stewart conducted a national poll on American attitudes towards biodiversity for the Biodiversity Project (Belden *et al.* 2002). Approximately four in ten Americans recognize the term "biodiversity" and can characterize it. Fifty-five percent of the total surveyed maintained that biodiversity was important to them and its loss was ranked as the most important environmental problem.

Table 14.1 Survey of American values regarding the natural world

Value	Definition	Function
Utilitarian	Practical and material exploitation of nature	Physical sustenance/security
Naturalistic	Direct experience and exploration of nature	Curiosity, discovery, recreation
Ecologistic-scientific	Systematic study of structure, function	Knowledge, understanding, observational skills
Aesthetic	Physical appeal and beauty of nature	Inspiration, harmony, security
Symbolic	Use of nature for language and thought	Communication, mental development
Humanistic	Strong emotional attachment and "love"	Bonding, sharing, cooperation, companionship
Moralistic	Spiritual reverence and ethical concern for nature	Order, meaning, kinship, altruism
Dominionistic	Mastery, physical control, dominance of nature	Mechanical skills, physical prowess, ability to subdue
Negativistic	Fear, aversion, alienation from nature	Security, protection, safety, awe

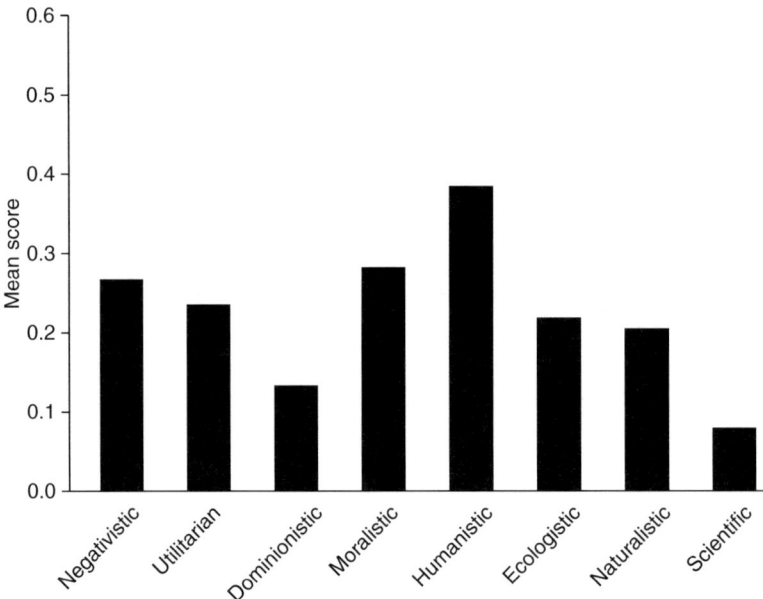

Figure 14.1 Survey of American attitudes to the natural world.

Sixty-nine percent claimed that we have a personal obligation to protect biodiversity, and 65 percent said it was a moral responsibility. Participants were asked, "Which is the most important reason for you personally to care about protecting the environment?" Responses are given in Fig. 14.2.

What is striking is that the most important reasons are responsibilities to future genera-tions (40%), nature is a product of God's creation (23%), and protecting the "balance of nature" (17%). The notion of "respect for nature" as environmental ethicists like Taylor have urged as paramount is suggested as fundamental by only 10 percent of Americans. Thus, Americans are concerned about the natural world, and biodiversity specifically, largely because of future gen-erations and as stewards of God's creation.

Leiserowitz *et al.* (2005) reported on a 2002 survey that seems to challenge the above; namely, it claims Americans "strongly agreed that nature has intrinsic value" (Leiserowitz *et al.* 2005: 25). They report that over 75 percent of Americans accept humans have moral duties and obligations to other animals, plants, and non-living nature. In fact, more than 75 percent of Americans agreed that "Nature has value within itself regardless of any value humans place on it" (Leiserowitz *et al.* 2005: 28). Moral duties and obligations to animals, plants, and non-living nature of course are consistent with an anthropocentrist approach. The claim that nature has value regardless of the values that humans place on it is problematically ambiguous. One way to understand it is that nature has intrinsic value. A second way is that nature's value, be it intrinsic or instrumental, is not due to our *placing* value on it. Thus, given this ambiguity, the Leiserowitz *et al.* (2005) approach is inconclusive.

Recently, Vucetich *et al.* (2015) have challenged the claim that "nature's intrinsic value cannot be used to justify conservation because nature's intrinsic value is not widely believed" (Vucetich *et al.* 2015: 5). They rightfully note that previous studies (Steel *et al.* 1994, Vaske and Donnelly 1999, Kaltenborn and Bjerke 2002), which tried to determine whether and to what extent people are non-anthropocentrists, are flawed. The questions these researchers asked participants

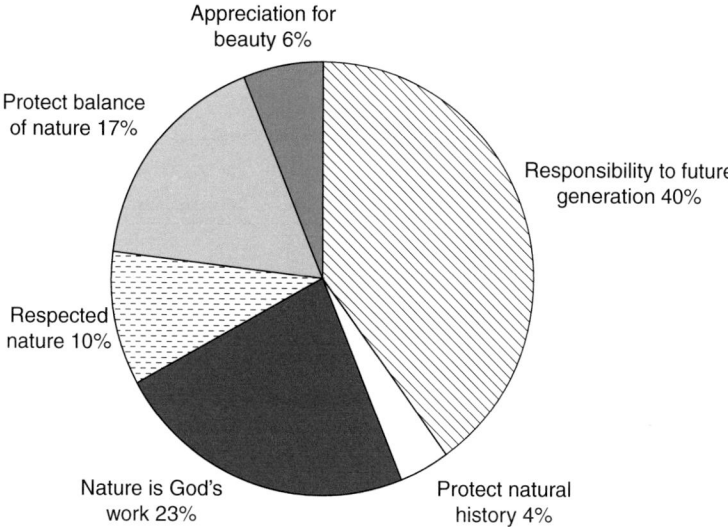

Figure 14.2   Most important reasons for Americans protecting the environment.

are too imprecise since anthropocentrists and non-anthropocentrists could answer "yes" to them. Consider these statements,

> Forests give us a sense of peace and well-being.
> Forests rejuvenate the human spirit.
> Forests let us feel close to nature.
> I need time in nature to be happy. (Vucetich *et al.* 2015: 6)

One could value forests instrumentally or intrinsically and agree with the above statements. Vucetich *et al.* surveyed a sample of Ohioans regarding their views on the value of wildlife. Specifically, they provided two statements and determined whether they agreed or not.

1. Wildlife have inherent value, above and beyond their utility to people.
2. Wildlife are only valuable if people get to utilize them in some way. (Vucetich *et al.* 2015: 10)

They found that 82 percent of >2,700 households accepted (1); namely, wildlife has inherent value. In fact, there was little difference between those that hunt, fish, trap, view wildlife, and those that don't, regarding their acceptance of (1). From this study and one other national survey, they write, "This suggests that conservationists who reject nature's intrinsic value are out of the mainstream of their peers." Vucetich *et al.*'s study is flawed, however. For Americans, especially ones that hunt, fish, trap, etc., "wildlife" typically is associated with deer, bears, birds, etc. A majority of Americans think that such animals matter morally for their own sake. But, this is consistent with the view that animals that can suffer deserve moral consideration, and does not extend moral consideration to non-sentient organisms (e.g. plants and microbes) and collectives (e.g. populations and species). Thus, their empirical study provides weak support that many accept "nature's intrinsic value."

At this point you might wonder, who cares why Americans say they should protect the natural world? What are *good* reasons for so doing? First, as we saw above, there are straightforward

good reasons for protecting biodiversity ("the environment") and they concern future generations of humans and sentient living things. Second, warranted reasons only matter for the practical decision-making if they motivate decision-makers. The Argument from Teleology did not
provide such warrant. But even if it did, it is not on the list of primary motivations of Americans.
Third, from the data described, we see reasons regarding emotional attachments, our legacy, and
religious injunctions are paramount in people's minds. Insofar as these considerations motivate
their pro-environmental behavior, we should strategically focus on them.[11]

Moral psychology and sentimentalism

In moral psychology, there is a debate regarding the nature of moral judgment. Sentimentalists
claim that moral judgments involve emotions. That is, they are necessarily affective. Rationalists
deny this. If sentimentalism is correct, then it has important implications for what motivates
conservation of biodiversity. Let's first examine the case for a sentimentalist moral psychology.
There is a great deal of evidence that emotions occur when we offer moral judgments. First,
neuroimaging studies show that the areas of the brain associated with emotion are active during moral judgment (Greene and Haidt 2002, Heekeren *et al.* 2003, Moll *et al.* 2003, Sanfey
et al. 2003, Singer *et al.* 2006). When subjects were asked to consider moral sentences versus
neutral sentences; were offered inequitable versus equitable pay-offs in the ultimatum games; or
violations of social rules like spitting food at dinner as opposed to spitting due to choking; the
parts of subjects' brains associated with emotion were far more active than in the non-moral
cases. Greene *et al.* (2001) showed using fMRI that emotions were involved in "trolley cases."
Consider the following two cases.

> Switch problem: a runaway trolley is headed for five people who will be killed if it
> continues on its present course. However, you can hit a switch, which turns the trolley
> onto a different track killing one bystander. Should you hit the switch?
> Footbridge problem: A trolley threatens to kill five people. You are standing next to
> a very large stranger on a footbridge above the track. However, the only way to save
> the five is to push the one onto the track. Should you push him?

Most people say "Yes" and "No," respectively. Why? The usual answer is that in the former
it involves letting someone die and the latter involves killing them. And, killing someone is
worse than letting them die. Greene *et al.*'s hypothesis regarding our conviction is that in switch
problems we are merely impersonally hitting a switch, in footbridge problems we are personally pushing a person, and the latter elicits emotions far more than the former (Greene *et al.*
2001: 2106). Those areas of the brain associated with emotion were significantly more active in
"moral-personal" than in "moral-impersonal" cases. Thus, Greene *et al.* conclude,

> How do people manage to conclude that it is acceptable to sacrifice one for the sake
> of five in one case but not in the other? We maintain that emotional response is likely
> to be the crucial difference between these two cases.
>
> (Greene *et al.* 2001: 2107)

Second, when a subject's disgust is aroused, they judge actions as morally worse. The
disgust aroused in subjects is independent of the actions evaluated and can even be aroused
unconsciously (Schnall *et al.* 2008). Schnall *et al.* asked subjects to morally evaluate stories
while sitting at a desk which was clean or one that was very dirty (e.g. has an old pizza box,

chewed pencils, and a dirty cup). Subjects' moral judgments are much more severe at the dirty desk compared to the tidy one. As another example, Wheatley and Haidt (2005) hypnotized subjects to feel disgust whenever the words "take" or "often" were heard. If one said that a congressperson "takes bribes" or "is often bribed" they would judge them far more harshly.

Third, some argue psychopaths cannot distinguish between moral and conventional violations due to emotional deficits (Blair 1995). Moral norms are ones that are authority-independent, general, and serious and conventional norms are authority-dependent, specific, and are not particularly bad to violate (Turiel 1983). Blair showed that psychopaths have extreme difficulty determining whether a norm is moral or conventional. Psychopaths suffer deficiencies in affect. Thus, Blair and others have hypothesized that the best explanation of psychopaths' failure to pass the moral/conventional task is that they lack certain emotions. As such, they lack the ability to make moral judgments (though see Kennett 2006, Aharoni *et al.* 2012).

Fourth, when morally dumbfounded, we do not change our opinions (Haidt 2003). Consider Haidt's cannibal story:

> Jennifer works in a medical school pathology lab as a research assistant. The lab prepares human cadavers that are used to teach medical students about anatomy. The cadavers come from people who had donated their body to science for research. One night Jennifer is leaving the lab when she sees a body that is going to be discarded the next day. Jennifer was a vegetarian, for moral reasons. She thought it was wrong to kill animals for food. But then, when she saw a body about to be cremated, she thought it was irrational to waste perfectly edible meat. So she cut off a piece of flesh, and took it home and cooked it. The person had died recently of a heart attack, and she cooked the meat thoroughly, so there was no risk of disease. Is there anything wrong with what she did?

Subjects claimed eating the cadaver was morally wrong. However, when asked why, subjects gave reasons inconsistent with the case. Nevertheless, they refused to give up their moral judgment. They were morally dumbfounded. Haidt claims moral judgments are more a product of emotion than reason.

One of the major debates in metaethics concerns the role of motivation and moral judgment (Smith 1994, Björnsson *et al.* 2015). Externalists claim that one can judge an action is wrong and be completely unmotivated to avoid it. For example, one could judge that "Anthropogenic species extinction is morally wrong" and not care one whit about it. Internalists claim that necessarily when one sincerely makes a moral judgment one is motivated accordingly. You cannot genuinely judge something is wrong and not be motivated somewhat. The evidence above suggests that internalism is correct. Moreover, if moral judgment involves affect and thus motivation, then non-motivating judgments are simply not moral judgments. It is thus even more crucial for environmental ethicists to concern themselves with the considerations that actually motivate people to pro-environmental behavior. Otherwise, they will be insincere, as well as ineffective.

Cognitive biases

Beginning with Herbert Simon and his work on bounded rationality, and then especially through that of Daniel Kahneman and Amos Tversky, psychologists have documented how in a blooming, buzzing, and confusing world we use heuristics to make decisions. However, our cognitive processes are subject to biases, and they can be very relevant to our environmental decision-making. Here I want to mention some of the biases that are relevant.

The *availability heuristic* concerns our tendency to consider only those alternatives that easily come to mind (Tversky and Kahneman 1974, Greenberg *et al.* 1989, Gardner and Stern 2002). For example, if we consider species extinction, it is easier to imagine the loss of charismatic megafauna like polar bears rather than, say, disruptions to a trophic cascade. Our consideration of environmental risks can thus be unrepresentative. The *anchor and adjustment heuristic* concerns how we anchor alternatives with an example and then rank the other alternatives in relation to it (Tversky and Kahneman 1974). For example, if we anchor protection of wolves with a complete ban on takings, then protecting wolves will seem extreme. Thus, allowing for the killing of wolves that threaten livestock will seem moderate. The extremity or moderateness of an action is dependent on our anchoring. The *loss aversion heuristic* concerns the fact that for some good or service, we are more strongly averse to losing it as opposed to gaining it (Kahneman and Tversky 1979). This is closely related to *the framing effect heuristic* in which the same information can be associated with aversion or preference depending on how it is framed (Tversky and Kahneman 1981). The *temporal discounting heuristic* concerns our tendency to prefer rewards that occur sooner rather than later (Hendrickx and Nicolaij 2004).

There are many cognitive heuristics and biases that are relevant to conserving biodiversity. Our pro-environmental behaviors are dependent on the alternatives we consider and how they are anchored, whether we consider the effects as losses or gains, and how temporally close or remote their effects are (Andreou 2007). Many philosophers assume that our failure to come to terms with ethical issues is due to an absence of information. If we improve the public's understanding of the relevant science, for example, they will make reasonable decisions. However, the cognitive biases go deep and our poor decisions are not simply due to an absence of information (Nisbet and Mooney 2009).

Our best account of moral psychology suggests several things. First, in order to motivate individuals to pro-environmental behavior, we must present considerations connected to what people care about. If we do not, then they will not be motivated by these considerations. As we have seen, the primary reasons people care about biodiversity are the impacts on future generations and religious commitments. Respect for nature per se appears to be a minority concern. Second, any attempt to motivate pro-environmental behavior ought to attend to cognitive biases. Whatever reasons are offered for caring about biodiversity, they should relate to what moves us in the short run, concern losses as opposed to gains, should be anchored in non-extreme ways, and should be framed appropriately. In the next section, I present reasons for preserving biodiversity which fit this bill.

Ecosystem services and biodiversity

I have considered the Argument from Teleology as offered by philosophers and we have found it wanting. Likewise, I have argued that a sentimentalist account of moral judgment is best supported by our current evidence. Coupled with an account of our cognitive biases given by cognitive psychologists, then we should look for ethical reasons that are motivationally effective and less subject to our biases. One such class of reasons concerns the *ecosystem services* biodiversity provides. We will first consider a more general discussion of ecosystems services and then consider a concrete example.

Biodiversity, including genes, species, ecosystems, and their variety, provides food, fuel, fiber, and medicine.[12] In one famous example, the rosy periwinkle (*Catharanthus roseus*) has been used for treating diseases including diabetes, malaria, and Hodgkin's lymphoma. The annual world fish catch is about 100 million metric tons valued between US$50 and US$100 billion and the commercial harvest of freshwater fish in 1990 was 14 million tons valued at $8.2 billion. We use about 7,000 plant species for food, but about 70,000 plants species are known to be edible. Of the top 150 prescription drugs used in the US, 118 are based on natural sources. Pharmaceuticals in the developed world are valued at $40 billion per year. The natural world sustains us.

Biodiversity also sustains us indirectly: purification of air and water, mitigation of floods and droughts, detoxification and decomposition of wastes, regeneration and renewal of soil and soil fertility, pollination of crops and natural vegetation, control of the vast majority of potential agricultural pests, dispersal of seeds and translocation of nutrients, moderation of temperature extremes and the force of winds and waves, and aesthetic experiences and intellectual stimulation. Consider three examples: soil, pollinators, and pests.

Soil shelters and supports seeds as they grow, retains and delivers nutrients to plants, and plays a central role in the decomposition of organic matter and wastes. It is crucial for regulating the Earth's carbon, nitrogen, and sulfur cycles. Soil degradation caused by humans affects nearly 20 percent of the Earth's vegetated surface. Most flowers require pollinators for reproduction (of 240,000 plants species, 200,000 require an animal pollinator). This includes 70 percent of the crop species that feed the world. Over 100,000 species of bats, bees, beetles, birds, butterflies, and flies provide these services. Approximately a third of our food is derived from plants pollinated by wild pollinators. Pests, our competitors, include insects, rodents, fungi, snails, nematodes, and viruses. They destroy between 25 and 50 percent of the world's crops and are especially important given how harmful pesticides can be; 99 percent of pests are controlled by natural enemies like birds, spiders, wasps, and ladybugs.

Some think of ecosystem services as narrowly "utilitarian" and "economic" (Daniel *et al.* 2012). This is not so. Ecosystems sustain us culturally too through recreation, appreciation of natural beauty, ecotourism, health, historical monuments, and spiritual experiences. In 2012, the US National Park System by itself had over 282 million recreational visits. These benefits are not just found in "wild" areas accessible only to REI members, but includes urban green spaces too. For example, children with ADHD do better after physical activities outside than with activities indoors (Taylor *et al.* 2001).

To make this discussion concrete, let's consider the ecosystem services that salmon provides.

Ecosystem services and salmon

Salmon has enormous value for fishermen, processors, distributors, restaurants, suppliers, boat-builders, tour operators, fishing guides, and charter boat operators.[13] As of 1988 there were an estimated 62,750 salmon-dependent jobs in the Pacific Northwest of the US, which generated about $1.25 billion to the regional economy. In the 1990s, the actual economic value of Columbia-based salmon fisheries dropped as low as $2 million. Salmon encourages recreation and tourism to the Pacific Northwest in the United States and Alaska. Additionally, they are important to sport fishing and angling. Salmon serve as a regional symbol and are found represented in art and souvenirs. They also serve as a flagship species for other species in the region. Salmon are incredibly important to Native American life and their ceremonial rituals. Additionally, young salmon are a rich source of food for fish and birds given their lipid content. Adults provide carbon, phosphorus, and nitrogen from the ocean to nutrient-poor lakes and streams. Their carcasses provide food for invertebrates like algae, fungi, and bacteria, and for vertebrates like bears, foxes, wolves, ravens, and eagles. In order to flesh out this ecosystem services argument, let's consider one very specific service: salmon's contribution to ecosystem productivity.

Spawning salmon serve as a food resource for other species, and when they die after spawning, their carcasses provide nutrients such as carbon, nitrogen, and phosphorus to freshwater systems. Recently, scientists have documented that these salmon-derived nutrient subsidies may have significant impacts on both freshwater and riparian communities. Adults return to freshwater late summer and fall where they cease feeding, spawn, and die. After some months, young emerge from gravel in early spring and spend up to two years in a freshwater habitat before migrating. The fish remain at sea for one to seven years, gaining over 90 percent of their biomass before returning. The nutrient flux of salmon biomass into a freshwater can be massive; 20 million sockeye can yield 5.4×10^7 kilograms (kg) of biomass which equates to 2.4×10^4 kg of phosphorus, 1.8×10^5 kg of nitrogen, and 2.7×10^5 kg of carbon. Stream systems serve as "conduits" for the input of ocean-derived material into freshwater and terrestrial systems. Bears move carcasses into riparian forest where they are partially consumed. Many stream insects have an aerial phase in which they fly far from streams. Avian scavengers remove chunks of salmon and sometimes leave them on land. Increased lake productivity is caused by salmon-nutrient inputs which increase phytoplankton and zooplankton. Salmon carcasses increase population size and growth rates of invertebrates; juvenile salmonids may grow faster feeding on the vertebrates which feed on the carcasses. Invertebrate scavengers increase in abundance due to salmon. Insectivorous riparian birds are found in greater numbers around salmon streams than non-salmon streams, suggesting that they are responding to the salmon "pulse." Bear populations are up to 80 percent larger in coastal areas where salmon are abundant rather than in interior areas. The fitness (growth rates, litter sizes, and reproductive success) of salmon consumers (birds, bears, etc.) are directly related to salmon availability. There is evidence that riparian shrubs and trees are positively affected by salmon. Finally, we also can provide an argument for the protection of *wild* salmon as opposed to just *hatchery* salmon. The diversity of salmon stocks also causes greater resilience by "spreading the risk" in the face of environmental fluctuations affecting their life history strategies. Hence, we should preserve not only hatchery salmon, but wild salmon stocks too.

Ecosystems sustain us directly and indirectly. As one example, I considered salmon in the Pacific Northwest. Those in the region love their forests and rivers. Salmon are key to love of place. Before we conclude, I would like to consider two objections raised to the ecosystem services argument I have just provided.

Two objections to the ecosystem services argument

First, consider a worry raised by environmental philosopher Eric Katz (1979) following Krieger (1973). He writes,

> Humanity could enjoy an artificial, plasticized world which produces more social utility than a world filled with natural objects and resources. As our space program has demonstrated, humans can even survive in an artificial environment. The simple fact of the matter is that the interests of humanity are not necessarily connected with the preservation of the natural environment. Any ethical theory which places its emphasis on the satisfaction of human needs can support a policy of preservation only on a contingent basis. Obligations to preserve natural objects and resources are overridden whenever a greater amount of human satisfaction can be attained by non-preservation.
>
> (Katz 1979: 362)

Katz claims the benefits provided by biodiversity justify protection only if artificial systems do not provide the same benefits. However, artificial systems can provide those benefits. Therefore, the benefits provided by biodiversity justify protection of biodiversity.

However, consider one of the most adventurous attempts to replace those services, Biosphere II in Arizona (Cohen and Tilman 1996). It was considered a complete disaster. Oxygen concentrations fell from 21 to 14 percent; nitrous oxide was very high, which can cause brain damage; nineteen of twenty-five species went extinct; all the pollinators went extinct; and in the ocean, there were excessive nutrients creating algal mats. As Joel Cohen and David Tilman write,

> The major retrospective conclusion that can be drawn is simple. At present there is no demonstrated alternative to maintaining the viability of Earth. No one yet knows how to engineer systems that provide humans with the life-supporting services that natural ecosystems produce for free.
>
> (Cohen and Tilman 1996: 1151)

We can cannot replace the ecosystems that sustain us.

A second argument comes from Katie McShane (2007). She argues that anthropocentrism is incompatible with love, respect, and awe regarding nature. Love is an "other-centered" emotion; to love another is to believe their value concerns something more than your own interests. Respect for another requires belief that their interests are of equal importance to your own. Awe for something requires that you believe that their greatness goes "beyond your needs, interests, or attitudes" (p. 176). She writes,

> If to love something is to think of it as having a kind of value that doesn't depend on us and our interests, then according to anthropocentrism, to love the natural world is to make a mistake about its value.
>
> (McShane 2007: 177)

McShane recognizes emotions are important in moral judgment, and is friendly to the sentimentalism discussed above. However, McShane assumes love requires the beloved has a value going beyond us. She raises deep issues and I can only scratch the surface here. Still, this seems incorrect.

I love Wayne Shorter's music. I love *Speak No Evil*, his work with Weather Report, his recent *Without a Net*, and the albums in between. I am partial to jazz and electronica, but people say they love music of all kinds. Nick Hornby eloquently writes about his favorite music,

> And mostly all I have to say about these songs is that I love them, and want to sing along to them, and force other people to listen to them, and get cross when these other people don't like them as much as I do.
>
> (Hornby 2003: 6)

The love of music is hard to make sense of as valuable outside of our experience of it. Music's value seems connected to our needs, interests, and attitudes. With regard to awe, is this not but a certain kind of perceptual-affective experience? In light of this, if one can adopt attitudes of love, respect, and awe to music, then one can do with regard to the natural world.

Conclusion

Ideally in environmental philosophy, we determine what the values of organisms and collectives are, articulate warranted reasons for those claims, and finally provide the reasons that motivate pro-environmental behavior. Environmental philosophers have spent much of their time asking whether organisms, species, and ecosystems have intrinsic value. I have claimed that this is the wrong question to be asking; more cautiously, it is not the *only* question we should be asking. First, the Argument from Teleology fails, and there is no other good argument to that conclusion. Second, data from the social sciences suggest that this claim is not something that most Americans take to be a reason to protect biodiversity. Third, given a sentimentalist moral psychology, if a consideration is not motivating, then a moral judgment regarding it is not in the offing. Fourth, given this psychology and our understanding of cognitive biases, I have provided what I take to be a more powerful argument for protecting biodiversity; namely, the ecosystem services it provides. Thus, we return to a very simple, pragmatic thought articulated by Aldo Leopold: "To keep every cog and wheel is the first precaution of intelligent tinkering" (Leopold 1989: 190).

Acknowledgments

I thank Justin Garson, Avram Hiller, Katie McShane, and Anya Plutynski for feedback on this essay.

Notes

1 I have formulated this argument to avoid Derek Parfit's (1984) "repopulation paradox." We are not asking of some particular group of future people, will *they* be better or worse off; rather, we are asking of *whomever* exists in the future, will those under one policy be better off than those under another policy.

2 This might sound odd. Are not species *just* their organisms? Not quite, since the difference between a species and a set of organisms are the intraspecific relationships between them. Thus, the well-being of species, if it has such a thing, consists in the constituent organisms and their intraspecific relationships. The same point applies to the other types of collectives.

3 Many take mind-independent value to suggest a form of moral realism. Expressivism can make sense of mind-independence too. If I say, "Coastal redwoods are valuable," nothing in what I say is contingent on my or anyone else approving of them. That is, my, or anyone's approving of them is not part of the content of what I have expressed (see Blackburn 1984 and Carter 2004 for a discussion). The question becomes what kind of mind-independence is relevant to issues of value.

4 Having a good is in effect the same thing as having interests.

5 I cannot do justice to his theory here. The attitude of Respect for Nature involves regarding wild living things as having inherent worth. We do this when we accept that they have a good, and whatever is so regarded should be preserved or preserved for its own sake. He argues that this attitude best fits a worldview called the Biocentric Outlook. Accepting the attitude and outlook thus invokes acceptance of duties, priorities, and virtues. As Dale Jamieson has noted, it seems that this attitude of Respect for Nature is an "invitation" rather than a requirement (Taylor 2011: x).

6 I can respect the male's ability without thinking that the world is better for it. This respect is not moral, but a type of fear.

7 We might add the proviso that these interests occur without reference to any other entity. However, as we found above, this restriction is problematic.

8 Interestingly, it turns out the actual clutch size measured is slightly less than the optimal one predicted by theory. The best explanation for this result is that there is a trade-off between maximizing the number of surviving young per brood and maximizing lifetime reproductive success (Visser and Lessells 2001).

9 The Argument from Teleology might very well apply to genes and genetic diversity, however. Genes are selected for and as such we could talk of their interests. But, here again, I think few of us would accept we have obligations to gene replication as such.

10 I do think talk of intrinsic value and our environment can be made intelligible. However, my approach to metaethics and normative ethics would be radically different from the above. First, we humans care about much the same things given what sorts of beings we are. Second, given the importance of coordinating our actions in situations of moderate scarcity and limited benevolence, intrapersonal and interpersonal consistency is something we attempt to achieve. Thus, we try to accommodate those who have biocentric sentiments and can even come to share them (Gibbard 1992, Blackburn 1993, Lenman 2007). Moral psychology and sentimentalism of the next sections combine easily with this approach.

11 I hypothesize that there are a variety of values associated with the natural world, many of which warrant pro-environmental behavior. However, I also hypothesize that not all of them motivate such behavior. Thus, we should focus on them if we hope to achieve progressive change. This is not a plea for dishonest advocacy, or whatever "works."

12 The information discussed here about ecosystem services is taken from Daily *et al.* (1997) and Daily (1997).

13 The information on salmon and their life history, along with the ecosystem services they provide, are taken from Gende *et al.* (2002), Oregon Trout (2001), Quinn (2011), and Woody *et al.* (2003).

References

Agar, N. 2001. *Life's Intrinsic Value: Science, Ethics, and Nature*. New York: Columbia University Press.

Aharoni, E., W. Sinnott-Armstrong, and K. A. Kiehl. 2012. "Can Psychopathic of-23 Fenders Discern Moral Wrongs? A New Look at the Moral/Conventional Distinction." *Journal of Abnormal Psychology* 121(2): 484.

Anderson, E. 1995. *Value in Ethics and Economics*. Cambridge, MA: Harvard University Press.

Andreou, C. 2007. "Environmental Preservation and Second-order Procrastination." *Philosophy & Public Affairs* 35(3): 233–248.

Belden, N., B. Russonello, and K. Stewart. 2002. *Americans and Biodiversity: New Perspectives*. Washington, DC: Biodiversity Project.

Bertram, B. C. 1975. "Social Factors Influencing Reproduction in Wild Lions." *Journal of Zoology* 177(4): 463–482.

Björnsson, G., F. Björklund, C. Strandberg, J. Eriksson, and R. F. Olinder. 2015. *Motivational Internalism*. Oxford: Oxford University Press.

Blackburn, S. 1984. *Spreading the Word*. Oxford: Clarendon Press.

Blackburn, S. 1993. *Essays in Quasi-realism*. Oxford: Oxford University Press.

Blair, R. J. R. 1995. "A Cognitive Developmental Approach to Morality: Investigating the Psychopath." *Cognition* 57(1): 1–29.

Boorse, C. 1976. "Wright on Functions." *The Philosophical Review* 85(1): 70–86.

Callicott, J. B. 1987. "The Conceptual Foundations of the Land Ethic." In *Companion to a Sand County Almanac: Interpretive and Critical Essays*. Madison: University of Wisconsin Press.

Carter, A. 2004. "Projectivism and the Last Person Argument." *American Philosophical Quarterly* 41: 51–62.

Casebeer, W. D. 2003. *Natural Ethical Facts: Evolution, Connectionism, and Moral Cognition*. Cambridge, MA: MIT Press.

Cohen, J. E., and D. Tilman. 1996. "Biosphere 2 and Biodiversity: The Lessons So Far." *Science* 274(5290): 1150–1151.

Cummins, R. C. 1975. "Functional Analysis." *The Journal of Philosophy* 72: 741–765.

Daily, G. 1997. *Nature's Services: Societal Dependence on Natural Ecosystems*. Washington, DC: Island Press.

Daily, G. C., S. Alex, P. R. Ehrlich, L. Goulder, P. A. Matson, H. A. Mooney, R. Postel, H. Schneider, D. Tilman, and G. M. Woodwell. 1997. "Ecosystem Services: Benefits Supplied to Human Societies by Natural Ecosystems." *Issues in Ecology* 1: 1–18.

Daniel, T. C., A. Muhar, A. Arnberger, O. Aznar, J. W. Boyd, K. M. Chan, R. Costanza, T. Elmqvist, C. G. Flint, P. H. Gobster, *et al.* 2012. "Contributions of Cultural Services to the Ecosystem Services Agenda." *Proceedings of the National Academy of Sciences* 109(23): 8812–8819.

Davies, P. S. 2001. *Norms of Nature: Naturalism and the Nature of Functions*. Cambridge, MA: MIT Press.

Elliott, R. 1997. *Faking Nature: The Ethics of Environmental Restoration*. New York: Routledge.

Foot, P. 2001. *Natural Goodness*. Oxford: Oxford University Press.

Gardner, G. T., and P. C. Stern. 2002. "Human Reactions to Environmental Hazards: Perceptual and Cognitive Processes." In *Environmental Problems and Human Behavior*, 205–252. Boston, MA: Allyn & Bacon.

Gende, S. M., R. T. Edwards, M. F. Willson, and M. S. Wipfli. 2002. "Pacific Salmon in Aquatic and Terrestrial Ecosystems Pacific Salmon Subsidize Freshwater and Terrestrial Ecosystems through Several Pathways, which Generates Unique Management and Conservation Issues But Also Provides Valuable Research Opportunities." *BioScience* 52(10): 917–928.

Gibbard, A. 1992. *Wise Choices, Apt Feelings: A Theory of Normative Judgment*. Cambridge, MA: Harvard University Press.

Godfrey-Smith, P. 1994. "A Modern History Theory of Functions." *Nous* 28(3): 344–362.

Goodpaster, K. E. 1978. "On Being Morally Considerable." *The Journal of Philosophy* 75: 308–325.

Greenberg, M. R., D. B. Sachsman, P. M. Sandman, and K. L. Salomone. 1989. "Network Evening News Coverage of Environmental Risk." *Risk Analysis* 9(1): 119–126.

Greene, J., and J. Haidt. 2002. "How (and Where) Does Moral Judgment Work?" *Trends in Cognitive Sciences* 6(12): 517–523.

Greene, J. D., R. B. Sommerville, L. E. Nystrom, J. M. Darley, and J. D. Cohen. 2001. "An FMRI Investigation of Emotional Engagement in Moral Judgment." *Science* 293(5537): 2105–2108.

Haidt, J. 2003. "The Emotional Dog Does Learn New Tricks: A Reply to Pizarro and Bloom (2003)." *Psychological Review* 110(1): 197–198.

Heekeren, H. R., I. Wartenburger, H. Schmidt, H.-P. Schwintowski, and A. Villringer. 2003. "An FMRI Study of Simple Ethical Decision-making." *Neuroreport* 14(9): 1215–1219.

Hendrickx, L., and S. Nicolaij. 2004. "Temporal Discounting and Environmental Risks: The Role of Ethical and Loss-related Concerns." *Journal of Environmental Psychology* 24(4): 409–422.

Hornby, N. 2003. *Songbook*. London: Penguin.

Jamieson, D. 1998. "Animal Liberation is an Environmental Ethic." *Environmental Values* 7(1): 41–57.

Johnson, L. E. 1993. *A Morally Deep World: An Essay on Moral Significance and Environmental Ethics.* Cambridge: Cambridge University Press.

Kahneman, D., and A. Tversky. 1979. "Prospect Theory: An Analysis of Decision under Risk." *Econometrica: Journal of the Econometric Society* 47: 263–291.

Kaltenborn, B. P., and T. Bjerke. 2002. "Associations between Environmental Value Orientations and Landscape Preferences." *Landscape and Urban Planning* 59(1): 1–11.

Katz, E. 1979. "Utilitarianism and Preservation." *Environmental Ethics* 1(4): 357–364.

Kellert, S. R. 1997. *The Value of Life: Biological Diversity and Human Society.* Washington, DC: Island Press.

Kennett, J. 2006. "Do Psychopaths Really Threaten Moral Rationalism?" *Philosophical Explorations* 9(1): 69–82.

Krieger, M. H. 1973. "What's Wrong with Plastic Trees? Rationales for Preserving Rare Natural Environments Involve Economic, Societal, and Political Factors." *Science* 179(4072): 446–455.

Lack, D., *et al.* 1954. *The Natural Regulation of Animal Numbers.* Oxford: Oxford University Press.

Leiserowitz, A. A., R. W. Kates, and T. M. Parris. 2005. "Do Global Attitudes and Behaviors Support Sustainable Development?" *Environment: Science and Policy for Sustainable Development* 47(9): 22–38.

Lenman, J. 2007. "What is Moral Inquiry?" *Proceedings of the Aristotelian Society, Supplementary Volumes*, 63–81.

Leopold, A. 1989. *A Sand County Almanac, and Sketches Here and There.* Oxford: Oxford University Press.

McShane, K. 2007. "Anthropocentrism vs. Nonanthropocentrism: Why Should We Care?" *Environmental Values* 16(2): 169–186.

Millikan, R. G. 1984. *Language, Thought, and Other Biological Categories: New Foundation for Realism.* Cambridge, MA: MIT Press.

Moll, J., R. de Oliveira-Souza, and P. J. Eslinger. 2003. "Morals and the Human Brain: A Working Model." *Neuroreport* 14(3): 299–305.

Neander, K. 1991. "Functions as Selected Effects: The Conceptual Analyst's Defense." *Philosophy of Science* 58: 168–184.

Nisbet, M. C., and C. Mooney. 2009. "Framing Science." In *Understanding and Communicating Science: New Agendas in Communication*, ed. L.A. Kahlor and P. Stout, 40. New York: Routledge.

Norton, B. G. 1987. *Why Preserve Natural Variety?* Princeton, NJ: Princeton University Press.

Odenbaugh, J. 2015. "Nothing in Ethics Makes Sense Except in the Light of Evolution? Natural Goodness, Normativity, and Naturalism." *Synthese*, 1–25.

O'Neill, J. 1992. "The Varieties of Intrinsic Value." *The Monist* 75: 119–137.

Oregon Trout. 2001. *Oregon Salmon: Essays on the State of the Fish at the Turn of the Millennium.* Oregon Trout.

Parfit, D. 1984. *Reasons and Persons.* Oxford: Oxford University Press.

Post, J. F. 2006. "Naturalism, Reduction and Normativity: Pressing from Below." *Philosophy and Phenomenological Research* 73(1): 1–27.

Quinn, T. P. 2011. *The Behavior and Ecology of Pacific Salmon and Trout.* Vancouver: UBC Press.

Regan, T. 1985. *The Case for Animal Rights.* Berkeley: University of California Press.

Rolston, H. 2012. *Environmental Ethics.* Philadelphia, PA: Temple University Press.

Routley, R. 1973. "Is there a Need for a New, an Environmental Ethic? In *Proceedings of the 15th World Congress of Philosophy*, 205–210. Varna: Sophia Press.

Sagoff, M. 1974. "On Preserving the Natural Environment." *Yale Law Journal* 84: 205–267.

Sanfey, A. G., J. K. Rilling, J. A. Aronson, L. E. Nystrom, and J. D. Cohen. 2003. "The Neural Basis of Economic Decision-making in the Ultimatum Game." *Science* 300(5626): 1755–1758.

Sarkar, S. 2005. *Biodiversity and Environmental Philosophy: An Introduction.* Cambridge: Cambridge University Press.

Schnall, S., J. Haidt, G. L. Clore, and A. H. Jordan. 2008. "Disgust as Embodied Moral Judgment." *Personality and Social Psychology Bulletin* 34: 1096–1109.

Singer, P. 1995. *Animal Liberation.* New York: Random House.

Singer, T., B. Seymour, J. P. O'Doherty, K. E. Stephan, R. J. Dolan, and C. D. Frith. 2006. "Empathic Neural Responses are Modulated by the Perceived Fairness of Others." *Nature* 439(7075): 466–469.

Smith, M. 1994. *The Moral Problem.* Oxford: Blackwell.

Sober, E., and D. S. Wilson. 1999. *Unto Others: The Evolution and Psychology of Unselfish Behavior.* Cambridge, MA: Harvard University Press.

Steel, B. S., P. List, and B. Shindler. 1994. "Conflicting Values about Federal Forests: A Comparison of National and Oregon Publics." *Society & Natural Resources* 7(2): 137–153.

Swenson, W., D. S. Wilson, and R. Elias. 2000. "Artificial Ecosystem Selection." *Proceedings of the National Academy of Sciences* 97(16): 9110–9114.

Taylor, A. F., F. E. Kuo, and W. C. Sullivan. 2001. "Coping with ADD: The Surprising Connection to Green Play Settings." *Environment and Behavior* 33(1): 54–77.

Taylor, P. W. 2011. *Respect for Nature: A Theory of Environmental Ethics.* Princeton, NJ: Princeton University Press.

Turiel, E. 1983. *The Development of Social Knowledge: Morality and Convention.* Cambridge: Cambridge University Press.

Tversky, A., and D. Kahneman. 1974. "Judgment under Uncertainty: Heuristics and Biases." *Science* 185(4157): 1124–1131.

Tversky, A., and D. Kahneman. 1981. "The Framing of Decisions and the Psychology of Choice." *Science* 211(4481): 453–458.

Varner, G. E. 1998. *In Nature's Interests?* Oxford: Oxford University Press.

Vaske, J. J., and M. P. Donnelly. 1999. "A Value–attitude–behavior Model Predicting Wildland Preservation Voting Intentions." *Society & Natural Resources* 12(6): 523–537.

Visser, M., and C. Lessells. 2001. "The Costs of Egg Production and Incubation in Great Tits (*Parus major*)." *Proceedings of the Royal Society of London. Series B: Biological Sciences* 268(1473): 1271–1277.

Vucetich, J. A., J. T. Bruskotter, and M. P. Nelson. 2015. "Evaluating Whether Nature's Intrinsic Value is an Axiom of or Anathema to Conservation." *Conservation Biology* 29(2): 321–332.

Wheatley, T., and J. Haidt. 2005. "Hypnotic Disgust Makes Moral Judgments More Severe." *Psychological Science* 16(10): 780–784.

Williams, G. C. 2008. *Adaptation and Natural Selection: A Critique of Some Current Evolutionary Thought.* Princeton, NJ: Princeton University Press.

Woody, E., E. C. Wolf, and S. Zuckerman. 2003. *Salmon Nation: People, Fish, and Our Common Home.* Corvallis: Oregon State University Press.

Wright, L. 1973. "Functions." *The Philosophical Review* 82(2): 139–168.

Wright, L. 1976. *Teleological Explanations: An Etiological Analysis of Goals and Functions.* Berkeley: University of California Press.

Wynne-Edwards, V. C. 1962. *Animal Dispersion in Relation to Social Behaviour.* Edinburgh: Oliver & Boyd.

15

WHAT WOULD LEOPOLD DO?

Considering assisted colonization as a conservation strategy

Ben A. Minteer

Introduction: the dilemma of conservation interventions

The reports are gloomy, and relentless. A recent analysis performed by the journal *Nature* found that, globally, 41 percent of all amphibians and 26 percent of mammal species are thought to currently face extinction, with considerably higher percentages at risk in the future if current threats, from climate change and habitat loss to species exploitation, continue unabated (Monastersky 2014). Similarly, the Audubon Society, in its widely cited 2014 *Birds and Climate Report*, projected that climate change would cause 314 of the 588 North American bird species studied to lose more than half their current climatic range by 2080, with extinction looming for those species unable to adapt to smaller or new spaces (http://climate.audubon.org/). Some ecologists have recently also suggested that we may be on the verge of a major extinction event in the planet's ocean species, underscoring the biotic scope of the crisis across the seas as well as the land (McCauley *et al.* 2015).

These and dozens of similar assessments have led many biodiversity scientists to proclaim that we are witnessing a sixth mass extinction episode on the planet, a period of destruction rivaling the disappearance of the dinosaurs (see e.g. Kolbert 2014, Ceballos *et al.* 2015). But it isn't only accelerating extinction trends that are troubling. The wider pattern of population declines – and decreases in abundance of individuals within them – is also cause for great concern, prompting some biologists to refer to the current situation as also a mass "defaunation," a term that captures the full sweep of human-driven decline in animal populations – as well as the impacts of these losses on the healthy functioning of ecosystems (Dirzo *et al.* 2014). Again, the numbers here are unsettling. The 2014 Living Planet Index (a collaboration between the World Wildlife Fund and the London Zoological Society) estimated that globally, on average, vertebrate species populations have declined 52 percent since 1970 (http://wwf.panda.org/about_our_earth/all_publications/).

Over the years, such scientific reports and predictions have profoundly shaped our understanding of the scope of our moral obligation to recover and conserve biodiversity from evolving anthropogenic threats. It's an understanding, I'd argue, that's increasingly defined by the tension between two competing impulses when assessing environmental responsibility in a time of rapid and likely unprecedented global change. On the one hand, there's a strong desire to "do whatever it takes" to recover and protect vulnerable species from any further losses. On the other, there's the sentiment that we should avoid taking risky and, some would say, reckless actions that may make things worse, including undermining other important environmental values (e.g. nature's wildness and autonomy) as we consider bolder efforts – from active relocation to genetic manipulation – to save species from plunging into the void (Minteer 2015a).

Biodiversity scientists, wildlife managers, naturalists, advocates, and philosophers have all been lured by the binding force of this loss-and-recovery narrative in conservation, a storyline that has driven efforts to curb species declines and extinctions via a clutch of scientific and policy strategies, from the traditional (e.g. regulations, referenda, and refugia) to the radical (e.g. the fast-evolving wizardry of genetic engineering and synthetic biology). Promising achievable solutions to the global extinction crisis, the more ambitious efforts raise thorny questions about the value of wild species and wildness – and our weighing of the moral wages of an interventionist conservation in a time of increasing human influence on the landscape.

In this chapter, I'll discuss a highly contested strategy to address certain challenges to species conservation under climate change, a proposal that seems to be caught between these opposing conservation impulses. I'll outline a broad philosophical framework for approaching this debate over how to save species from evolving anthropogenic threats when conventional conservation approaches seem to fail us – and when more radical interventions might need to be considered. In doing so, I'll appeal directly to the insights of Aldo Leopold, one of the pillars of American conservation thought and practice (not to mention environmental philosophy), though my reading will be somewhat different than many of the more familiar interpretations of his conservation project. I'll close the chapter with a few brief conclusions for thinking about the risks and rewards of aggressive environmental interventions in what some are calling the "age of humans."

Debating "assisted colonization"

The threat that global climate change (GCC) poses to biodiversity in this century has forced conservationists (including scientists, managers, activists, and philosophers) to wrestle with a set of novel and difficult scientific, strategic, and ethical questions concerning the management and value of species – as well as the meaning and significance of ecological integrity under rapidly changing and possibly unprecedented climatic and ecological conditions (Camacho *et al.* 2010, Sandler 2013). It's no small challenge: GCC has been linked to a range of species-level impacts, including physiological, phenological, and distributional changes that have significant implications for the viability of species over the long run (Root and Hughes 2005, Parmesan 2006, Bellard *et al.* 2012).

Although estimates of increased extinction risk due primarily to GCC are the most disconcerting to conservationists, they're often difficult to make given the tangle of environmental and societal variables at play. Still, one influential assessment placed up to a third of the world's species on a path to climate-driven extinction as a result of GCC (Thomas *et al.* 2004; see also Hannah 2012). By all accounts, global climate change is thus emerging as a very real and significant threat to biodiversity in this century (Maclean and Wilson 2011, Urban 2015).

But it's also a uniquely complex one. GCC, for example, can combine with and magnify other drivers of biodiversity loss, including land use change and the spread of invasive species and emerging infectious diseases (see e.g. Root *et al.* 2003, Barnosky 2009, Krosby *et al.* 2010, Altizer *et al.* 2013). The interaction of rapid climate change and habitat fragmentation has been of particular concern to conservationists, especially scenarios in which plant or animal populations vulnerable to climate shifts cannot adapt fast enough to a changing environment – and can't disperse naturally to more suitable habitats due to the growing presence of human barriers (e.g. subdivisions, office parks, and highways). Unable to move to higher ground (for example) on their own, the conservation of these plants and animals pose a strategic, policy, and philosophical challenge for conservationists used to responding to more direct and proximate threats (e.g. pavers, poachers, and pollutants, etc.) by keeping human activities at bay, that is, setting up protected areas to save habitat, enforcing strict no-take policies, and so on.

These scenarios have forced some biodiversity scientists and advocates to explore a range of anticipatory and managerially bold interventions into ecological systems before plant and wildlife populations thought to be at risk due to GCC start to enter the extinction vortex. One of the more controversial interventionist strategies is "assisted colonization" (or "managed relocation"), that is, the translocation (moving) of biological populations threatened by current or future climate change to locations outside their indigenous/historical range. The technique has already been performed for a number of species, including the *Torreya taxifolia*, a conifer with a small and diminishing range in Florida's panhandle, and butterfly species in the UK (Minteer and Collins 2010, Marris 2011). Not surprisingly, the assisted colonization (AC) idea has divided members of the conservation community, at times quite sharply (see e.g. McLachlan *et al.* 2007, Ricciardi and Simberloff 2009a, 2009b, Stone 2010, Webber *et al.* 2011, Neff and Larson 2014).

The primary objection appears to be AC's potential to disrupt populations and disturb the ecological integrity of the "receiving" systems (i.e. the new habitats for the translocated populations). It's a critique amplified by our own epistemic limitations: we simply lack the ability, critics claim, to accurately predict how a relocated species will act when transplanted in a new ecosystem outside its native range (see e.g. Davidson and Simkanin 2008, Ricciardi and Simberloff 2009a, 2009b, Webber *et al.* 2011). But there are other worries. These include animal welfare concerns in cases where further stress is placed on sentient animals during movement, and AC's potential to increase the risk of disease transmission. Some critics have also predicted that AC will likely fail to save relocated species given the spotty success record of past translocations – and the fact that relocated populations may be especially vulnerable to additional threats in their new habitat, especially if the introduced population size is, as expected, small (e.g. Huang 2008, Minteer and Collins 2012).

For their part, supporters of considering AC as a conservation strategy have countered that many of these risks are not deal-breakers because they're ultimately manageable. We can, they argue, develop the required knowledge via experimentation and adaptive conservation management – and develop the analytical techniques and risk assessment protocols to help shore up the evaluation of appropriate translocation candidates (and recipient ecosystems) (Hoegh-Guldberg *et al.* 2008, Richardson *et al.* 2009, Chauvenet *et al.* 2012, Schwartz *et al.* 2012, IUCN/SSC 2013). Some cautious supporters of AC also make the overtly moral argument that, even if ecological risks cannot always be minimized, our moral obligation to save species from anthropogenic threats like climate change requires that we consider radical conservation strategies such as AC even if doing so may be difficult, costly, or plagued by unpredictability and the potential for unwanted ecological consequences (Minteer and Collins 2010).

One of the aspects of AC that makes it so controversial as a conservation strategy is its apparent break with the philosophy of nature preservationism that has traditionally underpinned conservation efforts, especially in the United States. A core tenet of the preservationist tradition is that species should be protected within their historical habitats, that is, the indigenous geographical ranges and the evolutionary contexts in which they evolved. Whether we're talking about establishing parks and protected areas, habitat protection provisions in the US Endangered Species Act, or similar other conservation policy tools, the traditional preservationist model sets a species' historic range as the appropriate context for recovery, restoration, and conservation.

The proposed movement of populations to locations well outside their native habitat as part of a significantly altered understanding of "in-situ" field conservation effort thus seems to fall well outside the conventional conservation norm of saving-species-in-their-place.

AC clearly also upends longstanding preservationist values surrounding human intervention in and manipulation of ecological systems, particularly moral aversion to human meddling and control of nature that characterizes several strains of the contemporary ecocentric position in environmental philosophy (e.g. Elliot 1997, Katz 1997). Conservation scientists and wildlife managers, though, have long engaged in practices that could be characterized as "interventionist," such as captive breeding and the control and management of experimental populations. Even those areas administered under the stringent preservationist directives of the 1964 Wilderness Act, for example, are at times managed in a more active manner, including prescribed burning, species reintroductions (e.g. wolves), the stocking of non-native fish, pesticide spraying, the use of helicopters, and other activities that depart from a pure "hands off" philosophy (Long and Biber 2014). Nevertheless, for many critics, the degree of intervention suggested by assisted colonization falls well outside the lines of acceptable conservation practice.

Still others have questioned whether AC actually preserves the very values it claims to be protecting. Philosopher Ronald Sandler (2010, 2012), for example, has argued that the threat of ecological harm following AC outweighs any expected benefits to the translocated species because moving these populations outside their historical habitats reduces their conservation importance given the place-based nature of species value. Specifically, Sandler believes that such interventions disrupt the unique ecological and evolutionary relationship a species has with its native system, properties ultimately dependent on traditionally defined in-situ conditions.

This argument, though, rests on a fairly narrow understanding of the source of species value and the motives behind biological conservation. For example, it underplays the many aesthetic and cultural reasons for saving individual species (past and present), values that are not entirely dependent on a species' mooring in its native range but that are often significant drivers of conservation concern among the wider public. It also ignores the extensive empirical record, in the US and elsewhere, of successful conservation-driven translocations, including the introduction of species to novel environments to enhance their chance for survival. Australian and New Zealand wildlife managers, for instance, have long engaged in the translocation of populations for conservation purposes, even if it wasn't called "assisted colonization" – and even if global climate change wasn't the primary impetus (Seddon *et al.* 2015). And in the US, wildlife biologists have regularly discussed (and occasionally engaged in) conservation-driven translocations, including considering new habitats not part of a given species' historical range, since at least the early 1960s (Winston *et al.* 2014).

Skeptical arguments like Sandler's are, however, important because they remind us that normative considerations of place, evolutionary history, ecological integrity, and other traditional values in biodiversity conservation are not to be cavalierly brushed aside when making decisions about the movement of species for conservation purposes. These considerations may not always have "trumping power" in conservation decision-making, especially as managers cope with the novel conditions presented by rapid environmental change and attendant accelerating extinction risk, but they should not be taken lightly in deliberations over the merits of assisted colonization, especially given the high stakes involved (i.e. ecological disruption, extinction).

But there is another potential ethical concern here, a bigger worry that hangs not only over discussions about the risks and rewards of AC but also over other interventionist proposals that open the door to increased human manipulation and control of species and ecosystems. It's the objection that, although well intended, such efforts do not in the end address the deeper moral problem: the need to restrain ourselves on the landscape and especially to rein in our ecologically destructive activities on the planet. By putting us in a more commanding position in the natural world – which at the extreme end of the continuum promotes us to the role of "planetary manger" – strategies such as AC can appear to evade this deeper moral challenge of environmental forbearance, and possibly even exacerbate (albeit inadvertently) an already dysfunctional human–nature relationship (Minteer 2015b).

Searching for solid ground: Aldo Leopold's pragmatic preservationism

The controversy over assisted colonization, then, presents something of a moral conundrum for conservationists: to save species from emerging anthropogenic threats we may have to consider actions that entail more, rather than less, control of the natural environment. But in doing so, we run the risk of increasing other ecological vulnerabilities as a byproduct of the more aggressive conservation interventions – and undermining other important environmental values and goals that we can and should care about (e.g. preserving the integrity of ecosystems; maintaining a meaningful sense of the wild, etc.). It isn't clear how to weigh these concerns, how to navigate the apparently warring responsibilities of biodiversity conservation in the age of intervention. What, we might therefore ask, would Aldo Leopold do?

Although Leopold remains the most important figure in the history of environmental philosophy (and in the history of conservation biology), it seems an ill-advised question. For one thing, it's being asked more than sixty-five years after his death, and in a time characterized by a markedly different set of conservation challenges than those Leopold grappled with in the first half of the twentieth century. Moreover, Leopold's widely admired ethical project – his "land ethic" – might appear poorly suited to the "new normal" of conservation under climate change, an environment in which historical baselines and older notions of wilderness, "integrity," and other preservationist ideals seem to be losing much of their scientific and managerial relevance (Marris 2011). Indeed, the oft-cited "summary maxim" of the land ethic, "A thing is right when it tends to preserve the integrity, stability, and beauty of the biotic community. It is wrong when it tends otherwise," certainly can reinforce this reading (Leopold 1989: 224–225; see Norton 2005 for an alternative view). It's increasingly clear, however, that we need a philosophy of conservation today able to anchor a more experimental and activist approach to environmental management, one in which the goal is not to arrest ecological change or preserve some pre-disturbance ideal of ecological integrity, but rather to determine and guide rates of acceptable change in rapidly transforming socio-ecological systems (Minteer 2012).

So the preservationist Leopold might seem a poor choice to guide us in the current age of "intervention ecology" (Hobbs *et al.* 2011). I want to argue, however, that a wider analysis of Leopold's writing reveals a more versatile and nuanced picture of his conservation philosophy than the simple preservationist reading suggests, especially his views concerning conservation interventions. The land ethic is rightly elevated in discussions of Leopold's conservation thought, but placed within the broader context of his other writing (in *Sand County* and elsewhere) we can see that it's part of a wider vision of environmental management that doesn't preclude significant conservation interventions in plant and wildlife populations and in ecological systems (provided certain scientific and normative standards are met). Leopold, as we'll see, held an eminently practical and flexible view of the activity of nature preservation, one that allowed for significant ecological manipulation and experimentation – including pre-emptive actions – to save species vulnerable to extinction. But as we'll also see, his support for such activities and interventions was also leavened with a heady dose of humility and restraint, laced with key caveats and qualifications that make his prescriptions especially useful and timely for current conservation debates such as the one over assisted colonization.

One of the first objections to claiming that Leopold would have supported AC, at least in certain well-defined contexts, hinges on its rather obvious single-species emphasis. It's true that Leopold was a well-known (perhaps the most well-known) ecological holist, privileging the health and integrity of the biotic community over more individualistic – including single-species conservation – goals. But it's also clear that his moral, aesthetic, and scientific regard for the worth of individual wild species, especially (but not only) predators, was quite powerful and that it endured even as his thought and writing took on a pronounced ecological character in the 1930s and 1940s.

"It hardly seems necessary to say," he wrote in 1920, "that the wiping out of a species is wanton barbarism, especially species of high value, from both the sporting and esthetic points of view" (Meine and Knight 1999: 131). In one of his more lyrical essays in *A Sand County Almanac*, Leopold laments the loss of the passenger pigeon, an iconic extinction event in American conservation history: "We have erected a monument to commemorate the funeral of a species. It symbolizes our sorrow … There will always be pigeons in books and in museums, but these are effigies and images, dead to all hardships and to all delights" (Leopold 1989: 109). Elsewhere in *Sand County* he writes about the shooting of one of the last grizzly bears in Arizona on Escudilla Mountain during his early forestry career on the Apache National Forest. "The government trapper who took the grizzly knew he had made Escudilla safe for cows," Leopold wrote. But "he did not know he had toppled the spire off an edifice a-building since the morning stars sang together … Escudilla still hangs on the horizon, but when you see it you no longer think of bear. It's only a mountain now" (Leopold 1989: 137).

What's interesting in this passage is that it shows how Leopold, despite his well-known emphasis on the bigger (i.e. landscape-level) picture, never lost sight of the unique value of individual species – particularly those, such as the passenger pigeon, the wolf, and the grizzly, that had been victims of human avarice, short-sightedness, and/or fear. Top predators like the grizzly and the wolf (see his moving and influential essay, "Thinking Like a Mountain" in *Sand County*) take on great aesthetic and moral significance in Leopold's philosophical system, so much so that without its bear, Escudilla can "only" be a mountain. Leopold's regard for individual species and their conservation, of course, was not walled off from his broader ecological vision: just as the grizzly lent a special dimension to the high country of eastern Arizona, so, too, did the ecosystem reciprocate (after a fashion), conferring, he suggested, *de facto* value on its constituent elements. As he remarked in his 1938 essay, "Conservation," if the ecological system considered as a whole is good, "then every part is good, whether we understand it or not. If the biota, in the course of aeons, has built something we like but do not understand, then who but a fool would discard seemingly useless parts?" This led Leopold to make one of his more famous observations about the imperative of species preservation, an early statement of the presumptive value of species as part of a general principle of biotic insurance: "To keep every cog and wheel is the first precaution of intelligent tinkering" (Meine and Knight 1999: 141–142). Each species had a known or potential contributory value for the healthy functioning of the whole, especially given the limits of our ecological understanding.

We can say, therefore, that Leopold did not downplay the value of and efforts to care for single species within his conservation philosophy; rather, a complex mix of aesthetic, moral, and ecological values motivated his commitment to species protection. Leopold's vaunted ecological holism was sufficiently differentiated so that a full appreciation of both the biotic components and the wider ecological wholes and processes was possible in his complete axiological system. Among other things, this conclusion has implications for how Leopold's work might contribute to current debates over assisted colonization, which often seem to pit single species-centered considerations (typically in the "pro-AC" camp) against the more holistic concerns about ecological integrity (a marker of the "anti-AC" position).

Several additional features of Leopold's conservation philosophy are directly relevant to navigating the current debates over AC and species preservation under rapid environmental change more generally. One is his explicit support for experimental and manipulative approaches to environmental management, including for conservation purposes. Not surprisingly, these themes are quite strong in his earlier writing in wildlife (game) management, a more utilitarian context that found Leopold striking a strongly activist tone. But I also believe he never fully jettisoned his support for what we might call interventionist environmental management, provided it was done carefully and didn't violate what Leopold took to be key ecological and normative provisions (as we will see below).

In his 1930 essay, "The American Game Policy in a Nutshell," Leopold observed that conservationists always seem to have ideas about which course of action is the most desirable and lamented the fact that they often conflict. It was a recipe for a stalemate, Leopold believed, writing that, "We are in danger of pounding the table about them, instead of going out on the land and giving them a trial." He urged a more experimental approach: "[We should] quit arguing over abstract ideas, and instead go out and try them" (Meine 2013: 289). It's a pragmatic suggestion that Leopold would describe in more detail in his seminal 1933 textbook, *Game Management*. For example, discussing the restoration of bird populations, Leopold commends the "judicious use of those tools employed in gardening or landscaping or farming" to build environments able to attract desired species. It was, as he wrote, simply a process of "deliberately and intelligently reversing the processes which are destroying bird environments" (Leopold 1933: 405).

Leopold would push this activist line of thinking even further. That same year, in his landmark essay, "The Conservation Ethic," he explicitly linked this experimental and manipulative approach to wildlife management to the challenge presented by species extinction. In doing so, he unmistakably promoted a hands-on approach to ecological recovery and conservation:

> Why do species become extinct? Because they first become rare. Why do they become rare? Because of the shrinkage in the particular environments in which their particular adaptations enable them to inhabit. Can such shrinkage be controlled? Yes, once the specifications are known. How known? Through ecological research. How controlled? By *modifying the environment* with those same tools and skills already used in agriculture and forestry.
>
> (Flader and Callicott 1991: 190; emphasis added)

In step with Leopold's endorsement of a robustly experimental approach to wildlife conservation (including interventions that entailed a significant degree of habitat modification and manipulation) was his support for more preemptive forms of conservation practice. In fact, he regularly expressed frustration at the generally reactive and elegiac mode of conservation, as in the essay, "Post-War Prospects," which Leopold penned in 1944. "One defect in conservation," he noted, "is that it is so far an ex post facto effort. When we have nearly finished disrupting a fauna and flora, we develop a nostalgic regret about it, and a wish to save the remnants." Instead, Leopold wondered, "Why not do the regretting and saving in advance?" (Meine 2013: 492). It was a vision of anticipatory conservation that was itself ahead of its time.

But Leopold's conservation philosophy was progressive in a number of other ways, too, including, I'd argue, its attitude toward non-native species. As we've seen, one of the main objections to proposals such as AC is that it will amount to nothing more than "planned invasions."

That is, many critics describe AC as a recipe for ecological disaster given the translocated species' potential to disrupt native species and habitats when released into new environments. Invasion biologist Dan Simberloff has been one of the strongest critics of AC in scientific circles, calling it (in a widely cited paper co-authored with A. Ricciardi), "ecological roulette" (Ricciardi and Simberloff 2009a). Interestingly, Simberloff has elsewhere argued that Leopold similarly held an increasingly negative view toward non-native species over the years, an attitude that rested on the great conversationist's aesthetic and ecological aversion toward exotics as well as his steadfast commitment to the goals of ecological integrity and the protection of native flora and fauna (Simberloff 2012).

Simberloff is generally correct in noting Leopold's concern about the conservation implications of the spread of invasive species (or "pests" as he often called them) – as well as Leopold's prizing of both nativeness and a particular notion of ecological integrity (including in "The Land Ethic"). But I think he also overstates his case, on two counts. For one thing, while Leopold may have had a generally negative view of non-natives in conservation contexts, he wasn't nearly as categorical a critic of exotic species as Simberloff suggests. Indeed, Leopold seems to have subscribed to what could be described as a consequentialist view toward non-native plants and animals. "[N]o species is inherently a pest, and any species may become one," he remarked in his 1943 essay, "What is a Weed?" (Flader and Callicott 1991: 309). Although it's true that Leopold embraced a rosier view of exotic species in his earlier writing on game management in the 1930s than he did in his later years, he was far from an ideologue about the necessity of maintaining conditions of strict nativeness in conservation.

We can see this more nuanced view of non-native species and ecological integrity emerge in Leopold's mature thinking about ecological restoration. In the 1930s, Leopold was appointed research director of the University of Wisconsin's new arboretum (at the time he was also serving at the university as Professor of Game Management). His original plan for the arboretum was for it to serve as a research site for university students, but it was also envisioned as a restoration effort, emphasizing the region's native plant and animal species (Newton 2006: 267). Yet, as Bill Jordan and George Lubick (2011: 91) note in their insightful study of the history of the science and practice of ecological restoration in America, Leopold would soon temper his restoration goals as he realized that the re-creation of strict historical assemblages and communities was not always possible. This found Leopold slowly warming to the view that, at least in some cases, novel associations of plants and animals that could be maintained effectively within a mixed environmental and human matrix formed a realistic and valid conservation target.

The ultimate ecological objective for Leopold, in other words, was not a rigid, all-or-nothing recreation of past ecosystems and species compositions, but rather the maintenance of what he called "land health," defined as the persistence of the self-renewing capacities of the ecosystem. As Leopold put it in his 1944 essay, "Conservation: In Whole or in Part?":

> Conservation is a state of health in the land … It is a state of vigorous self-renewal … Such collective functioning of interdependent parts for the maintenance of the whole is characteristic of an organism. In this sense land is an organism, and conservation deals with its *functional integrity*, or health.
>
> (Flader and Callicott 1991: 310; emphasis added)

So, despite Simberloff's (2012: 504) assertion that Leopold's fidelity to "integrity" (especially in an aesthetic sense) was so powerful that it "could no more be maintained by adding a nonnative species than could the integrity of the Mona Lisa be maintained by adding a moustache or a necklace, even a pretty necklace," Leopold in fact held a more plastic and pragmatic understanding of ecological integrity. It was *functional* integrity, not historical integrity that Leopold was after, and although his working hypothesis was that native species were key to maintaining land health, he was no purist on the issue. If, for example, non-native species introduced for conservation purposes (i.e. to reduce the threat of extinction) could be accommodated by ecological systems such that they didn't reduce the diversity and fertility of the system – the "yardsticks" for land health Leopold advanced in his important essay, "Biotic Land Use," written in the early 1940s (Leopold 1999) – then their presence wasn't objectionable *a priori*. It's a view, I believe, that tracks nicely with some of the current revisionist thinking about exotic species in ecology and biodiversity science, where the emphasis on a species' origin (i.e. whether it was introduced by humans) is less important than whether an introduced species is producing benefits or harm to native biodiversity, ecological services, and/or other goods (e.g. Davis *et al.* 2011).

Two key provisos

In sum, then, we can conclude the following: (1) Leopold's conservation philosophy, while anchored in ecological holism, was nuanced and sensitive enough to allow for a strong attachment to the value and preservation of *individual* species; (2) Leopold clearly supported (early and late in his career) *experimental and preemptive conservation efforts*, including those requiring a significant degree of manipulation of wildlife populations and their habitats; and (3) Leopold held a *consequentialist and broadly pragmatic attitude toward non-native species*, focusing on the objective of *conserving functional (rather than purely historical) ecological integrity* and the wider goal of promoting land health.

Therefore, with respect to current discussions of assisted colonization, I'd argue that a more discerning reading of Leopold's conservation philosophy suggests a cautious and provisional acceptance of the practice, under certain conditions. *If* conservation-driven translocation of species (including to systems outside their indigenous range) is determined to be the only way to save a species at risk of extinction, and *if* doing so does not undermine ecological "integrity" understood functionally (rather than historically) as the maintenance of land health (i.e. the self-renewal capacity of the landscape), then assisted colonization could be deemed acceptable within Leopold's conservation philosophy.

At the same time, however, I believe there are two major caveats to this conclusion that emerge from Leopold's writing. They're significant stipulations, in no small part because they introduce critical ecological and moral constraints on human interventions in populations and ecosystems, even in situations where such efforts are potentially supportable according to many of the other managerial norms contained within Leopold's conservation philosophy.

The first is a clear preference in Leopold's writing for stretching traditional species conservation strategies to their breaking point before adopting more radical techniques. This view can be summarized in the context of current debates over AC with a simple directive: *try native habitat expansion first.* "The combined evidence of history and ecology," Leopold wrote in 1939, "seems to support one general deduction: the less violent the man-made changes [in the land], the greater the probability of successful readjustment in the [biotic] pyramid ..." (Flader and Callicott 1991: 270). In *Sand County*, Leopold's understanding of the structure and function of the biotic community translated into a specified strategy for managing wild species in situations where native habitat proved insufficient. "The most feasible way to enlarge the area available for wilderness fauna is for the wilder parts of the National Forests, which usually surround the [National] Parks, to function as parks in respect of threatened species," he wrote in his essay "Wilderness" (Leopold 1989: 198).

In other words, I think Leopold would have encouraged the exhaustion of traditional, in-situ approaches to species conservation (including the expansion of protected areas, the construction of wildlife corridors, and native habitat modification to increase resilience) before making the decision to preemptively move species into new environments, especially those habitats well outside their indigenous ranges. Such a view fits nicely within his broader normative commitment to the standard of land health and also provides a bridge to the preservationist reading of Simberloff and others who stress Leopold's abiding concern with ecological integrity and nativeness in his conservation philosophy.

The second proviso is explicitly moral in character: *don't let our tools and techniques run the show.* It was a powerful and enduring preoccupation of Leopold's, namely, the concern that our technological prowess would outstrip our ecological humility, caution, and self-possession. "Our tools are better than we are, and grow faster than we do," he wrote in 1938. "They suffice to crack the atom, to command the tides. But they do not suffice for the oldest task in human history: to live on a piece of land without spoiling it" (Flader and Callicott 1991: 254). It's a reminder that efforts such as AC will prove meaningless – and maybe perversely counterproductive – if they convince us that we've solved the underlying conservation problem "simply" by moving populations to more hospitable environments. That is, if proposed strategies such as AC end up taking the place of more serious attempts to control and mitigate our environmental destructiveness, then we will have failed to meet Leopold's "oldest task," even if we believe otherwise. Without addressing the deeper moral and cultural drivers of our environmental pathologies, the translocation of species to help them adapt to climate change will be tantamount to moving deck chairs on the *Titanic*.

It's an especially critical proviso in Leopold's conservation philosophy because it serves as a kind of moral safeguard, which keeps the machinery of conservation from spinning so far out of control that it runs roughshod over values other than keeping species alive in a changing world (such as, for example, wildness and ecological restraint in a fast humanizing landscape). And it's why I'd argue that, even though Leopold's conservation system would permit the cautious consideration of AC when deemed necessary to save species from newer anthropogenic threats (and if doing so was not reasonably expected to disrupt land health), it doesn't justify some of the more radical technocentric ideas traveling under the banner of "conservation" that have emerged in recent years.

I don't believe the Leopoldian position, for example, would support aggressively manipulative proposals to employ the latest tools of genetic engineering and synthetic biology to try to create facsimiles of extinct species such as the passenger pigeon and the thylacine (aka Tasmanian tiger) with the goal of returning these lost species to the landscape. Such "de-extinction" schemes, which have garnered considerable media attention (e.g. Zimmer 2013, Rich 2014), have at times been lumped with AC and other novel conservation approaches by conservation scientists (e.g. Seddon *et al.* 2014). I would argue, however, that de-extinction is neither a prudent nor a morally defensible conservation strategy, not least because it violates Leopold's critical proviso to not rely on technological fixes to solve what are ultimately foundational moral and cultural problems surrounding the human relationship to nature (Minteer 2015).

All this is to say, then, that even though Leopold may accurately be described as a "preservationist" (both in the moral and historical sense), his version was considerably more activist, pragmatic, and philosophically nuanced than one-dimensional varieties minimizing human intervention into nature. Leopold's conservation thought is not exactly what you'd call programmatic, but I think it's possible to derive from it a set of general "Leopoldian prescriptions" for species conservation under rapid environmental change. Broad normative directives, they provide a philosophical anchor point for current discussions about conservation interventions, one that departs from the traditional laissez-faire preservationist model but that importantly does not grant us complete carte blanche in environmental systems:

(1) Pursue *mitigation strategies* that address the deep drivers of underlying "land sickness" (e.g. the societal and ecological forces driving habitat loss/fragmentation/change, species exploitation and decline, etc.).
(2) Simultaneously pursue *adaptive conservation strategies* to protect species and safeguard valued ecological features and functions (including strategies that may even relax commitments to historical ecosystems and assemblages, such as AC), *while not further degrading land health.*
(3) Support and conduct *experimental research* into novel conservation interventions to inform and guide wise and effective *adaptive conservation management under global change.*
(4) Perform 1–3 while maintaining an attitude of *humility and precaution* in environmental management, particularly a respect for those environmental values (e.g. wildness) that might be imperiled by more intensive modes of conservation intervention.

Conclusion: boldness and restraint in the Anthropocene

Extinction casts a large shadow. The legacy of biotic loss in the moral imagination – the nagging reminders of the myopia and avarice that led to the disappearance of species such as the passenger pigeon, the near extinction of the bison and California condor, and so on – continues to provoke a range of responses among conservationists. For some, it prompts a profound sense of regret, which in its darkest moments can lead to a posture of hopeless resignation. For others, this reminder of the biological stakes of our societal choices is simply evidence of the need for greater precaution, not just in how we develop and consume on the planet, but in how we prioritize and conduct our conservation efforts. For still others it stokes a fiery rededication to the cause, a determination to combat the elegiac conservation narrative by doing whatever it takes to save species from extinction in an era of increased human influence.

As we've seen, however, the more extreme forms of intervention for biodiversity conservation purposes raise serious questions about the scope of human responsibility for wild species and places in this century. And they're some of the most difficult questions conservationists, philosophers, and advocates can grapple with as we seek to square the various demands of holding a principled and consistent environmental ethic in a rapidly changing world, one able to maintain a fidelity to preservationist norms even as the strategies for preserving nature demand that we consider activities that appear to depart from the storied tradition of John Muir, Leopold, and David Brower (Minteer and Pyne 2015).

As we contemplate going down this more interventionist path we need to make sure that the more aggressive (though well-intended) efforts to conserve biodiversity do not become driven by a "save species at any cost" mantra. If we end up sacrificing other important environmental and moral values, such as the respect for nature's wildness and a sense of human proportion on the landscape, I think we'll have lost something quite profound in our environmental ethos: that sense of forbearance and moral restraint that we think of as being the best of the American nature preservationist tradition, but that is also increasingly undermined by the growing recognition of our unrelenting pressure on wild populations and places. Conservation in the Anthropocene (what many are now calling our current age of human influence on the planet) must therefore be a balancing act between the pragmatic need for action and the moral wisdom of ecological self-control in an age of increasing human influence (Minteer 2015a).

"I have purposely presented the land ethic as a product of social evolution because nothing so important as an ethic is ever 'written' … It evolves in the minds of a thinking community," Leopold wrote at the end of "The Land Ethic" (Leopold 1989: 225). It's a powerful reminder that although Leopold can point us in certain directions, underscoring certain values and concerns that we should weigh carefully in our management decisions, in the end, developing a responsible and "evolved" environmental ethic for conservation under environmental change will likely hinge on our own answers to a set of tough questions. Will we be able to develop a responsible "ethics of acceptable intervention" in rapidly changing ecological systems? Can we retain a Leopoldian spirit of humility toward the environment and rein in our more destructive technological endeavors, even as our planetary influence inevitably grows? Will we be able to keep alive a vital sense of the wild as we further transform the natural world in new and previously unimaginable ways (including in the name of conservation)? These are the questions our own "thinking community" will and must continue to grapple with as we calibrate conservation in the age of humans.

References

Altizer, S., *et al.* 2013. "Climate Change and Infectious Diseases: From Evidence to a Predictive Framework." *Science* 341: 514–519.

Barnosky, A. D. 2009. *Heatstroke: Nature in an Age of Global Warming*. Washington, DC: Shearwater/Island Press.

Bellard, C., *et al.* 2012. "Impacts of Climate Change on the Future of Biodiversity." *Ecology Letters* 15: 365–377.

Camacho, A. E., *et al.* 2010. "Reassessing Conservation Goals in a Changing Climate." *Issues in Science and Technology* 26: 21–26.

Ceballos, G., *et al.* 2015. "Accelerated Modern Human-induced Species Losses: Entering the Sixth Mass Extinction." *Science Advances* 1: 1–5.

Chauvenet, A. L. M., *et al.* 2013. "Maximizing the Success of Assisted Colonizations." *Animal Conservation* 16: 161–169.

Davidson, I., and C. Simkanin. 2008. "Skeptical of Assisted Colonization." *Science* 322: 1048–1049.

Davis, M. A., *et al.* 2011. "Don't Judge Species on Their Origins." *Nature* 474: 153–154.

Dirzo, R. *et al.* 2014. "Defaunation in the Anthropocene." *Science* 345: 401–406.

Elliot, R. 1997. *Faking Nature. The Ethics of Environmental Restoration*. London: Routledge.

Flader, S. L., and J. B. Callicott. 1991. *The River of the Mother of God and Other Essays by Aldo Leopold*. Madison: University of Wisconsin Press.

Hannah, L., ed. 2012. *Saving a Million Species: Extinction Risk from Climate Change*. Washington, DC: Island Press.

Hobbs, R. J., *et al.* 2011. "Intervention Ecology: Applying Ecological Science in the Twenty-first Century." *BioScience* 61: 442–450.

Hoegh-Guldberg, O., *et al.* 2008. "Assisted Colonization and Rapid Climate Change." *Science* 321: 345–346.

Huang, D. 2008. "Assisted Colonization Won't Help Rare Species." *Science* 322: 1049.

IUCN/SSC. 2013. *Guidelines for Reintroductions and Other Conservation Translocations*. Version 1.0. Gland: International Union for the Conservation of Nature.

Jordan, W. R., and G. M. Lubick. 2011. *Making Nature Whole: A History of Ecological Restoration*. Washington, DC: Island Press.

Katz, E. 1997. *Nature as Subject: Human Obligation and Natural Community*. Lanham, MD: Rowman & Littlefield.

Kolbert, E. 2014. *The Sixth Extinction: An Unnatural History*. New York: Henry Holt & Co.

Krosby, M., *et al.* 2010. "Ecological Connectivity for a Changing Climate." *Conservation Biology* 24: 1686–1689.

Leopold, A. 1933. *Game Management*. New York: Charles Scribner's Sons.

Leopold, A. 1989 (orig. 1949). *A Sand County Almanac*. Oxford: Oxford University Press.

Leopold, A. 1999. *For the Health of the Land*, ed. J. B. Callicott and E. T. Freyfogle. Washington, DC: Island Press.

Long, E., and E. Biber. 2014. "The Wilderness Act and Climate Change Adaptation." *Environmental Law* 44: 623–694.

Marris, E. 2011. *Rambunctious Garden: Saving Nature in a Post-Wild World*. New York: Bloomsbury.

Maclean, I. M. D., and R. J. Wilson. 2011. "Recent Ecological Responses to Climate Change Support Predictions of High Extinction Risk." *PNAS* 108: 12337–12342.

McCauley, D. J., *et al.* 2015. "Marine Defaunation: Animal Loss in the Global Ocean." *Science* 347: 6219.

McLachlan, J. S., J. J. Hellmann, and M. W. Schwartz. 2007. "A Framework for Debate of Assisted Migration in an Era of Climate Change." *Conservation Biology* 21: 297–302.

Meine, C., ed. 2013. *Aldo Leopold: A Sand County Almanac and Other Writings on Ecology and Conservation*. New York: The Library of America.

Meine, C., and R. L. Knight, ed. 1999. *The Essential Aldo Leopold: Quotations and Commentaries*. Madison: University of Wisconsin Press.

Minteer, B. A. 2012. *Refounding Environmental Ethics: Pragmatism, Principle, and Practice*. Philadelphia, PA: Temple University Press.

Minteer, B. A. 2015a. "The Perils of De-Extinction." *Minding Nature* 8(1): 11–17.

Minteer, B. A. 2015b. "When Extinction is a Virtue." In *After Preservation: Saving American Nature in the Age of Humans*, ed. B.A. Minteer and S.J. Pyne, 96–104. Chicago, IL: University of Chicago Press.

Minteer, B. A., and J. P. Collins. 2010. "Move It or Lose It? The Ecological Ethics of Relocating Species under Climate Change." *Ecological Applications* 20: 1801–1804.

Minteer, B. A., and J. P. Collins. 2012. "Species Conservation, Rapid Environmental Change, and Ecological Ethics." *Nature Education Knowledge* 3: 14.

Minteer, B. A., and S. J. Pyne. 2015. *After Preservation: Saving American Nature in the Age of Humans*. Chicago, IL: University of Chicago Press.

Monastersky, R. 2014. "Life – a Status Report." *Nature* 516: 158–61.

Neff, M. W., and B. M. H. Larson. 2014. "Scientists, Managers, and Assisted Colonization: Four Contrasting Perspectives Entangle Science and Policy." *Biological Conservation* 172: 1–7.

Newton, J. L. 2006. *Aldo Leopold's Odyssey*. Washington, DC: Island Press/Shearwater.

Norton, B. G. 2005. *Sustainability: A Philosophy of Adaptive Ecosystem Management*. Chicago, IL: University of Chicago Press.

Parmesan, C. 2006. "Ecological and Evolutionary Responses to Recent Climate Change." *Annual Review of Ecology, Evolution, and Systematics* 37: 637–669.

Ricciardi, A., and D. Simberloff. 2009a. "Assisted Colonization is Not a Viable Conservation Strategy." *Trends in Ecology & Evolution* 24: 248–253.

Ricciardi, A., and D. Simberloff. 2009b. "Assisted Colonization: Good Intentions and Dubious Risk Assessment." *Trends in Ecology & Evolution* 24: 476–477.

Rich, N. 2014. "The Mammoth Cometh." *New York Times*, February 27. Available online at www.nytimes.com/2014/03/02/magazine/the-mammoth-cometh.html.

Richardson, D. M., *et al.* 2009. "Multidimensional Evaluation of Managed Relocation." *PNAS* 106: 9721–9724.

Root, T. L., and L. Hughes. 2005. "Present and Future Phenological Changes in Wild Plants and Animals." In *Climate Change and Biodiversity*, ed. T. E. Lovejoy and L. J. Hannah, 61–88. New Haven, CT: Yale University Press.

Root, T. L., *et al.* 2003. "Fingerprints of Global Warming on Wild Animals and Plants." *Nature* 421: 57–60.

Sandler, R. 2010. "The Value of Species and the Ethical Foundations of Assisted Colonization." *Conservation Biology* 24: 424–431.

Sandler, R. 2012. *The Ethics of Species: An Introduction*. Cambridge: Cambridge University Press.

Sandler, R. 2013. "Climate Change and Ecosystem Management." *Ethics, Policy & Environment* 16: 1–15.

Schwartz, M. W., *et al.* 2012. "Managed Relocation: Integrating the Scientific, Regulatory and Ethical Challenges." *BioScience* 62: 732–743.

Seddon, P. J., *et al.* 2014. "Reintroducing Resurrected Species: Selecting DeExtinction Candidates." *Trends in Ecology and Evolution* 29: 140–147.

Seddon, P. J., *et al.* 2015. "Proactive Conservation or Planned Invasion? Past, Current and Future Use of Assisted Colonisation." In *Advances in Reintroduction Biology of Australian and New Zealand Fauna*, ed. D. Armstrong *et al.*, 105–126. Clayton, Vic: CSIRO Publishing.

Simberloff, D. 2012. "Integrity, Stability, and Beauty: Aldo Leopold's Evolving View of Nonnative Species." *Environmental History* 17: 487–511.

Stone, R. 2010. "Home, Home Outside the Range?" *Science* 329: 1592–1594.

Thomas, C. D., *et al.* 2004. "Extinction Risk from Climate Change." *Nature* 427: 145–148.

Urban, M. C. 2015. "Accelerating Extinction Risk from Climate Change." *Science* 348: 571–573.

Webber, B. L., *et al.* 2011. "Translocation or Bust! A New Acclimatization Agenda for the 21st Century?" *Trends in Ecology & Evolution* 26: 495–496.

Winston, J., *et al.* 2014. "Old Wine, New Bottles? Using History to Inform the Assisted Colonization Debate." *Oryx* 48: 186–194.

Zimmer, C. 2013. "Bringing Them Back to Life." *National Geographic*, April. Available online at http://ngm.nationalgeographic.com/2013/04/125-species-revival/zimmer-text.

PART IV

Measurement and methodology

16

BIODIVERSITY INDICATORS NEED TO BE FIT FOR PURPOSE

Kerrie Wilson, Jacqueline England, and Shaun Cunningham

Background

The International Union for Conservation of Nature (IUCN) estimates that roughly 22,000 species worldwide are threatened with extinction (IUCN 2014). This number is predicted to increase due to the impacts of human activities (Sala *et al.* 2000), but also as more species are described and have their extinction risk assessed. As a consequence of the extinction of species, intergovernmental policies and agreements have been developed to conserve biological diversity, including the Convention on Biological Diversity (CBD), which has three main objectives: (1) conservation of biological diversity; (2) sustainable use of the components of biological diversity; and (3) fair and equitable sharing of the benefits arising out of the utilization of genetic resources (Convention on Biological Diversity 2003).

What is biodiversity? The term "biodiversity" is a contraction of "biological diversity" and is meant to capture the structural, functional, or taxonomic heterogeneity of biology (Sarkar 2008). The term was coined by Walter G. Rosen in 1986 (Sarkar 2002) and its global legal framework defined by the CBD in 1992 (United Nations Environment Programme 1992). The CBD define biodiversity as "the variability among living organisms from all sources including, *inter alia*, terrestrial, marine, and other aquatic ecosystems and the ecological complexes of which they are part; this includes diversity within species, between species and of ecosystems" (Convention on Biological Diversity 2003).

The term "biodiversity" is inherently connected to the term "conservation biology", which encompasses a social goal, that is, to conserve biodiversity. The concept of biodiversity thus has a normative component, along with scientific and descriptive components (Norton 1987, Callicott *et al.* 1999, Sarkar 2008). There are also normative decisions surrounding what components of biodiversity should be conserved (Sarkar 2002). Justifications for conserving biodiversity range from biodiversity possessing intrinsic value (i.e. biodiversity has value "in and for itself") to instrumental values (i.e. considered valuable by humans such as harvested forest products or species with medicinal values; Norton 1987, Maguire and Justus 2008, Justus *et al.* 2009, Minteer and Miller 2011).

Biodiversity indicators are defined by the CBD to be "information tools that summarise data on complex and sometimes conflicting issues to indicate the overall status and trends of biodiversity" (Convention on Biological Diversity 2003). Such indicators are employed to motivate changes in land use and management, and to assess the response of biodiversity to policy inter-ventions and other actions (Butchart *et al.* 2010). Consequently, biodiversity indicators are a key component of efforts to mitigate biodiversity loss. Given the multiplicity of biodiversity defini-tions, it can be measured in numerous ways (Sarkar and Margules 2002, Caillon and Degeorges 2007) and therefore no single indicator will likely describe biodiversity as a whole.

A range of criteria have been developed for appraising indicators (Heink and Kowarik 2010b) including their feasibility, efficiency, responsiveness, timeliness, relevance, and effective-ness (Gregory *et al.* 2005, Lamb 2009, Benedek 2014). One of the key criteria for indicators is that they are *fit for purpose* (Mace and Baillie 2007, Jones *et al.* 2011, Vačkářa *et al.* 2012). Therefore, biodiversity indicators must measure the right things, as determined by the objective of the indicator or the purpose for which it will be applied (Mace and Baillie 2007). Biodiversity indicators should also reflect the norms, values, and goals of society and the ethical motivations for conservation (Robertson and Hull 2001, Minteer and Miller 2011). However, such norma-tive components are inherently difficult to measure objectively (Heink and Kowarik 2010a).

In this chapter we: (a) identify a core set of criteria for reviewing indicators, (b) critically review a suite of popular indicators according to these criteria, and (c) identify ways to improve the extent to which indicators are fit for purpose.

We conducted an extensive literature search to compile a representative sample of the bio-diversity indicators that are in current use. The literature search was performed using ISI Web of Knowledge and Google Scholar in March 2014. Search terms included "biodiversity assess-ment", "biodiversity indicator", "biodiversity index", and "biodiversity tool". An effort was made to search for biodiversity indicators from all continents by combining the above search terms with country names. This exhaustive process revealed that there was a small set of dis-tinct biodiversity indicators, which have been applied across numerous continents. The five indicators chosen for evaluation (Table 16.1) were selected because they: (a) are focused on biodiversity assessment; (b) are measured at a national scale or above; (c) are documented in the peer-reviewed literature; (d) cover a wide range of biodiversity components among them (e.g. particular taxa through to whole ecosystems); and (e) have distinct characteristics to the other indicators selected for review.

The biodiversity indicators were reviewed according to their: (1) objective and utility; (2) rep-resentativeness of biodiversity; and (3) quality of information. Given the multiple conceptual-izations, perceptions, and possible measurements of biodiversity, appraising the level of "fit" of

Table 16.1 The biodiversity indicators reviewed, with example applications and key references

Indicator	Example applications	Key references
Red List Index (RLI)	Status and trends of extinction risk of bird species	Butchart *et al.* 2004, 2007 Koyanagi and Furukawa 2013
	Status of grasslands in Japan	
Living Planet Index (LPI)	Global vertebrate abundance	Collen *et al.* 2009, Loh *et al.* 2005
Nature Index (NI)	Ecosystems of Norway	Certain *et al.* 2011 Skarpaas *et al.* 2012
Natural Capital Index (NCI)	Vegetation of Hungary	Czúcz *et al.* 2012
Wild Bird Index (WBI)	Farmland birds in United Kingdom	Gregory *et al.* 2004 Gregory and van Strien 2010
	Birds of European countries	

indicators to their intended purpose requires appraisal from diverse viewpoints. We framed a specific perspective to reflect that commonly held by ecological scientists, who typically seek to accurately quantify changes in biodiversity. In contrast, we framed the general perspective to reflect that of policy-makers and the general public who might be interested in only the general trajectory of biodiversity change.

Indicators have diverse objectives, but largely unknown utility

All of the indicators were established with normative objectives in mind: to communicate changes to policy-makers and the general public in order to raise awareness of biodiversity loss. The NCI is also noted to have the capacity to achieve a specific normative objective at a local scale of "improving the sense of place" (Czúcz *et al.* 2012). Typically, reference is made to international policy, such as the CBD, multi-country policies such as the European Union Common Agriculture Policy (Gregory *et al.* 2005), and national-level policies. To be policy-relevant and meaningful, indicators ideally should be related to policy targets (Dennis *et al.* 2009). For example, the LPI, WBI, and the RLI are connected to CBD Target 4,[1] the RLI is connected to CBD Target 12,[2] and a national application of the RLI to Australian birds demonstrated potential utility for reporting on CBD Target 13[3] (Szabo *et al.* 2012).

One of the key attributes of effective biodiversity indicators is that they are easily understood and amenable to clear presentation (Gregory *et al.* 2005, Normander *et al.* 2012). The WBI is an example of an indicator that has influenced national-level policy (Aebischer *et al.* 2000, Chamberlain *et al.* 2000, Gregory *et al.* 2004) and this has been attributed to the synthesis of scientific data into a simple, understandable, and meaningful presentation (Gregory *et al.* 2004). Typically, the indicators are reported as a proportional change or on an ordinal scale (e.g. a dimensionless quantity between 0 and 100) to enhance communication. The "apparency" of taxa used in an indicator (i.e. whether their decline is noticed by the public) and responsiveness to changes in management are also important for effective communication. The extent that indicators are understood can thus be enhanced through engaging the public in the collation of information underpinning an indicator. This so-called "citizen science", where volunteers collect and/or process data as part of a scientific enquiry (Silvertown 2009, Tulloch *et al.* 2013), is particularly helpful for addressing questions that have a large spatial or temporal scope, such as biodiversity loss (Bonney *et al.* 2009). Citizen scientists have contributed to bird species population monitoring across extensive regions in Europe allowing the development of indicators for birds, including the WBI (Gregory *et al.* 2005). The WBI is an example of how the communication to the public and the media of the changes observed was enhanced through the process of engaging citizens in the collation of data.

To be easily communicated, indicators must simplify complex patterns, processes, and phenomena. From an ecological perspective, the objective of the five biodiversity indicators reviewed can be classified in three ways: (a) to measure change over time only (RLI); (b) to measure the current "state" relative to a baseline (NI and NCI); or (c) to measure change both over time and relative to a baseline (WBI and LPI). An important aspect of the utility of an indicator is whether this information can then be disaggregated (that is, separated into sub-categories or component data for further evaluation), particularly to allow the underlying drivers of change or impacts of policy and management to be evaluated.

All of the indicators reviewed have the stated capacity to be disaggregated. Data on habitat extent and quality is used to populate the NI and NCI and thus the results for these indicators are inherently connected to habitat conversion and degradation. The documentation for the NI suggests disaggregation be to the level of ecosystems rather than particular locations

(Certain *et al*. 2011). The WBI can also be disaggregated to particular habitat types and has been used to evaluate agricultural policy in the United Kingdom (Gregory *et al*. 2004). Subsets of species for which extinction risk has changed most rapidly according to the RLI can be identified and hence important threatening processes can be inferred (Butchart *et al*. 2004, Szabo *et al*. 2012). The RLI has been used for assessing the effectiveness of the Convention on International Trade in Endangered Species of Wild Fauna and Flora (CITES) (Bubb *et al*. 2009) and for differentiating the impacts of African protected areas policies (Nicholson *et al*. 2012). The LPI has been tested by assessing the impact of fisheries policies and was found to exhibit counterintuitive behaviour due to over-representation of some taxonomic and functional groups in the indicator (Collen *et al*. 2013), and contrasting impacts of policies on different groups caused by trophic interactions (Nicholson *et al*. 2012). As only a handful of analyses such as these have been undertaken, it is possible that the stated capacity for disaggregation (and hence the utility of the indicators) might be different from the realized capacity.

A key criterion of an indicator is the capacity for users to assess the significance of changes relative to a baseline (OECD 2003). All of the indicators reviewed, with the exception of the RLI, measure a change in state or trends relative to a baseline. For aggregated measures of biodiversity (i.e. the NI and NCI), there are challenges associated with choosing accurate baselines for all the components of biodiversity being evaluated.

The choice of baseline is important to facilitate comparisons and to obtain an accurate assessment of change (Bull *et al*. 2014). In terms of reference year, the length of the temporal sequence and when the observation commences relative to fluctuations in the data will be important. For example, for a given hypothetical indicator, if the reference year was prior to agricultural industrialization, then it may show a decline to the present time (Fig. 16.1a), if it was at the height of human-induced modification, it may show no change to the present (Fig. 16.1b), and if it was at a period of slowing modifications it may show a small increase to the present (Fig. 16.1c). Thus, the choice of reference year ultimately determines whether changes are observed and therefore the usefulness of the indicator for informing or assessing policy. Across and within all indicators, there is, however, no common reference year as the choice of

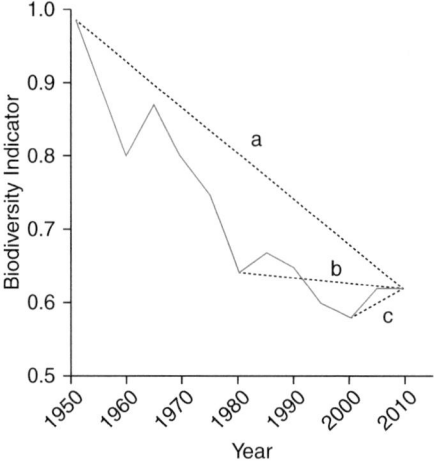

Figure 16.1 Impact of the reference year used to calculate a biodiversity index. Using an historical reference is likely to show a decline in biodiversity (a), whereas more recent years are likely to show smaller decreases (b), or increases (c) due to minor variations among years.

baseline is determined by data availability (e.g. 1970 for the LPI) or by experts (as is the case for the NCI). This limits the extent to which comparisons can be made among areas, with the documentation for the NCI explicitly expressing caution in relation to comparing regions with different baselines (Czúcz *et al.* 2012).

The choice of reference state for data on habitat quantity and quality is heavily reliant on expert assessment, and as such there is potential for ambiguity. For the NI and NCI, the reference state is assumed to be a "pristine" or "intact" state (e.g. for the NI it is a hypothetical undisturbed or sustainably managed ecosystem). These notions are dynamic in relation to societal preferences and values and what is feasible to achieve (Hobbs and Norton 1996). Important questions in relation to choice of reference state is "How does the 'ideal' state of an ecosystem differ among parts of society?" and "Is it feasible to return to the reference state?"

It is likely that there will be variation in the response of biodiversity to different drivers of change, with some species declining, some improving, and some species displaying an intermediate response. Indicators should be sufficiently sensitive to detect these changes and thus explicitly account for and distinguish natural fluctuations that can arise due to seasonal variations or longer-term environmental phenomena (e.g. due to changes in the Southern Oscillation Index; Gregory *et al.* 2005). The WBI notes procedures for controlling natural fluctuations and this is achieved coarsely by smoothing the trend lines. However, mean trends (as delivered by the LPI, WBI, and RLI) can hide substantial declines for some species if the majority of species are not declining. In the UK application of the WBI, a mean index was created by combining the species trends on a geometric scale, so that doubling of one species' index is balanced by halving of another (Gregory *et al.* 2004). As a result, the WBI will detect no change when half of the species increase, while the other half decrease, and upward trends could mask substantial declines in several species. In the case of the RLI, slow declines of common species are not well captured and recently evaluated species may introduce bias (Szabo *et al.* 2012). While the impacts of using mean trends is not a stated limitation of the aggregated indicators (i.e. NI and NCI), it is possible that the impact is greatest for these types of indicators.

Only a limited subset of biodiversity is measured

Biodiversity is a complex and multidimensional concept that has to be estimated using surrogates, which are assumed to represent the distributions and trends in other components of biodiversity. The choice of biodiversity surrogate is driven by the availability of appropriate monitoring data. Consequently, there is a potential for bias, either explicitly or implicitly, towards commonly surveyed taxa (e.g. birds, mammals, and butterflies; Boakes *et al.* 2010). Indeed, the biodiversity indicators reviewed that are focused on specific taxa are biased towards birds and mammals (RLI, WBI, and LPI). There is also variation in the relative abundance of species that are included, ranging from threatened species (for the RLI), common species (for the WBI), or a combination (for the LPI and NCI). For the WBI, common birds were employed as they were identified as being diverse, widespread and mobile, high in the food chain, and responsive to environmental change (Gregory and van Strien 2010). For the LPI, it is asserted that population trends for species may act as a proxy indicator for the state of the ecosystem that the species inhabits and by inference wider biodiversity (Loh *et al.* 2005). By assessing the ecosystem as a whole (in terms of habitat quantity and quality), the NCI and NI may be more representative of biodiversity than species-based indicators.

The indicators reviewed mainly focus on terrestrial birds and mammals and therefore account for only a small proportion of total biodiversity (Fig. 16.2). Ecologists have long debated the issue of whether single taxa should be used to make conservation decisions (Franklin 1993).

Different types of species have been theorized including "ecological indicators" that respond to stressors, "keystones" upon which a large part of a community's diversity depends, "umbrellas" that if protected will conserve a suite of species, "flagships" that are charismatic and can motivate political and societal change, and "vulnerables" that are prone to extinction in human-dominated landscapes (Noss 1990). However, there is limited evidence that a small subset of taxa can capture the breadth of biodiversity and may do no better than a random subset (Andelman and Fagan 2000). For example, vertebrates have been found to be poor predictors of the distribution of plants and plants to be poor predictors of invertebrates (Oliver *et al.* 1998). Basing trend assessments on the abundance of common species may mask changes for vulnerable species. General rules for surrogacy are yet to be found (Margules and Sarkar 2007, Grantham *et al.* 2010, Lewandowski *et al.* 2010) but existing evidence suggests that birds and mammals alone are unlikely to provide useful surrogates for all of biodiversity. Ecosystem indicators, which measure habitat extent and quality, may overcome some surrogacy issues since habitat area is a well-established predictor of biodiversity (MacArthur and Wilson 1967). However, the relationship between biodiversity and habitat quality is still poorly defined and understood (Fischer and Lindenmayer 2007).

There is little evidence that other cultural, ecological, or evolutionary processes are embodied in the indicators, the exception being the NI, which when applied in Norway included

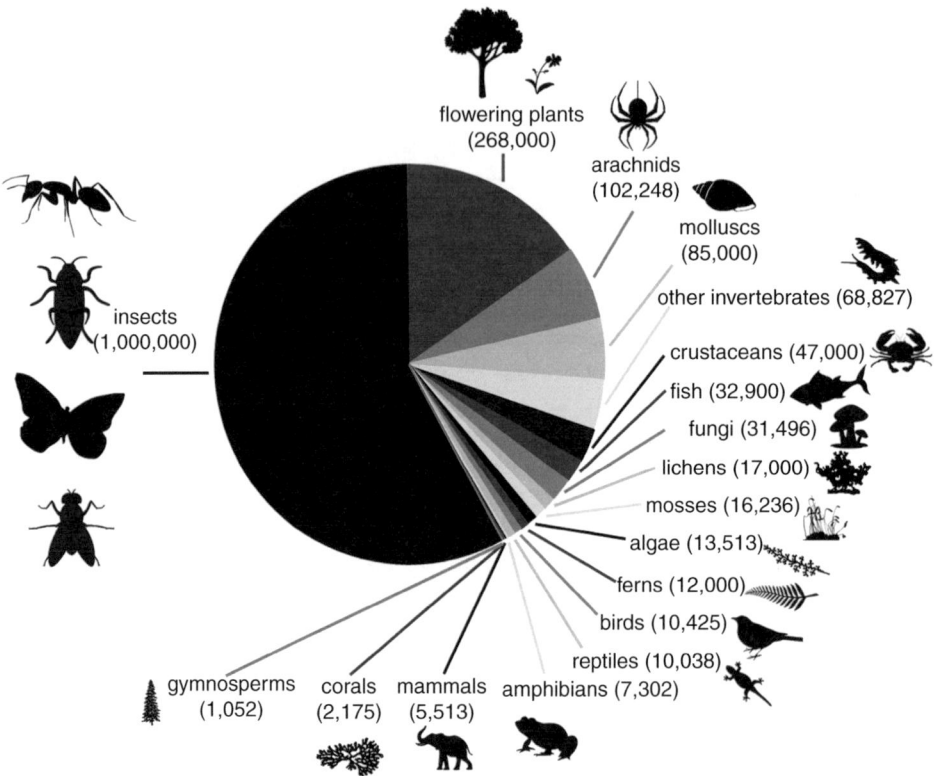

Figure 16.2 The number of described species for different taxa (number shown in brackets) (IUCN 2014). The IUCN Red List, on which the RLI is derived, has assessed 76,000 species and the goal is to extend this to 160,000 species by 2020.

ecosystem services (Certain *et al.* 2011). Abundant species are noted in the LPI to influence provisioning and regulating services provided by ecosystems (Collen *et al.* 2013), although this relationship was not quantified. For several of the indicators (LPI, NI, and NCI), biodiverse ecosystems are assumed to be more resistant (e.g. to invasion by introduced species) and resilient (i.e. will recover after disturbance) (Certain *et al.* 2011, Czúcz *et al.* 2012, Collen *et al.* 2013). However, the role that biodiversity plays in delivering these values is poorly understood (Folke *et al.* 2004, Hooper *et al.* 2012) and it is argued by some that these values could be provided by simplified ecosystems (Worm *et al.* 2006, Kareiva *et al.* 2007, Isbell *et al.* 2011).

A broader ecological view would be to survey change in population abundance for a representative range of taxa from marine, freshwater, and terrestrial environments. This is the approach taken for the Sampled Red List Index (Butchart *et al.* 2007). The sampled approach to the Red List Index (Butchart *et al.* 2007) was developed in order to determine the threat status and trends of lesser-known and less charismatic species groups in an attempt to provide a more broadly representative picture of biodiversity change. The index is based on a representative sample of 1,500 species selected for a number of taxonomic groups within vertebrates, invertebrates, plants, and fungi. Furthermore, although indicators are driven by current data availability, they could be more targeted in the future as this is the rationale for the development of a consolidated set of variables referred to as the Essential Biodiversity Variables (EBVs; Pereira *et al.* 2013). The EBVs align with CBD targets and comprise measures of genetic composition, species composition and traits, community composition, and ecosystem structure and function. The indicators reviewed here are largely relevant to the candidate EBV classes of species population (RLI, LPI, WBI), and ecosystem structure and function (NCI, NI), with some relevance to genetic composition in the case of the RLI.

There is a tension between the quality and coverage of information

A desired quality of biodiversity indicators is that they are scientifically sound, being based on verifiable data that have been collected using standard methods, with known accuracy and precision (Dennis *et al.* 2009). Furthermore, accurate population trend analysis (as required by the RLI, WBI, and LPI) relies on systematic monitoring programs with adequate repeat surveys (ideally annually) and extensive spatial coverage (OECD 2003, Gregory *et al.* 2005, Rhodes *et al.* 2006). The limited availability of data on biodiversity that meet such criteria presents obstacles for achieving equitable coverage, particularly for data-poor areas.

Data-poor areas typically coincide with the deforestation/degradation frontier and thus are where information is urgently needed to facilitate quick responses to mitigate the loss of biodiversity. Tropical regions are an epicentre of the global biodiversity crisis because they are the most species-rich and the most rapidly changing. In tropical regions monitoring data are typically sparse, along with general biological knowledge (Lawler *et al.* 2006), the availability of experts, and of how well land-use impacts are understood (Law and Wilson 2015). Data coverage for the LPI, for example, is explicitly noted to be better in temperate than tropical regions (Collen *et al.* 2013). Future development of biodiversity indicators must prioritize the facilitation of timely and efficient assessment in biodiverse but information-poor regions.

The NI and NCI use data on habitat extent and quality, and these data types may be more readily obtained in data-poor areas through remote-sensing technology or land-use inventories (Kerr and Ostrovsky 2003). However, for the NCI, there is vagueness in relation to the metric that should be used to assess habitat quality (Czúcz *et al.* 2012) and even indirect proxies such as data on threatening processes (e.g. logging) can be included. Furthermore, relatively fine-resolution (e.g. < 1 km^2) ground measures or field validation of habitat quality data are required

(Czúcz *et al.* 2012). The NI employs a range of information sources, which may be useful in data-poor areas, although care must be taken when combining data from different sources to ensure currencies are commensurable and to minimize duplication of information. Regardless of the type of input data, data that are measured in a consistent way (e.g. using comparable techniques at similar resolutions) are required to ensure valid comparisons among areas and between times. However, all of the indicators, with the exception of the WBI, are not underpinned by widely accepted survey techniques.

While the rationale for the construction of the indicators and the process required to populate these with data is quite well documented, there is subjectivity in relation to the choice of weightings and how these are applied, particularly for the RLI, NCI, and the LPI. For the LPI, all declines, regardless of how close they bring a species to extinction, are considered equal (Collen *et al.* 2013). For the NI, all major ecosystems are weighted equally while some component indicators are given more weight. By combining quantity and quality into one indicator, the NCI relies on a hypothetical equivalence in terms of ecological value between smaller intact patches of habitat and patches that are larger, yet degraded. This might be acceptable for species with generalist habitat requirements but not for species with specialist requirements such as those requiring intact habitat over expansive areas. All weightings, even equal weighting, represent a choice that requires clear justification. An important question is "who should choose weightings?" Should it be scientific experts, policy-makers, or stakeholders that are impacted by biodiversity change?

All of the indicators are affected by error in the underlying data, sampling bias, sampling intensity, and the timeframe assessed. The RLI, LPI, and WBI allow for the estimation of confidence intervals on the rate of decline (and for the WBI trends lines are smoothed to control for sampling error). The NI allows confidence intervals to be generated for each ecosystem type based on three sources of data uncertainty: numerical uncertainty, data uncertainty, and uncertainty because of lack of knowledge (Certain *et al.* 2011). In comparison, the NCI does not provide a measure of error, although the results have been demonstrated to be sensitive to data and weightings employed for the quality measure (Czúcz *et al.* 2012). The establishment of systematic monitoring approaches would reduce uncertainty associated with reported values, although strict standards for data collection and analysis could further limit the spatial and temporal coverage of indicators.

Conclusions

All of the indicators reviewed here have been developed with a normative objective in mind (mostly to inform international biodiversity policy and associated targets or for communication). Ideally, biodiversity indicators should connect the field of ecological science with policy-making or science communication. There are four main points of concern that ecologists and policy-makers should keep in mind while assessing indicators. First, there is a tension between the needs of ecologists and those who form policy. Second, the indicator species that are often used are unlikely to be representative of biodiversity. Third, the spatial and temporal coverage of information is uneven. Finally, the indicators have been rarely used to evaluate the impacts of policy interventions.

Ecologists are generally concerned with the quantitative attributes of an indicator, such as whether an indicator includes accurate data with an assessment of error or whether the indicator is representative. In terms of its development, the LPI represents a special case. The LPI was developed initially as a communications tool for a World Wildlife Fund campaign. However, the increasing role of the LPI as a policy tool for monitoring progress toward the

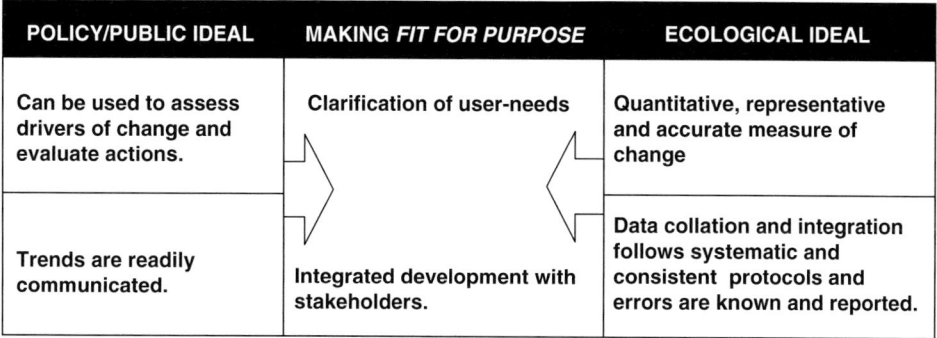

POLICY/PUBLIC IDEAL	MAKING *FIT FOR PURPOSE*	ECOLOGICAL IDEAL
Can be used to assess drivers of change and evaluate actions.	**Clarification of user-needs**	**Quantitative, representative and accurate measure of change**
Trends are readily communicated.	**Integrated development with stakeholders.**	**Data collation and integration follows systematic and consistent protocols and errors are known and reported.**

Figure 16.3 An integrated approach to improve the extent to which biodiversity indicators are fit for purpose.

CBD and other biodiversity targets initiated efforts to improve the robustness, sensitivity, and representativeness of the indicator (Collen *et al.* 2009, 2013). While the accuracy of indicators is important to scientists, it is potentially less important for the normative purposes of biodiversity indicators. The WBI is the only indicator that uses survey data of a defined standard and all indicators report mean trends, but does this matter? If this question is resolved in the early stages of indicator development then the level of fit of the indicator will be enhanced. Ultimately, indicators are intended to influence political decisions about the allocation of resources to an issue that society deems important (Robertson and Hull 2001). However, the utility of the indicators to identify the underlying drivers of change and the impacts of policy is yet to be extensively tested. Instead, a large amount of research and development has been focused on the accuracy and representativeness of indicators. Ultimately, to be fit for purpose, the design and testing of indicators should be user-driven (Fig. 16.3), involving policy-makers, scientific experts, environmental groups, citizens, and other important stakeholders (Dennis *et al.* 2009).

Notes

1 By 2020, at the latest, governments, business and stakeholders at all levels have taken steps to achieve or have implemented plans for sustainable production and consumption and have kept the impacts of use of natural resources well within safe ecological limits.

2 By 2020, the extinction of known threatened species has been prevented and their conservation status, particularly of those most in decline, has been improved and sustained.

3 By 2020, the genetic diversity of cultivated plants and farmed and domesticated animals and of wild relatives, including other socio-economically as well as culturally valuable species, is maintained, and strategies have been developed and implemented for minimizing genetic erosion and safeguarding their genetic diversity.

References

Aebischer, N. J., R. E. Green, and A. D. Evans. 2000. "From Science to Recovery: Four Case Studies of How Research Has Been Translated into Conservation Action in the UK." In *Ecology and Conservation of Lowland Farmland Birds*, ed. N. J. Aebischer, A. D. Evans, P. V. Grice, and J. A. Vickery. Tring: British Ornithologists' Union.

Andelman, S. J., and W. F. Fagan. 2000. "Umbrellas and Flagships: Efficient Conservation Surrogates or Expensive Mistakes?" *Proceedings of the National Academy of Sciences of the United States of America* 97: 5954–5959.

Benedek, Z. 2014. "On the Potential Policy Use of Some Selected Biodiversity Indicators: Limitations and Recommendations for Improvements - Short Communication." *Journal of Forest Science* 60: 84–88.

Boakes, E. H., P. J. K. McGowan, R. A. Fuller, C. Q. Ding, N. E. Clark, K. O'Connor, and G. M. Mace. 2010. "Distorted Views of Biodiversity: Spatial and Temporal Bias in Species Occurrence Data." *PLoS Biology* 8, e1000385.

Bonney, R., C. B. Cooper, J. Dickinson, S. Kelling, T. Phillips, K. V. Rosenberg, and J. Shirk. 2009. "Citizen Science: A Developing Tool for Expanding Science Knowledge and Scientific Literacy." *Bioscience* 59: 977–984.

Bubb, P. J., S. H. M. Butchart, B. Collen, H. Dublin, V. Kapos, C. Pollock, S. N. Stuart, and J.-C. Vié. 2009. *IUCN Red List Index: Guidance for National and Regional Use.* Gland, Switzerland: IUCN.

Bull, J., A. Gordon, E. Law, K. Suttle, and E. Milner-Gulland. 2014. "Importance of Baseline Specification in Evaluating Conservation Interventions and Achieving No Net Loss of Biodiversity." *Conservation Biology* 28: 799–809.

Butchart, S. H. M., H. Resit Akçakaya, J. Chanson, J. E. M. Baillie, B. Collen, S. Quader, W. R. Turner, R. Amin, S. N. Stuart, and C. Hilton-Taylor. 2007. "Improvements to the Red List Index." *PLoS ONE* 2: e140.

Butchart, S. H. M., A. J. Stattersfield, L. A. Bennun, S. M. Shutes, H. R. Akcakaya, J. E. M. Baillie, S. N. Stuart, C. Hilton-Taylor, and G. M. Mace. 2004. "Measuring Global Trends in the Status of Biodiversity: Red List Indices for Birds." *PLoS Biology* 2: 2294–2304.

Butchart, S. H. M., M. Walpole, B. Collen, A. Van Strien, J. P. W. Scharlemann, R. E. A. Almond, J. E. M. Baillie, B. Bomhard, C. Brown, J. Bruno, K. E. Carpenter, G. M. Carr, J. Chanson, A. M. Chenery, J. Csirke, N. C. Davidson, F. Dentener, M. Foster, A. Galli, J. N. Galloway, P. Genovesi, R. D. Gregory, M. Hockings, V. Kapos, J.-F. Lamarque, F. Leverington, J. Loh, M. A. McGeoch, L. McRae, A. Minasyan, M. H. Morcillo, T. E. E. Oldfield, D. Pauly, S. Quader, C. Revenga, J. R. Sauer, B. Skolnik, D. Spear, D. Stanwell-Smith, S. N. Stuart, A. Symes, M. Tierney, T. D. Tyrrell, J.-C. Vie, and R. Watson. 2010. "Global Biodiversity: Indicators of Recent Declines." *Science* 328: 1164–1168.

Caillon, S., and P. Degeorges. 2007. "Biodiversity: Negotiating the Border between Nature and Culture." *Biodiversity and Conservation* 16: 2919–2931.

Callicott, J. B., L. B. Crowder, and K. Mumford. 1999. "Current Normative Concepts in Conservation." *Conservation Biology* 13: 22–35.

Certain, G., O. Skarpaas, J.-W. Bjerke, E. Framstad, M. Lindholm, J.-E. Nilsen, A. Norderhaug, E. Oug, H.-C. Pedersen, A.-K. Schartau, G. I. Van Der Meeren, I. Aslaksen, S. Engen, P.-A. Garnåsjordet, P. Kvaløy, M. Lillegård, N. G. Yoccoz, and S. Nybø. 2011. "The Nature Index: A General Framework for Synthesizing Knowledge on the State of Biodiversity." *PLoS ONE* 6: e18930.

Chamberlain, D. E., R. J. Fuller, R. G. H. Bunce, J. C. Duckworth, and M. Shrubb. 2000. "Changes in the Abundance of Farmland Birds in Relation to the Timing of Agricultural Intensification in England and Wales." *Journal of Applied Ecology* 37: 771–788.

Collen, B., J. Loh, S. Whitmee, L. McRae, R. Amin, and J. E. M. Baillie. 2009. "Monitoring Change in Vertebrate Abundance: the Living Planet Index." *Conservation Biology* 3: 317–327.

Collen, B., L. McRae, J. Loh, S. Deinet, A. De Palma, R. Manley, and J. E. M. Baillie. 2013. "Tracking Change in Abundance: The Living Planet Index." In *Biodiversity Monitoring and Conservation.* Oxford: Wiley-Blackwell.

Convention on Biological Diversity. 2003. *Handbook of the Convention on Biological Diversity.* London: Earthscan.

Czúcz, B., Z. Molnár, F. Horváth, G. G. Nagy, Z. Botta-Dukát, and K. Török. 2012. "Using the Natural Capital Index Framework as a Scalable Aggregation Methodology for Regional Biodiversity Indicators." *Journal of Nature Conservation* 20: 144–152.

Dennis, P., M. Arndorfer, K. Balázs, R. G. H. Bunce, A. Centeri, A. Corporaal, D. Cuming, M. Deconchat, W. Dramstad, B. Elyakime, E. Falusi, W. Fjellstad, M. D. Fraser, B. Freyer, J. K. Friedel, I. Geijzendorffer, R. Jongman, M. Kainz, G. M. Marcos, T. Gomiero, S. Grausgruber-Gröger, E. Kelemen, S. R. Moakes, P. Nicholas, M. G. Paoletti, L. Podmaniczky, P. Pointereau, J.-P. Sarthou, N. Siebrecht, D. Sommaggio, S. D. Stoyanova, N. Teufelbauer, D. Viaggi, A. Vialatte, and S. Wolfrum. 2009. "Conceptual Foundations for Biodiversity Indicator Selection for Organic and Low-input Farming Systems." Available online at www.biobio-indicator.org/deliverables/D21.pdf.

Fischer, J., and D. B. Lindenmayer. 2007. "Landscape Modification and Habitat Fragmentation: A Synthesis." *Global Ecology and Biogeography* 16: 265–280.

Folke, C., S. Carpenter, B. Walker, M. Scheffer, T. Elmqvist, L. Gunderson, and C. S. Holling. 2004. "Regime Shifts, Resilience, and Biodiversity in Ecosystem Management." *Annual Review of Ecology Evolution and Systematics* 35: 557–581.

Franklin, J. F. 1993. "Preserving Biodiversity: Species, Ecosystems or Landscapes?" *Ecological Applications* 3: 202–205.

Grantham, H. S., R. L. Pressey, J. A. Wells, and A. J. Beattie. 2010. "Effectiveness of Biodiversity Surrogates for Conservation Planning: Different Measures of Effectiveness Generate a Kaleidoscope of Variation." *PLoS ONE* 5(7): e11430.

Gregory, R. D., and A. van Strien. 2010. "Wild Bird Indicators: Using Composite Population Trends of Birds as Measures of Environmental Health." *Ornithological Science* 9: 3–22.

Gregory, R. D., G. Noble, and J. Custance. 2004. "The State of Play of Farmland Birds: Population Trends and Conservation Status of Lowland Farmland Birds in the United Kingdom." *Ibis* 146: 1–13.

Gregory, R. D., A. V. Strien, P. Vorisek, A. W. G. Meyling, D. G. Noble, R. P. B. Foppen, and D. W. Gibbons. 2005. "Developing Indicators for European Birds." *Philosophical Transactions of the Royal Society of London B* 360: 269–288.

Heink, U., and I. Kowarik. 2010a. "What are Indicators? On the Definition of Indicators in Ecology and Environmental Planning." *Ecological Indicators* 10: 584–593.

Heink, U., and I. Kowarik. 2010b. "What Criteria Should Be Used to Select Biodiversity Indicators?" *Biodiversity and Conservation* 19: 3769–3797.

Hobbs, R., and D. Norton. 1996. "Towards a Conceptual Framework for Restoration Ecology." *Restoration Ecology* 4: 93–110.

Hooper, D. U., E. C. Adair, B. J. Cardinale, J. E. K. Byrnes, B. A. Hungate, K. L. Matulich, A. Gonzalez, J. E. Duffy, L. Gamfeldt, and M. I. O'Connor. 2012. "A Global Synthesis Reveals Biodiversity Loss as a Major Driver of Ecosystem Change." *Nature* 486: 105–U129.

Isbell, F., V. Calcagno, A. Hector, J. Connolly, W. S. Harpole, P. B. Reich, M. Scherer-Lorenzen, B. Schmid, D. Tilman, J. Van Ruijven, A. Weigelt, B. J. Wilsey, E. S. Zavaleta, and M. Loreau. 2011. "High Plant Diversity is Needed to Maintain Ecosystem Services." *Nature* 477(7363): 199–202.

IUCN. 2014. "The IUCN Red List of Threatened Species." Version 2014.3. Available online at www.iucnredlist.org (accessed 8 December 2014), International Union for Conservation of Nature and Natural Resources.

Jones, J. P. G., B. E. N. Collen, G. Atkinson, P. W. J. Baxter, P. Bubb, J. B. Illian, T. E. Katzner, A. Keane, J. Loh, E. V. E. McDonald-Madden, E. Nicholson, H. M. Pereira, H. P. Possingham, A. S. Pullin, A. S. L. Rodrigues, V. Ruiz-Gutierrez, M. Sommerville and E. J. Milner-Gulland. 2011. "The Why, What, and How of Global Biodiversity Indicators Beyond the 2010 Target/El Porqué, Qué y Cómo de los Indicadores Globales de Biodiversidad Más Allá de la Meta 2010." *Conservation Biology* 25: 450–457.

Justus, J., M. Colyvan, H. Regan, and L. Maguire. 2009. "Buying into Conservation: Intrinsic Versus Instrumental Value." *Trends in Ecology and Evolution* 24: 191–197.

Kareiva, P., S. Watts, R. McDonald, and T. Boucher. 2007. "Domesticated Nature: Shaping Landscapes and Ecosystems for Human Welfare." *Science* 316: 1866–1869.

Kerr, J. T., and M. Ostrovsky. 2003. "From Space to Species: Ecological Applications for Remote Sensing." *Trends in Ecology & Evolution* 18: 299–305.

Koyanagi, T. F., and T. Furukawa. 2013. "Nationwide Agrarian Depopulation Threatens Semi-natural Grassland Species in Japan: Sub-national Application of the Red List Index." *Biological Conservation* 167: 1–8.

Lamb, E. G., E. Bayne, G. Holloway, J. Schieck, S. Boutin, J. Herbers, and D. L. Haughland. 2009. "Indices for Monitoring Biodiversity Change: Are Some More Effective Than Others?" *Ecological Indicators* 9: 432–444.

Law, E. A., and K. A. Wilson. 2015. "Providing Context for the Land-sharing and Land-sparing Debate." *Conservation Letters* 8: 404–413.

Lawler, J. J., J. E. Aukema, J. B. Grant, B. S. Halpern, P. Kareiva, C. R. Nelson, K. Ohleth, J. D. Olden, M. A. Schlaepfer, B. R. Silliman, and P. Zaradic. 2006. "Conservation Science: A 20-year Report Card." *Frontiers in Ecology & the Environment* 4: 473–480.

Lewandowski, A. S., R. F. Noss, and D. R. Parsons. 2010. "The Effectiveness of Surrogate Taxa for the Representation of Biodiversity." *Conservation Biology* 24: 1367–1377.

Loh, J., R. E. Green, T. Ricketts, J. Lamoreux, M. Jenkins, V. Kapos and J. Randers. 2005. "The Living Planet Index: Using Species Population Time Series to Track Trends in Biodiversity." *Philosophical Transactions of the Royal Society B − Biological Sciences* 360: 289–295.

Macarthur, R. H., and E. O. Wilson. 1967. *The Theory of Island Biogeography*. Princeton, NJ: Princeton University Press.

Mace, G. M., and J. E. M. Baillie. 2007. "The 2010 Biodiversity Indicators: Challenges for Science and Policy." *Conservation Biology* 21: 1406–1413.

Maguire, L. A., and J. Justus. 2008. "Why Intrinsic Value is a Poor Basis for Conservation Decisions." *BioScience* 58: 910–911.

Margules, C., and S. Sarkar. 2007. *Systematic Conservation Planning*. Cambridge: Cambridge University Press.

Minteer, B. A., and T. R. Miller. 2011. "The New Conservation Debate: Ethical Foundations, Strategic Trade-offs, and Policy Opportunities." *Biological Conservation* 144: 945–947.

Nicholson, E., B. Collen, A. Barausse, J. L. Blanchard, B. T. Costelloe, K. M. E. Sullivan, F. M. Underwood, R. W. Burn, S. Fritz, J. P. G. Jones, L. McRae, H. P. Possingham, and E. J. Milner-Gulland. 2012. "Making Robust Policy Decisions Using Global Biodiversity Indicators." *PLoS ONE* 7: e41128.

Normander, B., G. Levin, A.-P. Auvinen, H. Bratli, O. Stabbetorp, M. Hedblom, A. Glimskär, and G. A. Gudmundsson. 2012. "Indicator Framework for Measuring Quantity and Quality of Biodiversity – Exemplified in the Nordic Countries." *Ecological Indicators* 13: 104–116.

Norton, D. 1987. *Why Preserve Natural Variety?* Princeton, NJ: Princeton University Press.

Noss, R. F. 1990. "Indicators for Monitoring Biodiversity: A Hierarchical Approach." *Conservation Biology* 4: 355–364.

OECD. 2003. *OECD Environmental Indicators. Development, Measurement and Use*. Paris: OECD.

Oliver, I., A. J. Beattie, and A. York. 1998. "Spatial Fidelity of Plant, Vertebrate, and Invertebrate Assemblages in Multiple-use Forest in Eastern Australia." *Conservation Biology* 12: 822–835.

Pereira, H. M., S. Ferrier, M. Walters, G. N. Geller, R. H. G. Jongman, R. J. Scholes, M. W. Bruford, N. Brummitt, S. H. M. Butchart, A. C. Cardoso, N. C. Coops, E. Dulloo, D. P. Faith, J. Freyhof, R. D. Gregory, C. Heip, R. Hoeft, G. Hurtt, W. Jetz, D. S. Karp, M. A. McGeoch, D. Obura, Y. Onoda, N. Pettorelli, B. Reyers, R. Sayre, J. P. W. Scharlemann, S. N. Stuart, E. Turak, M. Walpole, and M. Wegmann. 2013. "Essential Biodiversity Variables." *Science* 339: 277–278.

Rhodes, J. R., A. J. Tyre, N. Jonzen, C. A. McAlpine, and H. P. Possingham. 2006. "Optimizing Presence–Absence Surveys for Detecting Population Trends." *Journal of Wildlife Management* 70: 8–18.

Robertson, D. P., and R. B. Hull. 2001. "Beyond Biology: Toward a More Public Ecology for Conservation." *Conservation Biology* 15: 970–979.

Sala, O. E., F. S. Chapin, J. J. Armesto, E. Berlow, J. Bloomfield, R. Dirzo, E. Huber-Sanwald, L. F. Huenneke, R. B. Jackson, A. Kinzig, R. Leemans, D. M. Lodge, H. A. Mooney, M. Oesterheld, N. L. Poff, M. T. Sykes, B. H. Walker, M. Walker, and D. H. Wall. 2000. "Biodiversity – Global Biodiversity Scenarios for the Year 2100." *Science* 287: 1770–1774.

Sarkar, S. 2002. "Defining Biodiversity; Assessing Biodiversity." *The Monist* 85: 131–155.

Sarkar, S. 2008. "Norms and the Conservation of Biodiversity." *Resonance* 13: 627–637.

Sarkar, S., and C. Margules. 2002. "Operationalizing Biodiversity for Conservation Planning." *Journal of Bioscience* 27: 299–308.

Silvertown, J. 2009. "A New Dawn for Citizen Science." *Trends in Ecology & Evolution* 24: 467–471.

Skarpaas, O., G. Certain, and S. Nybo. 2012. "The Norwegian Nature Index – Conceptual Framework and Methodology." *Norsk Geografisk Tidsskrift/Norwegian Journal of Geography* 66: 250–256.

Szabo, J. K., S. H. M. Butchart, H. P. Possingham, and S. T. Garnett. 2012. "Adapting Global Biodiversity Indicators to the National Scale: A Red List Index for Australian Birds." *Biological Conservation* 148: 61–68.

Tulloch, A. I. T., H. P. Possingham, L. N. Joseph, J. Szabo, and T. G. Martin. 2013. "Realising the Full Potential of Citizen Science Monitoring Programs." *Biological Conservation* 165: 128–138.

United Nations Environment Programme. 1992. Rio Declaration on Environment and Development. UN Doc. A/CONF.151/26 (vol. I) / 31 ILM 874 (1992). Rio de Janeiro, Brazil.

Vačkářa, D., B. Ten Brink, J. Loh, J. E. M. Baillie, and B. Reyers. 2012. "Review of Multispecies Indices for Monitoring Human Impacts on Biodiversity." *Ecological Indicators* 17: 58–67.

Worm, B., E. B. Barbier, N. Beaumont, J. E. Duffy, C. Folke, B. S. Halpern, J. B. C. Jackson, H. K. Lotze, F. Micheli, S. R. Palumbi, E. Sala, K. A. Selkoe, J. J. Stachowicz, and R. Watson. 2006. "Impacts of Biodiversity Loss on Ocean Ecosystem Services." *Science* 314: 787–790.

USING CONCEPTS OF BIODIVERSITY VALUE IN STRUCTURED DECISION-MAKING

Lynn A. Maguire

Structured decision-making

Structured decision-making augments human intuition by organizing decisions that are difficult because of trade-offs among multiple goals or because of uncertainty about outcomes, or both. Breaking complex decisions into simpler parts makes it is easier for human decision-makers to make choices that are (a) consistent with their underlying beliefs and values, (b) transparent to participants and to wider audiences, and (c) repeatable and correctable, in case of new information becoming available. In this chapter, I am going to focus on problems involving trade-offs among multiple goals; such problems are ubiquitous in conservation decision-making. Some examples include (a) public land management agencies struggling to satisfy the disparate interests of recreationists, conservationists, and extractive users such as mining companies; (b) conservation organizations wondering how to allocate limited resources among many candidate sites or species; and (c) individuals trying to align their housing, transportation, consumer purchasing, and other personal choices with their conservation aspirations.

There are many formats for addressing such problems through structured decision-making (Moffett and Sarkar 2006, Huang *et al.* 2011), but some common elements include (a) assuming that human decision-makers are rational in the sense that they would like to make choices that are consistent with their underlying beliefs and values; (b) recognizing that cognitive and emotional characteristics of human decision-making limit the reliability of unaided intuition in making such choices; and (c) providing procedures and structures that break complex decisions into smaller parts that are easier for humans to deal with consistently and without error. Structured decision-making is referred to as a "prescriptive" approach to analyzing decisions in the sense of recommending to human decision-makers how to make choices that are more likely to accomplish their goals than unaided intuition, rather than a "descriptive" approach in the sense of analyzing how human decision-makers operate intuitively (Slovic *et al.* 1977). Structured analysis is a decision aid, not a decision dictator; human decision-makers retain the authority and responsibility for making a decision.

One type of structured decision-making to help make trade-offs among multiple goals is multi-attribute utility analysis (Keeney and Raiffa 1976). (Others, such as Analytic Hierarchy

Process and outranking procedures, are nicely summarized in Department of Communities and Local Government 2009). A simple example of this type of analysis is a choice among three sites that might be acquired to promote conservation of a threatened species. An ideal site would already support a population of the species in question, would be surrounded by already protected land, and would be inexpensive, but no such site is available. One site already supports a small population of the species in question, but is surrounded by rapidly developing residential areas, and is quite expensive ($1 million). Another site has apparently suitable habitat but no known population of the threatened species, is surrounded by already developed residential areas, and is inexpensive ($500,000). Another site would require restoration to provide suitable habitat, is surrounded by already protected areas, and is intermediate in cost ($800,000). The best choice among the three depends on two components of value: (1) how much improvement to conservation is expected from moving from one level of performance to another for each of the three criteria used to evaluate the sites (e.g. how much relative value might improve by moving from a site requiring restoration to a site with suitable but unoccupied habitat or to a site that is already occupied), and (2) how much priority is given to each of the three criteria in composing the overall value of a site for conservation. The latter captures willingness to trade anticipated improvements in one criterion for decrements in another. Multi-attribute utility analysis provides procedures for articulating these components of value in quantitative form (even for qualitative criteria such as whether or not the site is occupied or needs restoration) and for forming an overall estimate of the comparative merits of the three sites. Estimates of the change in relative value obtained by moving from one level of performance to another and estimates of the priority accorded to different types of value in a multi-attribute problem are context-dependent. In different circumstances a human decision-maker might assign different priorities to cost versus habitat suitability versus presence of a threatened species.

Structured decision-making takes as given that human decision-makers, and what they value, are at the center of the decision process and that different human decision-makers might assign different changes in value to moving from one level of performance on a criterion to another level. For example, a regulatory agency charged with overseeing water quality improvements in a stream might consider improving impaired waters to the regulatory standard for turbidity to be nearly as satisfactory as reducing turbidity to half that specified by the standard, whereas a water quality watchdog group might consider meeting the regulatory standard to be nearly as unsatisfactory as the impaired waters. Similarly, different decision-makers might have different willingness to trade off performance on one criterion compared to others. For example, an industry holding an effluent discharge permit might put a much higher priority on treatment cost relative to water quality improvement than the water quality watchdog group. Differing assessments of components of value are often at the heart of disagreements about what course of action to take. Articulating those assessments clearly using the tools of structured decision-making can help disputing parties understand each other's positions and, sometimes, can help them arrive at a plan for coordinated action (Maguire and Boiney 1994, Gregory *et al.* 2012).

Also central to analyses of decisions with multiple goals is the observation that taking action almost always entails trade-offs among values that cannot all be maximized with any available alternative. Thus action in pursuit of conservation necessarily requires implicit or, less commonly, explicit exchanges among different kinds of value. Sometimes these different kinds of value pit aspects of biodiversity conservation, such as protection of a threatened species, against other types of value, such as financial efficiency or human comfort. Sometimes different kinds of conservation value at least appear to be in competition with each other, as in schemes to protect endangered caribou by culling wolves that prey on them (Hervieux *et al.* 2014).

How does this structured approach to the pursuit of value through conservation action relate to ongoing debates about which concepts of value should inform biodiversity conservation? Are the values used in structured decision-making for conservation intrinsic, instrumental, existential, biocentric, anthropocentric, measurable, infinite, incommensurable, monetary, market-based, or various combinations of these? Given the heated exchanges in the conservation literature and in popular media about sources of conservation value and the right ways to pursue conservation of biodiversity (e.g. Maguire and Justus 2008, Justus *et al.* 2009, Soulé 2013, Marvier 2014, Max 2014, Tallis *et al.* 2014, Silvertown 2015, Vucetich *et al.* 2015), this question has practical, as well as theoretical, importance. I will begin with the latter issue by examining arguments that have been raised in the conservation literature against various concepts of value (and methods of valuation) that have been used in reference to biodiversity.

Criticisms of concepts of biodiversity value used in structured decision-making

One of the tasks of a multi-attribute analysis such as that described above is to put unlike measures of value on a common footing to permit a synthesis of the overall merits of a conservation alternative. One possible common footing is to express all types of concerns in monetary terms, which can then be summed in a cost–benefit or cost-effectiveness analysis. Monetization of value for things not ordinarily bought and sold in markets and not ordinarily assigned prices has been a particular target of criticism (e.g. Chee 2004). A mild form of this criticism is that it is difficult to do a good job of monetizing non-market values (e.g. Hajkowicz 2007). A more thoroughgoing criticism is that monetization is inappropriate, and perhaps morally wrong, because monetary expression demeans the underlying value, implying that it might be alright to give it up in favor of a monetary payment or exchange it for something else deemed to have equivalent or greater monetary value (e.g. Winthrop 2014, Silvertown 2015). Another expression of this criticism, casting back to Kant's early distinction between intrinsic and instrumental value (1785/1959, as quoted in Callicott, this volume), is that monetization cheapens what is priceless, where "priceless" could mean that something is extremely valuable or it could mean that it is not possible, and perhaps not appropriate, to assign a monetary price to it. In these critiques, monetization is not merely a convenient way to express willingness to exchange more of one kind of value for less of another, but a particularly crass way of reducing everything to something that could be bought or sold. In conservation decisions involving indigenous tribes, monetizing kinds of value with traditional significance has been regarded as particularly offensive (Winthrop 2014).

As a way around the particular criticisms that monetization has provoked, Hajkowicz (2007) and others have pointed out that non-monetary units, such as utility, a unitless measure of relative satisfaction used in multi-attribute utility analysis, could be used instead. Implementing these non-monetary expressions of value has its own pitfalls, such as the potential for bias engendered by the way questions used to elicit relative satisfaction are framed (Tversky and Kahneman 1981). These can be mitigated, at least in part, by following well-structured protocols for elicitation (von Winterfeldt and Edwards 1986). In some decision-making processes, the issue of a common unit for different kinds of value might be finessed by making trade-offs through negotiation among disputing parties (e.g. some of the case studies in Gregory *et al.* 2012), where willingness to trade more of one kind of value for less of another might be agreed without quantifying explicitly the relative value of the items being traded off. Such decision processes have some structure, but do not include all the articulations of relative value and relative priority described above.

Using non-monetary units, or avoiding common units altogether, does not quell all criticisms, however. Some such criticisms are directed at exchanges that decrease the amount of something that has unique qualities, such as a particular endangered species, in order to have more of something else, such as improved water quality, which might benefit many species, including some that are endangered, and including humans. But, because improved water quality might be achieved in a variety of ways (e.g. engineered solutions versus "green" infrastructure), the values it provides are not unique in the same way that a particular species is unique. Some of the arguments that have been levied against basing conservation decisions on valuation of the ecosystem services provided by biodiversity raise this objection, worrying that such valuations overlook the unique loss of particular species, even if their ecosystem roles are still fulfilled (e.g. Silvertown 2015).

A more extreme form of this type of criticism has been termed "taboo" trade-offs, or sacred values, where even conceptualizing a decision as one involving trade-offs is considered morally repugnant (Hanselmann and Tanner 2008). Often the sacred values at stake are human lives or human longevity, but sometimes they are unique elements of non-human biodiversity – species, ecological communities, evolutionary processes, or evolutionary phenomena such as migration behavior of monarch butterflies. Winthrop (2014) relates an instance where an American Indian tribe placed a unique value on a traditional food source occurring in a particular place threatened by a pipeline. Tribal members could not accept the same food in a different place or a restoration to the same place of stocks of the same food obtained elsewhere, an example of sacred values and unwillingness to place these values in a trade-off framework: "From the standpoint of *practice*, assessing people's engagement with their environments through the language of trade-offs, which is fundamental to ecosystem services valuation, seems in many situations to be both methodologically and ethically inappropriate" (Winthrop 2014: 209).

Those replying to this form of criticism point out that it is difficult to function in the world at all without making trade-offs, even those involving sacred values. Few people are willing to forego automobile transportation entirely or drive at 5 mph, although those actions would save thousands of human lives and many more life-changing injuries. In the public arena of conservation decisions, the ubiquity of trade-offs seems equally compelling. Even Winthrop's (2014) traditional food source example can be recast in the form of implicit trade-offs, given that the tribe was willing to accept the pipeline if the topsoil from the construction, along with roots of the food plant, were reserved and replaced after the pipeline was installed. In this scheme, it might be acceptable to make trade-offs, but perhaps not to participate in a structured analysis framed explicitly in the language of trade-offs. On the other hand, in northwestern Canada in particular, there have been many successful negotiations of energy and water projects entailing trade-offs of traditional values of First Nation tribes using the type of deliberative structure described in the opening section of this chapter (see examples in Gregory *et al.* 2012).

This review of criticisms that have been leveled against measures of value used to articulate trade-offs has alluded to the sometimes competing, sometimes complementary concepts of value that have been used to describe elements of biodiversity. Now I will turn to an examination of whether and how these concepts of value might be included in, or at least might inform, structured decision-making for conservation.

Which concepts of biodiversity value are most compatible with structured decision-making?

Discussion of concepts of value underlying conservation decisions often rely on the dyadic distinctions of valuing biodiversity anthropocentrically versus biocentrically and of valuing biodiversity instrumentally versus intrinsically, but perhaps these concepts are not so dichotomous and exclusive as they might at first appear. Structured decision-making, whether fully quantified, as in multi-attribute utility analysis, or more qualitative, as in deliberative multicriteria analysis, is an unavoidably human-centered activity. Humans decide what types of value will be considered, how those types of value will be expressed and how different types of value will be weighted in situations involving trade-offs. Nevertheless, some participants in decision processes may attempt to "speak for" non-human interests, such as endangered species; and, some participants may rely on biocentric or ecocentric concepts of value, ascribing value to non-human entities in a manner independent of the usefulness or meaning of those entities to humans.

One often-discussed distinction among concepts of value for biodiversity is instrumental versus intrinsic value (e.g. Callicott 2006). Instrumental value refers to utility to a user, usually, but not always, a human user [Rolston (1994), for example, has argued that non-humans might also value things instrumentally, as a predator might its prey]. With respect to human users, this concept of instrumental value fits easily into structured decision-making, using either monetary or non-monetary representations of the relative value of something (e.g. a fisheries stock) for a particular decision context or relying on deliberative processes to implicitly assign human use-values to elements of biodiversity. This concept of biodiversity value is typically human-centered, variable among human users and perhaps variable for the same human user among different decision contexts.

Consumptive use of some element of biodiversity (e.g. eating fish) is the most obvious kind of instrumental value, but instrumental value has been construed broadly to encompass many kinds of value, including "non-use" values such as aesthetic value, educational value, bequest value (value to future generations), and existence value (e.g. a national park holding various elements of biodiversity has existence value even if no one visits it). Both use-values for things that are not ordinarily exchanged in markets and non-use values can be measured using dollar values through methods such as contingent valuation or using non-monetary units through structured methods of utility elicitation. These measures of value fit easily into structured decision-making, allowing many kinds of biodiversity value to be included in decision processes (e.g. Justus *et al.* 2009, table 1).

Intrinsic value, on the other hand, ascribes value to the thing itself (e.g. a species, an ecosystem, an individual animal or plant) without regard for its use by humans (or non-humans). Concepts of intrinsic value applied to elements of biodiversity are very heterogeneous, making it difficult to see where intrinsic value does and does not fit into a structured decision framework. In traditional Western philosophy, intrinsic value has often been limited to individual human life and well-being (as reviewed in Callicott 2006: 36). Some ethicists extend intrinsic value to individual non-human beings by a variety of criteria (e.g. Regan 2004). A differently grounded concept of intrinsic value from deep ecology accords intrinsic value holistically to the living world or to components such as entire species or ecosystems (e.g. Naess 1989, Callicott 1985, Rolston 1994). Usually, instrumental value and intrinsic value are viewed as distinct components of value, with the former reliant on a user and the latter not. However, Norton (2000) has suggested that instrumental and intrinsic value might better be viewed as a continuum rather than a dichotomy and that separating these concepts cleanly may be neither possible nor helpful in motivating conservation of biodiversity.

There is no barrier in structured decision-making to including different qualities of the same entity as different objectives for management, such as a fish population as a food source and the same fish population as an aspect of biodiversity, with different measurement criteria evaluating how well a particular management scheme addresses each of these aspects of value. It is, however, a feature of structured decision-making to avoid "double-counting" the same kinds of value in multiple criteria (Keeney and Raiffa 1976), so that the value of fish as a food and the value of fish as elements of biodiversity are understood to partition the overall value of the fish population into non-overlapping components. However, this sort of partitioning of value may be offensive to those who believe that the value of biodiversity is a holistic property that cannot, or at least ought not, to be separated into component parts (e.g. Norton 2000).

As far back as Kant's original articulation of intrinsic versus instrumental value (1785/1959, as quoted in Callicott, this volume), philosophers have recognized that the same thing (e.g. a fish species that is harvested for human consumption) might have both instrumental value (as a human food) and intrinsic value (as an element of biodiversity valuable for its own sake). Just as structured decision-making easily incorporates different kinds of instrumental value, including different kinds of instrumental value assigned to the same element of biodiversity, it also can accommodate both instrumental and intrinsic values of biodiversity, provided that all of those kinds of value can be assigned non-overlapping measures of relative merit in at least qualitative terms. This formulation permits analysis of trade-offs among kinds of value. In an attempt to de-escalate the rancor in recent debates about the types of value that should inform conservation action, some participants have pleaded for a pluralistic view, incorporating both instrumental and intrinsic forms of value in conservation decision-making (Hunter *et al.* 2014, Tallis *et al.* 2014).

Two barriers could prevent including multiple concepts of value, including intrinsic value, in structured decision-making for conservation: (1) inability to partition different types of value attributed to particular elements of biodiversity (to avoid double-counting when aggregating measures of overall value across all of its constituent parts), and (2) inability to measure, at least qualitatively, all the types of value relevant to a particular decision. As mentioned above, the first barrier, incorporating multiple kinds of value for the same element of biodiversity, is not a problem for structured decision-making in principle, although it could be difficult to do in practice. However, it could be perceived as inappropriate, or even offensive, by those for whom value is a holistic property and not to be rendered piecemeal.

The second barrier, measuring all types of biodiversity value, could be more problematic. Some concepts of intrinsic value describe it as "immeasurable" (e.g. McCauley 2006). This view seems to take Kant's (1785/1959, as quoted in Callicott, this volume) description of things with intrinsic value as priceless a step further, to having value that cannot be expressed in any metric, monetary, or otherwise. This concept does not fit well into a structured decision framework that depend on expressions of relative value and of willingness to trade off different kinds of value in order to achieve an overall conservation goal. There is a variation of this concept, however, that characterizes intrinsic value as being immeasurable in the sense of inability to quantify relative value, but exchangeable in the sense that some things have more intrinsic value than others and could be traded off for one another. Callicott (2006: 45) argues that "[i]ntrinsic values are objectified through public debate" rather than through quantification in monetary or non-monetary terms. Some provisions of the US Endangered Species Act (e.g. section 6, describing what constitutes "taking" of endangered species) objectify values associated with endangered species relative to human activities, particularly on privately owned land (Callicott 2006). This formulation of intrinsic value, although not quantified, does permit trading off one kind of value against another. Similarly, Callicott (2006) argues that the Endangered Species Act offers some guidance on how to weigh the intrinsic values of endangered and threatened species against utilitarian values such as economic benefit, putting the burden of proof on those advocating utilitarian values. Naess (1989) has argued for priorities among kinds of intrinsic biodiversity value that could guide trade-offs where not all entities having intrinsic value can receive equal protection.

Other authors conceive of intrinsic value quite differently, focusing on independence from any human valuer as an essential characteristic of intrinsic value (e.g. Rolston 1994, Redford and Richter 1999). This concept of value does not fit into the structured decision-making framework because, absent a human intermediary who might be interrogated about the relative values of different items all having intrinsic value, it is hard to see how to express intrinsic values except perhaps as constraints external to an analysis of options in a conservation management decision. For example, one might omit from consideration any options with adverse consequences for any endangered species, or any other element of biodiversity, but these constraints seem so demanding as to be unworkable.

In a similar vein, some authors declare intrinsic value to be not only immeasurable, but infinite (e.g. McCauley 2006). This view also does not fit well into structured decision-making because the well-being of anything with intrinsic value would then trump consideration of any other kinds of value. If there were more than one item of intrinsic value (e.g. more than one species with incompatible requirements), there would be no way to adjudicate how much to benefit one such species compared to another, a problem that afflicts the very controversial use of culling of one species to benefit another (e.g. wolves versus endangered caribou; Hervieux *et al.* 2014). This also seems unworkable in a decision-making context. Colyvan *et al.* (2010) explicitly reject the claim that biodiversity is infinitely valuable, in part on grounds that such a concept of value would preclude setting conservation priorities and would likely produce decisions counter to the underlying conservation goals being pursued.

"Behind-the-scenes" roles for less compatible concepts of value

If concepts of intrinsic value that are independent of human valuers, immeasurable, and possibly infinite, do not fit well into structured decision-making, might they nevertheless perform some behind-the-scenes functions? One possibility is that the measures used in structured decision-making do not capture value, but rather express preferences (Mumford and Callicott 2003). In this formulation, values are underlying motivators of action for individuals and for higher levels of human social organization, such as political or resource management institutions. Preferences are subjective and personal, usually ascribed to individuals, but also, and especially in the case of public decision-making for conservation, ascribed to organizations participating in collaborative decision processes (e.g. resource management agencies, non-governmental conservation organizations). Multi-attribute utility analysis, one implementation of structured decision-making, uses utility functions to express relative satisfaction at different levels of performance on some valued criterion (e.g. extent of an ecological community) and weights to express priorities among valued criteria (e.g. at what rate hectares of that ecological community might be exchanged for increased population size of an at-risk species). These utility functions and weights reflect decision-maker, and perhaps also stakeholder, preferences, which are, in turn, underpinned by their values.

The concept of preference as "constructed" rather than innate (e.g. Fischhoff 1991) supports the idea that values function behind the scenes in structured decision-making. Fischhoff (1991), and many others since, find that people have well-formed concepts of value for only a few deep-seated qualities (e.g. love of family members, loyalty to friends, lack of regard for those seen as unlike oneself) and for a relatively few items encountered regularly in everyday professional or personal life (e.g. goods regularly bought and sold in markets). Otherwise, preferences prove to be quite labile and subject to bias, depending on how the questions in elicitation protocols are framed [e.g. expressing the benefits of a vaccine program in terms of lives saved versus lives lost; Tversky and Kahneman (1981)]. Values might underlie expressions of preference used in structured decision-making by providing motivation and commitment toward actions that benefit valued entities. The intrinsic value of species or other ecological and evolutionary entities might inform preferences in specific decision situations in a manner similar to the way that broad statements of vision or mission set aspirational goals that are then translated into more specific objectives to be pursued in daily activities.

Another role for concepts of value might be as motivators for conservation action (where the course of action might or might not be chosen through some sort of structured decision process). Concepts of the value of biodiversity that do not seem to fit well in a structured decision framework might "bring parties to the table" to discuss conservation alternatives. Immeasurable intrinsic values might be used to move a conservation agenda forward. Soulé (quoting C. P. Snow in Max 2014) referred to blurring the distinction between what is intrinsically, and immeasurably, valuable about biodiversity and what is instrumentally valuable about biodiversity for human use as a "moral escalator" (perhaps the counterpart of a slippery slope) from which it might be hard to get off. Soulé may well recognize both intrinsic and instrumental values of biodiversity, but he, and many others (e.g. Vucetich *et al.* 2015), argue on moral grounds for using the former, rather than the latter, to motivate conservation action.

In addition to moral arguments recommending that only certain kinds of value be used to motivate conservation, there may be strategic reasons to rely on one, or many, concepts of value, including those that do and do not fit easily into structured decision frameworks. In a contentious decision, sticking firmly to the moral high ground of intrinsic and immeasurable value might be the best place to start in order to wind up with the most favorable result for conservation goals. A different strategic use of concepts of value comes from those arguing for a plurality of concepts and measures – intrinsic and instrumental, monetary and non-monetary – as a pragmatic approach to conservation, allowing parties with differing motivations to work together to accomplish conservation goals (Hunter *et al.* 2014, Tallis *et al.* 2014). Using the tools of structured decision-making to incorporate different concepts of the value of biodiversity, including those that appear to fit easily and explicitly into this framework as well as those that do not fit as well and function mainly behind the scenes, can enhance the ability of disputing parties to understand each other's positions and underlying motivations and work together fruitfully.

Summary

Structured decision-making for biodiversity conservation relies on measurement, usually quantitative, of how well different conservation actions advance underlying goals and of willingness of decision participants to trade off progress toward some goals against losses in others. The methods for achieving this measurement of what is valuable are quite flexible and permit use of qualitative as well as quantitative criteria. Concepts of the value of biodiversity that are human-centered, such as the many forms of instrumental value including non-use values (e.g. aesthetic value, existence value, scientific value), and some versions of the concept of intrinsic value that allow quantification or prioritization of some type, fit neatly into structured decision-making. Concepts of intrinsic value that eschew measurement of any kind, or declare intrinsic value to be infinite or independent of any human valuer, do not fit neatly into structured decision-making. Nevertheless, even these concepts of value may inform structured decisions by motivating human preferences and commitments to conservation action.

Acknowledgments

This chapter was inspired by the Decision Making for Complex Environmental Problems Working Group supported by the National Center for Ecological Analysis and Synthesis (NCEAS), a center funded by the NSF (grant DEB-0553768), the University of California–Santa Barbara and the State of California. I thank Baird Callicott, Jack Justus, Michael Nelson, and Sahotra Sarkar for their generous comments; no doubt some disagreements remain.

References

Callicott, J. 1985. "Intrinsic Value, Quantum Theory, and Environmental Ethics." *Environmental Ethics* 7: 257–275.

Callicott, J. 2006. "Explicit and Implicit Values." In *The Endangered Species Act at Thirty: Conserving Biodiversity in Human-Dominated Landscapes*, ed. J. Scott, D. Goble, and F. Davis, 36–48. Washington, DC: Island Press.

Chee, Y. 2004. "An Ecological Perspective on the Valuation of Ecosystem Services." *Biological Conservation* 120: 549–565.

Colyvan, M., J. Justus, and H. Regan. 2010. "The Natural Environment is Valuable but Not Infinitely Valuable." *Conservation Letters* 3: 224–228.

Department of Communities and Local Government. 2009. *Multi-criteria Analysis: A Manual* (ISBN 978-1-4098-1023-0). Available online at www.gov.uk/government/publications/multi-criteria-analysis-manual-for-making-government-policy (accessed 2 February 2016).

Fischhoff, B. 1991. "Value Elicitation: Is There Anything in There?" *American Psychologist* 46: 835–847.

Gregory, R., L. Failing, M. Harstone, G. Long, T. McDaniels, and D. Ohlson. 2012. *Structured Decision Making: A Practical Guide to Environmental Management Choices*. Oxford: Wiley-Blackwell.

Hajkowicz, S. 2007. "Can We Put a Price Tag on Nature? Rethinking Approaches to Environmental Valuation." *Australasian Journal of Environmental Management* 14: 6–10.

Hanselmann, M., and C. Tanner. 2008. "Taboos and Conflicts in Decision Making: Sacred Values, Decision Difficulty, and Emotions." *Judgment and Decision Making* 3: 51–63.

Hervieux, D., M. Hebblewhite, D. Stepnisky, M. Bacon, and S. Boutin. 2014. "Managing Wolves (*Canis lupus*) to Recover Threatened Woodland Caribou (*Rangifer tarandus caribou*) in Alberta." *Canadian Journal of Zoology* 92: 1029–1037.

Huang, I., J. Keisler, and I. Linkov. 2011. "Multi-criteria Decision Analysis in Environmental Sciences: Ten Years of Applications and Trends." *Science of The Total Environment* 409: 3578–3594.

Hunter, M., K. Redford, and D. Lindenmayer. 2014. "The Complementary Niches of Anthropocentric and Biocentric Conservationists." *Conservation Biology* 28: 641–645.

Justus, J., M. Colyvan, H. Regan, and L. Maguire. 2009. "Buying into Conservation: Intrinsic vs. Instrumental Value." *Trends in Ecology and Evolution* 24: 187–191.

Keeney, R., and H. Raiffa. 1976. *Decisions with Multiple Objectives: Preferences and Value Tradeoffs*. New York: John Wiley and Sons.

Maguire, L., and L. Boiney. 1994. "Resolving Environmental Disputes: A Framework Incorporating Decision Analysis and Dispute Resolution Techniques." *Journal of Environmental Management* 42: 31–48.

Maguire, L., and J. Justus. 2008. "Why Intrinsic Value is a Poor Basis for Conservation Decisions." *Bioscience* 58: 910–911.

Marvier, M. 2014. "New Conservation is True Conservation." *Conservation Biology* 28: 1–3.

Max, D. 2014. "Green is Good." *The New Yorker*, May 12, 54–63.

McCauley, D. 2006. "Selling Out on Nature." *Nature* 443: 27–28.

Moffett, A., and S. Sarkar. 2006. "Incorporating Multiple Criteria into the Design of Conservation Networks: A Mini-review." *Diversity and Distributions* 12: 125–137.

Mumford, K., and J. Callicott. 2003. "A Hierarchical Theory of Value Applied to the Great Lakes and Their Fishes." In *Values at Sea: Ethics for the Marine Environment*, ed. D. Dallmeyer, 50–74. Athens: The University of Georgia Press.

Naess, A. 1989. *Ecology, Community, Life Style: An Outline of an Ecosophy*. D. Rothenberg, trans. Cambridge: Cambridge University Press.

Norton, B. 2000. "Biodiversity and Environmental Values: In Search of a Universal Earth Ethic." *Biodiversity and Conservation* 9: 1029–1044.

Redford, K., and B. Richter. 1999. "Conservation of Biodiversity in a World of Use." *Conservation Biology* 13: 1246–1256.

Regan, T. 2004. *The Case for Animal Rights*, 2nd edn. Berkeley and Los Angeles: University of California Press.

Rolston, H. 1994. *Conserving Natural Value*. New York: Columbia University Press.

Silvertown, J. 2015. "Have Ecosystem Services Been Oversold?" *Trends in Ecology and Evolution* 30: 641–648.

Slovic, P., B. Fischhoff, and S. Lichtenstein. 1977. "Behavioral Decision Theory." *Annual Review of Psychology* 28: 1–39.

Soulé, M. 2013. "The 'New' Conservation." *Conservation Biology* 27: 895–897.

Tallis, H., J. Lubchenco and 238 co-authors. 2014. "A Call for Inclusive Conservation." *Nature* 515: 27–28.

Tversky, A., and D. Kahneman. 1981. "The Framing of Decisions and the Psychology of Choice." *Science* 211: 453–458.

Vucetich, J., J. Bruskotter, and M. Nelson. 2015. "Evaluating Whether Nature's Intrinsic Value is an Axiom of or Anathema to Conservation." *Conservation Biology* 29: 321–332.

Winthrop, R. 2014. "The Strange Case of Cultural Services: Limits of the Ecosystem Services Paradigm." *Ecological Economics* 108: 208–214.

von Winterfeldt, D., and W. Edwards. 1986. *Decision Analysis and Behavioral Research*. Cambridge: Cambridge University Press.

18

MEASURING BIODIVERSITY AND MONITORING ECOLOGICAL AND EVOLUTIONARY PROCESSES WITH GENETIC AND GENOMIC TOOLS

Alan R. Templeton

Introduction

Genetics plays a dual role in biodiversity. First, genetic diversity is a type of biodiversity in itself. Indeed, genetic diversity is the most fundamental and basic level of biodiversity because it contributes to all other levels of biodiversity. Without genetic diversity, there is no evolution. Genetic diversity is the raw material for the adaptation of organisms to their environment through the process of natural selection. Adaptations in turn shape ecological communities and ecosystem functioning. Genetic divergence is also necessary for the origin of new species. Hence, species diversity is ultimately derived from genetic diversity. Because genetic diversity is necessary for the evolutionary process, it is not surprising that the maintenance of genetic diversity is often a goal in conservation plans. Moreover, the genetic processes of adaptation and speciation are often needed to protect against extinction when environments are changing. Given the global impact of humans on the environments of the Earth and the rapid pace of human-induced global climate change, it is patent that we must preserve genetic diversity in order to maintain the potential for evolutionary change.

Before this infusion of genetic diversity into conservation biology, conservation plans generally were concerned with preserving species, habitats, etc., as static units. The realization that genetic diversity plays a major role in conservation through a dynamic process (evolution) led to a major philosophical shift in conservation biology. Increasingly conservation plans focus on preserving and restoring the dynamic processes (both genetic and non-genetic) that shape adaptation, speciation, communities, and landscapes.

Second, genetic diversity and its measurement often play a critical role in the design and monitoring of conservation programs even when genetic diversity is not the primary focus of the program. Quite often, measurements of genetic diversity and their spatial and temporal patterns are the only practical ways of inferring critical parameters for conservation management at higher levels of biodiversity.

The importance of genetic measurement and monitoring in conservation has increased dramatically due to developments in the fields of genetics and genomics. The newer techniques of molecular genetics and genomics are more portable than ever across species, allowing them to be applied to virtually any species of conservation interest. Moreover, the newer techniques can perform genetic surveys from extremely small samples, and even indirect samples such as feces, hair, and DNA naturally released by organisms into the environment (environmental DNA, or eDNA). This has revolutionized the study and management of rare and elusive species, and has revealed whole realms of biodiversity that were previously hidden from our view.

Genetic biodiversity

Conservation management for genetic diversity is primarily directed at the species level and below because the genetic diversity that makes adaptation and speciation possible lies within the gene pools of species and their local populations. The most straightforward measurement of genetic diversity in a gene pool is the number of alleles (alternative forms of homologous genes) at a locus. New alleles are the direct product of mutation, the ultimate source of new genetic diversity. Recent theory shows that evolvability, the potential for new mutations to produce adaptive alleles in the future, is influenced by many genetic variables (Wagner 2007, Hu *et al.* 2014, Payne and Wagner 2014, Griswold 2015). Of these, the simplest to measure is the number of alleles existing at a locus, with evolvability increasing as the number of alleles increases (Wagner 2007). Unfortunately, the evolutionary properties of allele number are difficult to study analytically. Although there is a rich theory for how evolution operates within species (Templeton 2006), allele number plays at most only a minor role in this theory. Instead, much of this theory uses different types of heterozygosity (when the two alleles at a diploid gene locus are different) as measures of genetic diversity. Allele number has rarely been used in conservation biology, whereas various heterozygosity measures have been used to establish many "rules" for conservation management. For example, the "one migrant per generation" rule (OMPG) states that local populations can maintain high levels of expected heterozygosity (the probability that two homologous genes drawn at random from the population's gene pool are different alleles) by exchanging one reproductive individual per generation (Wang 2004). This rule is important because human activity frequently causes habitat fragmentation that prevents the movement from one patch of livable habitat to another. Therefore, a major objective in many conservation programs is to either establish dispersal corridors between habitat patches or to alter the landscape matrix separating the habitat patches in a manner that would allow some gene flow (the movement of genes between local populations). Although allele number is not as analytically tractable as expected heterozygosity, the properties of allele number are amenable to study through computer simulations. Recent simulations show that the OMPG rule frequently fails to preserve the number of alleles (Greenbaum *et al.* 2014). Expected heterozygosity measures are mostly determined by common alleles. However, rare alleles are the ones most likely to be lost under habitat fragmentation, and are therefore the most sensitive indicators of the amount of gene flow. These simulations highlight the importance of distinguishing between heterozygosity and allele number as measures of genetic diversity, and show that a general and frequently used rule in conservation biology is highly dependent upon the exact measure of genetic diversity being employed.

Another example showing how different genetic diversity measures can yield different results is provided by the genetic monitoring of a translocated population of eastern collared lizards (*Crotophytus collaris collaris*) in the Missouri Ozarks (Neuwald and Templeton 2013). In the Ozarks, collared lizards live in glades – open rocky habitats characterized by a desert-like microclimate imbedded in a woodland matrix. Glades are fire-maintained habitats, and European settlement of the Ozarks eventually led to forest-fire suppression, particularly after World War II. Fire suppression allowed the red "cedar" (*Juniperus virginiana*) to invade the glades and for the forest matrix to transition from an open woodland to a forest with a thick, woody understory. These changes in the woodland matrix prevent dispersal between glade populations separated by as little as 50 meters of unburned forest (Templeton *et al.* 2011). Fire suppression therefore fragmented collared lizards into ever smaller populations with little to no gene flow. Fire suppression also greatly reduced recolonization potential when local extinction occurred due to habitat degradation. As a result, more than 75 percent of the local glade populations in the Northeastern Ozarks had gone extinct by 1980 (Templeton *et al.* 2011). The Missouri Department of Conservation therefore embarked on a program of glade restoration, mainly through burning former glade habitat, followed by translocation of lizards from existing glade populations onto restored glades. Neuwald and Templeton (2013) genetically monitored a set of populations over twenty-four years that were established from three populations translocated in 1984, 1987, and 1989 (a total of twenty-eight translocated lizards) onto Stegall Mountain in the lower Ozarks. Major management shifts were made over these twenty-four years, and the yearly genetic monitoring of microsatellite variation (a type of hypervariable DNA regions) allowed a direct assessment of different genetic diversity measures as indicators of these management shifts. Allele number was roughly stable throughout these twenty-four years, showing no significant temporal trends. In contrast, the average within-glade expected heterozygosity (upper part of Fig. 18.1) was a sensitive indicator of management shifts. Between 1984 and 1993, management was restricted to just clearing glades and burning of the woodland matrix was forbidden. The ten years between 1984 and 1993 are the pre-burn phase in Fig. 18.1, and during this time there was almost no dispersal of lizards. The expected heterozygosity steadily declined, showing the expected pattern of erosion of genetic diversity under habitat fragmentation and

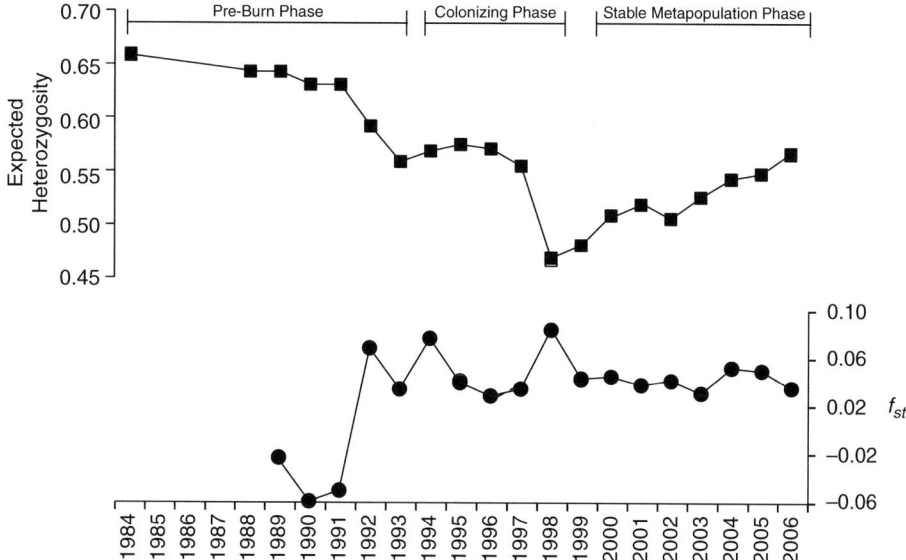

Figure 18.1 The expected heterozygosity within glade populations and the f_{st} among glade populations in collared lizards introduced onto Stegall Mountain from 1984 to 1989 under changing management and demographic regimes.

isolation. Burning of the glades and woodlands started in 1994, with many subsequent burns. The lizards immediately began dispersing through the burned woodland. The expected heterozygosity briefly went up as gene flow among the glades was established, but then decreased during a period of rapid colonization of new glades (Fig. 18.1). This colonization period was characterized by just a handful of lizards usually colonizing the unoccupied glades, and these founder events decreased average within-glade heterozygosity. By the year 2000, with continuing prescribed burns, the lizard population on Stegall Mountain reached an equilibrium size of 370 lizards with forty-four to fifty-five glades being occupied per year with some local extinction followed by recolonization. Since this demographically stable phase began, the expected heterozygosity within glades has steadily risen due to ongoing gene flow (Fig. 18.1).

The failure of allele number to measure management shifts in the collared lizards was probably due to the nature of the founder populations. Earlier genetic surveys had revealed that the collared lizards in the Northeastern Ozarks were an anomaly relative to the remainder of the species using another measure of genetic diversity commonly symbolized by f_{st} (Hutchison and Templeton 1999). f_{st} contrasts the expected heterozygosity found in the total population to the average expected heterozygosity within local populations. This measure is standardized such that $f_{st} = 0$ means that all the species' expected heterozygosity is found within each local population and with no genetic differences between local populations, and $f_{st} = 1$ means that there is no heterozygosity within local populations and the species' genetic diversity is found exclusively as genetic differences between local populations. The genetic surveys of the surviving Ozark populations revealed extreme genetic differentiation, often approaching an f_{st} close to 1 for even nearby glade populations (Hutchison and Templeton 1999). There was still much genetic variation in the Ozark lizards, but mainly as fixed genetic differences between glade populations with very little heterozygosity within glade populations. Therefore, it was decided to take the source lizards for translocation from many different glades, thereby establishing a mixed founder population with high levels of expected heterozygosity (the initial values in Fig. 18.1). Moreover, because few of the surviving natural glade populations were healthy, the same source glades were used repeatedly in all three translocated founder populations, making the initial founder populations extremely similar genetically, as shown by the initial low f_{st} value of the three translocated populations in the lower part of Fig. 18.1. All of these conditions meant that all alleles would have a frequency of about 10 percent or more in these founder populations, a frequency that is not considered rare in population genetics. Without several truly rare alleles, allele number would not be expected to be, nor was, a sensitive indicator of management shifts in this case. The contrast of Fig. 18.1 with the simulation results discussed earlier indicate that no one measure of genetic diversity is optimal in all situations. Attention must be directed to the specific context being monitored and the goals of the specific program. All too often, heterozygosity measures are used regardless of context and goals simply because they have been the standard in most evolutionary theory.

The decision to mix lizards from many source populations was controversial. The local populations of many species are distributed over heterogeneous environments in terms of climatic variables or other ecologically significant variation. Under these circumstances, there is the possibility that genetically differentiated local populations are also adapted to their local environments. Mixing individuals from these populations could break down these local adaptations and thereby have deleterious consequences for population viability – a phenomenon known as "outbreeding depression" (Templeton 1986). The high f_{st} values indicated much genetic differentiation among Ozark glade populations, and this could be an indicator of local adaptation. However, the pattern of genetic variation showed no correlation with geographic position or glade type (based on the underlying type of bedrock) within the Ozarks, and instead was consistent with a history of sudden fragmentation of a widespread, genetically uniform population followed by extreme loss of genetic variation within local populations and extreme differentiation between due to the evolutionary force of genetic drift (Hutchison and Templeton 1999). Genetic drift refers to the random changes in a population's gene pool that are induced by sampling a finite number of gametes to produce the next generation. All populations are subject to genetic drift, but small populations are particularly susceptible. Such small populations are often the focus of conservation efforts, so the impact of genetic drift is of great concern in conservation biology. Genetic drift causes a loss of genetic variation in isolated local populations (lower allele numbers and heterozygosity) and genetic differentiation among isolates (increases in f_{st}). Indeed, such a rapid increase in f_{st} is shown in Fig. 18.1 during the pre-burn phase, followed by a stabilization of f_{st} to a low value after gene flow was established due to prescribed burns.

Because the analysis of Hutchison and Templeton (1999) indicated that the high f_{st}'s found in the natural Ozark glade populations could be explained by demographic history, the mixed translocation policy was implemented. However, local adaptation was still a possibility, so monitoring of these three and many other translocated populations throughout the Ozarks was continued after the initial translocations. Most translocated populations thrived, and the ten-year success rate was 60 percent, compared to 20 percent for reptile and amphibian translocations in general (Templeton *et al.* 2007). Hence, outbreeding depression, if it occurred at all, was at most minor. Note that by monitoring, this conservation program was also a scientific experiment. Often in conservation we need to act with incomplete knowledge, and by monitoring we can learn from both our successes and our failures.

Genomics and biodiversity

With the advent of genomics, the inference of local adaptation can be made with greater rigor (Funk *et al.* 2012). Tens of thousands to millions of genetic polymorphisms can be surveyed, usually at the level of single nucleotide polymorphisms (SNPs), to reveal patterns of genetic differentiation over the entire genome. f_{st} can be measured on all of these SNPs, and typically one finds that the vast majority of SNPs share very similar f_{st} values. This background f_{st} is regarded as being the product of neutral variation responding to the species' balance of gene flow versus genetic drift. Often, a few markers show high outlier values of f_{st}, and these outliers are potential markers of local adaptation. The inference of local adaptation is strengthened when these outliers can be related to some important environmental variable. For example, the threespine stickleback (*Gasterosteus aculeatus*) is one of the few fish species that has populations inhabiting both freshwater and ocean habitats. A survey of 12,648 SNPs from three lake and two ocean populations revealed a pattern in which pairwise f_{st} values (that is, f_{st} calculated just between two populations) for 12,016 non-outlier loci grouped one of the lake populations with the two ocean populations, but the 632 outlier loci grouped the two ocean populations together and the three lake populations together (Funk *et al.* 2012). These observations imply that there are local adaptations to fresh- and to saltwaters.

As indicated by the collared lizard example, genetic drift is an important force acting upon genetic diversity. One commonly used, and abused, rule in managing genetic drift is the "50/500" rule (Franklin 1980) that reveals a different set of measurement problems. Genetic drift leads not only to the loss of alleles and heterozygosity from the gene pool, but also to inbreeding (when both parents of an individual share one or more common ancestors). Inbreeding can have deleterious consequences for survival and reproduction known as inbreeding depression. The "50" in this rule states that to avoid a deleterious rate of accumulation of inbreeding, a population should be started with no fewer than fifty individuals. The theoretical rate of accumulation of inbreeding per generation due to genetic drift is $1/(2N)$, so $N=50$ keeps this rate down to 1 percent per generation. Franklin (1980) justified this 1 percent rate by referring to unreferenced and unnamed "animal breeders" who felt that a 1 percent rate was tolerable in domestic animals. The "500" in this rule states that the population should be quickly grown in size to no fewer than 500 individuals to avoid a loss of genetic diversity. This number is based by Franklin (1980) on a single experiment on the rate of production of new variation by mutation that is needed to balance the effects of genetic drift for the trait of sternal–pleural bristle number in homozygous lines of the fruit fly *Drosophila melanogaster* (Lande 1976).

There has been much controversy about these numbers (for a recent example, see Frankham *et al.* 2014, Franklin *et al.* 2014), in part because the empirical basis for the 500 number is drawn from a single experiment on a single trait on laboratory strains of a single species, and the basis for the 50 number is completely anecdotal. In reality, the response to inbreeding is highly variable across species (Ralls *et al.* 1988, Templeton 2006), can evolve very rapidly (Templeton and Read 1983, 1984), and is actually adaptive in many species, including such extreme inbreeding as self-mating (Holsinger 1991). Thus, the 50/500 rule implicitly denies the importance of biodiversity by proclaiming one rule for all species. There are many examples of populations that seriously violate the 50/500 rule and that are highly successful (Templeton 1994). It is obviously far better to treat each population according to its individual characteristics and conservation context. However, in many cases the information that takes into account the individual circumstances of a particular population is not available and is not likely to be obtained. In such cases, a general rule provides at least some guidance.

Another serious problem with the 50/500 rule is how to measure the two numbers. Many managers treat the 50 and 500 as the census number of animals or the number of breeding individuals. However, both of these numbers are effective population sizes (Franklin 1980). In deriving the mathematical theory for the evolutionary impact of genetic drift, it was convenient to deal with an idealized population that is constant in size, has no subdivision ($f_{st} = 0$), has random mating among self-compatible hermaphrodites (that is, an individual is as likely to mate with itself as with any other individual), and the number of offspring from a mating is distributed as a Poisson distribution (Templeton 2006). Very few, if any, real populations satisfy these idealized conditions. Therefore, Wright (1931) came up with the idea of an effective population size in which the impact of genetic drift upon some genetic parameter of interest in the real population behaves as if it were a particular size in an idealized population. Effective sizes can differ dramatically from census sizes, so measuring the two sizes of the 50/500 rule is not straightforward. To make matters worse, different genetic parameters define different effective sizes. The two genetic parameters of interest here are the amount of inbreeding (the "50" part of the 50/500 rule) and the amount of genetic variation in the gene pool (the "500" part of the 50/500 rule). In particular, the "50" refers to an inbreeding effective size, and the "500" to a variance effective size (Franklin 1980). The measurement of these sizes requires much demographic information and/or genetic survey results and is not a simple census.

Table 18.1 illustrates these difficulties by giving the census, inbreeding effective, and variance effective sizes for three African rhinoceros populations (Braude and Templeton 2009). As can be seen, these three size measurements can be drastically different. Table 18.1 also undermines another widespread myth that effective sizes are always much smaller than census size. This is indeed the case for the southern white rhinoceros, but the opposite is true for the black and the northern white rhinoceros (Table 18.1). There is no mystery here because many factors

Table 18.1 Wild African rhinoceros census and effective population sizes

	Census size, 1997	*Inbreeding effective size*	*Variance effective size*
Black rhinoceros *Diceros bicornis*	2,600	18,840	4,189
Southern white rhinoceros *Ceratotherium simum simum*	8,440	106	240
Northern white rhinoceros *Ceratotherium simum cottoni*	23	69	41

common for endangered species can inflate an effective size well beyond the census size. For example, the inbreeding effective size is a function of the number of recent ancestors and not the current population size (Crow and Kimura 1970). In a declining population, a common situation with endangered species, the number of ancestors can be larger than the current census size, resulting in an inbreeding effective size that is larger than the current number of breeding individuals. Another example is that the rate of loss of genetic variability is reduced by population subdivision (Wright 1943, Chesser *et al.* 1993). For a highly subdivided or fragmented species – another common situation in endangered species – the variance effective size (which depends on the rate of loss of genetic variability) can be larger than the census size. Indeed, Chesser *et al.* (1980) recommend that population subdivision be considered as a management strategy to preserve genetic variation because it can greatly inflate the variance effective size. Unfortunately, the calculation of both inbreeding and variance effective sizes is almost never done in the conservation literature, so there is no way of knowing how often the 50/500 rule is truly implemented even in those management plans that claim to use it.

Genetic surveys can be used to measure genetic drift directly rather than through the much misunderstood effective size. The primary effect of genetic drift is to induce variation in allele frequencies over time, typically measured by the variance of allele frequency across sampling periods. The trouble with measuring drift through the variance of allele frequency is that the variance is theoretically a function of the allele frequency itself as well the number of gametes sampled, and this greatly complicates the statistics. Neuwald and Templeton (2013) eliminated this statistical difficulty by studying the arcsin, square root of the allele frequency. This mathematical transformation makes the variance over a unit of time, $(\Delta a)^2$, only a function of the number of gametes sampled and not allele frequency. This is therefore a direct measure of the force of genetic drift that avoids the complexities and potential for misunderstanding of using an effective size. Fig. 18.2 shows a plot of the yearly transformed variances for the collared lizards on Stegall Mountain. When there was only a single, small glade population (the year 1986), Fig. 18.2 reveals that genetic drift was very strong. With the addition of each new translocated population, genetic drift for the total population diminishes greatly. This drop clearly shows the impact of subdivided population structure in reducing the impact of genetic drift on the variance of allele frequencies (and hence loss from the total population). The force of genetic drift increased in the initial periods of the colonizing phase as population subdivision was reduced, but soon began to diminish to a very small level as the population increased and progressed to the stable metapopulation phase. Fig. 18.2 therefore provides a direct profile of the power of genetic drift throughout this conservation program, showing that the program was successful in ultimately minimizing its impact.

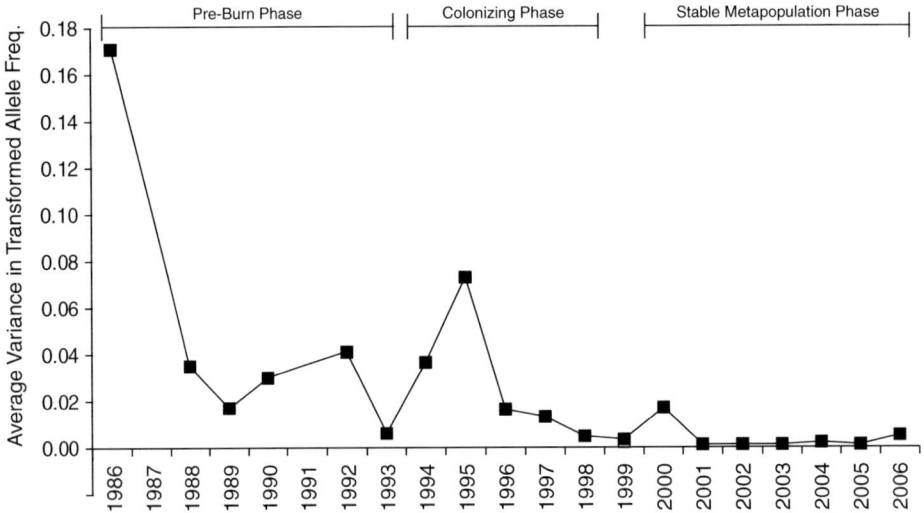

Figure 18.2 The strength of genetic drift as measured by the variance of the arcsin, square root transformed allele frequencies in the total population of collared lizards introduced onto Stegall Mountain from 1984 to 1989 under changing management and demographic regimes.

Genetics and genomics as a general biodiversity tool

Many conservation programs focus on species, habitat types, communities, etc., rather than genetic diversity. Even when genetic diversity is a goal of a conservation program, this goal cannot be realized unless there are viable populations in appropriate habitats, communities, and ecosystems. Genetics and genomics are an increasingly important tool for designing and managing biodiversity at these higher levels. For example, the genetic monitoring of the collared lizards clearly revealed that the suppression of woodland fires was the major cause of habitat fragmentation in this species and that it could be reversed by prescribed burns. This insight into management policy was necessary to have viable lizard populations (and other species as well) in the Ozarks. Genetics and genomics are now commonly used to infer optimal management strategies or policies in many species (e.g. Chakraborty *et al.* 2014, Feist *et al.* 2014, McMahon *et al.* 2014).

One common use of genetics and genomics is to identify or prioritize the population units in need of protection. The three major types of intraspecific populations in conservation biology are management units (MUs), evolutionarily significant units (ESUs), and subspecies. MUs are isolated populations that receive no or very little gene flow from other populations in the species (Funk *et al.* 2012). In the past, MUs were often identified as geographical isolates. However, geography is not always the best criterion for inferring MUs. For example, two species of moray eels, *Gymnothorax undulatus* and *G. flavimarginatus*, are distributed throughout the Indo-Pacific Ocean but are confined to reefs as reproductive adults (Reece *et al.* 2010). Different clusters of reefs are often separated by thousands of kilometers from the nearest outside reef cluster. Reece *et al.* (2010) surveyed both nuclear and mitochondrial DNA at the sequence level in these two species with samples throughout the entire Indo-Pacific. The null hypothesis of no genetic differentiation was tested with AMOVA (Analysis of MOlecular VAriation). In the traditional f_{st} analysis, one calculates the probability that two genes, randomly drawn from the appropriate populations, are different. However, with sequence data, one can quantify how different two genes are. For example, two genes might differ at a single nucleotide out of 500 nucleotides sequenced, whereas another pair of genes might differ at 10. Both pairs would be scored as heterozygous in an f_{st} analysis, but obviously the latter pair is more different than the first pair. This additional information can be incorporated into a statistic that quantifies how different two genes are, either by simply counting nucleotide differences or by using a model of mutation to estimate a genetic distance between the two pieces of sequenced DNA. In this case, the AMOVA indicated no significant genetic differentiation. Reece *et al.* (2010) also performed a statistical test to see if geographical distance resulted in any genetic isolation, and found no significant isolation by distance across the entire range of 22,000 km. Hence, for these two eel species, the entire Indo-Pacific (about two-thirds of the world) defines a single management unit. The reason for this may be that moray eels have the longest pelagic larval stages among reef fishes, which provides opportunities to disperse over great distances. This example shows the danger of defining MUs just by geography. Whenever possible, MUs should be defined by genetic surveys to ensure that they are indeed genetically differentiated from other populations (Funk *et al.* 2012).

ESUs and subspecies play a more important role in conservation because they have legal protection in many countries (Funk *et al.* 2012). Similar to MUs, both ESUs and subspecies are regarded as geographically circumscribed populations with sharp boundaries within a species that have strongly reduced gene flow. In addition, ESUs are also ecologically distinctive (Crandall *et al.* 2000), and traditionally subspecies are morphologically distinctive (Mayr 1970). The same types of genetic surveys and analyses used to define MUs are equally applicable as a first step in defining ESUs and subspecies. The sharp geographical boundaries of ESUs and subspecies are also critical here. Isolation by distance, mentioned above in the eel example, is a common phenomenon that yields genetically differentiated populations but without sharp geographical boundaries. However, if one samples populations located far apart but without sampling geographically intermediate populations, it is impossible to distinguish between a sharp geographical boundary and a gradual cline (Templeton 2013). For example, geographical clustering of human samples results in the appearance of sharply differentiated populations, whereas geographically uniform sampling of humans reveals no sharp genetic boundaries (Templeton 2013). Consequently, genetic differentiation among populations is necessary but not sufficient to define ESUs or subspecies.

Given a population with sharp genetic boundaries, the inference of an ESU also requires evidence of ecological distinctiveness. Often, this inference is made on the basis of ecological variables, such as habitat differences. For example, genetic surveys on African elephants reveal two major geographically circumscribed populations with sharp genetic boundaries despite contacting ranges. These two types of elephants live in two distinct habitat types – open savannas and closed forests (Roca *et al.* 2001). Given these highly distinct habitats and the fact that these two populations display morphological differences that appear adaptive (e.g. smaller size in the forest elephants), these two African elephant populations are minimally ESUs (more about their taxonomic status later). Increasingly, genome surveys are being used in addition to habitat differences for detecting local adaptation. For example, the stickleback fish survey discussed earlier indicated that the freshwater and ocean populations are adaptively differentiated, and hence are distinct ESUs (Funk *et al.* 2012). The ESUs in turn could contain multiple MUs. The same genetic survey that revealed adaptive outlier loci for freshwater versus ocean populations of sticklebacks also revealed by using the non-outlier genes that the three lake populations sampled are genetically distinct, and these lakes are geographically disjoint. Hence, each of the three lake populations would be MUs within the freshwater ESU of sticklebacks.

Subspecies were traditionally defined as morphologically distinct, geographically circumscribed populations with sharp boundaries. However, many taxonomists and evolutionary biologists have long questioned the biological significance and utility of the subspecies concept on the basis of morphological criteria alone (Mayr 1970). Genetic surveys have therefore become critical in modern inferences of subspecies. Given a geographically circumscribed population with sharp genetic boundaries, one criterion for elevating such a population to subspecies status is to demand that its level of genetic differentiation from the remainder of the species exceed some high threshold, such as an $f_{st} \geq 0.25$ or 0.30 (Smith *et al.* 1997). For example, the common chimpanzee, *Pan troglodytes*, has been split into five subspecies on geographical/morphological criteria under traditional taxonomy. However, a microsatellite genetic survey using an f_{st}-like statistic that includes a mutational model for microsatellites reveals that two of the traditional subspecies satisfy the $f_{st} \geq 0.25$ criterion. The other three, all found in tropical central Africa, show only modest genetic differentiation among themselves (0.03 to 0.07), but do have $f_{st} \geq 0.25$ with respect to the other two subspecies (Gonder *et al.* 2011). Hence, by the f_{st}-threshold definition, there are three, not five, subspecies of common chimpanzees (Templeton 2013). The trouble with this definition is that it depends on an arbitrary threshold that has no special biological meaning. For example, if the criterion $f_{st} \geq 0.30$ had been used instead, there would only be two subspecies of chimpanzees, not three. Another example is the Ozark collared lizards, for which many local glade populations easily satisfy either f_{st} threshold resulting in hundreds of "subspecies"; yet, this extreme genetic differentiation is a recent artifact of human suppression of forest fires in this region after World War II and is being reversed by proscribed burning in the Ozarks.

A more modern definition of subspecies embraced by many evolutionary biologists is to elevate any geographical population with sharp boundaries to subspecies status if it also represents a significant evolutionary lineage within the species; that is, the population has an evolutionary history of long-term restricted gene flow from the remainder of the species and traces back primarily to a single ancestral population (Templeton 2013). Statistical tests for inferring lineages from molecular genetic data are available, and these tests show that there are three evolutionary lineages within the common chimpanzee, which correspond to the three subspecies identified by the $f_{st} \geq 0.25$ criterion (Templeton 2013). Similarly, the forest and savanna African elephant populations also represent significant evolutionary lineages despite low levels of past and ongoing gene flow (Templeton 2009). Hence, the forest and savanna African elephant populations are not only ESUs but are also minimally subspecies.

Subspecies are not always identical to ESUs. For example, the habitat requirements of two of the common chimpanzee subspecies discussed above revealed three distinct ecological populations, with one of the subspecies subdivided into two different and geographically separate populations or ESUs (Clee *et al.* 2015). In contrast, the non-outlier genes in the stickleback study clearly show that the lake ESU populations do not define an evolutionary lineage coming primarily from a single ancestral population, and therefore the lake ESU is not a subspecies under the lineage definition and indeed encompasses multiple lineages.

The primary focus of many conservation programs is upon species. Despite the central role that species play in evolutionary and conservation biology, there is no consensus on a species definition (Templeton 1989). For example, Roca *et al.* (2001, 2007) concluded that the forest and savanna elephants were species and not just subspecies. Using the same data, Debruyne (2005) concluded that they were not different species. The problem here is the species concept being used (implicitly in this case by both sets of authors). If one accepts the biological species concept that species are reproductively isolated from one another, then forest and savanna elephants are not different species since they can and do hybridize under natural conditions, the hybrids are viable and fertile, and gene flow does occur. If one accepts the cohesion species concept that species are significantly different evolutionary lineages that display either significant ecological differentiation *and/or* significant reproductive isolation (with "significance" determined through statistical testing, and therefore not necessarily complete isolation) (Templeton 2001), then forest and savanna elephants are different species. Torstrom *et al.* (2014) reviewed the reptile literature, and found that genetic analyses are causing major changes in reptile species status, especially the elevation of subspecies to species, and that nine different species concepts have been used to justify changes in taxonomic status. Given the centrality of species in many conservation programs, this controversy and lack of consensus on what is or is not a species is producing major problems. Torstrom *et al.* (2014) argue for a standard method to determine the species–subspecies boundary, but no standard is currently agreed upon.

Even when there is a consensus on which groups are species, genetics is often used in species identification. Sibling species are morphologically nearly identical populations, even when there is total reproductive isolation between them. Sibling species are common, and often their identification was only possible by highly trained taxonomic specialists. As the priorities of science have changed, the pool of these taxonomic specialists has declined, making the identification of sibling species on morphological grounds more and more difficult. Fortunately, sibling species are often genetically distinct. DNA barcoding takes advantage of this fact by sequencing a specific DNA region that evolves rapidly in the relevant higher-order taxa and uses the sequence as a "barcode" to identify the species of an unknown specimen, sometimes discovering a new cryptic species (Kekkonen and Hebert 2014, Weitschek *et al.* 2014). DNA barcoding occasionally yields results that are discrepant with classical taxonomy, and has therefore been controversial (Waugh 2007). Improvements on the molecular genetics and data analysis of DNA barcoding are greatly improving the results, so barcoding is now a major tool in conservation biology.

Non-intrusive genetic and genomic sampling

Many of the modern genomic techniques can sequence extremely small quantities of DNA, even highly degraded DNA. This ability is revolutionizing the study of biodiversity. As shown in the previous section, genetics has many applications to conservation biology. However, often the main problem is obtaining the samples. Many populations and species are elusive, occur at low densities, live in habitats that make sampling difficult, or cannot be handled for sampling without endangering their well-being. Now it is possible to obtain DNA samples from feces, hair, antlers, museum and herbarium specimens, etc., and often in a manner that does not intrude in any way upon the individuals being sampled. Such indirect, non-intrusive sampling has exploded in conservation biology, and can be used for genetic monitoring, estimating population size, inferring social structure and behavior, kinship patterns, dispersal and gene flow patterns, habitat fragmentation, and other important parameters for conservation design and management (e.g. Eggert *et al.* 2014, Gray *et al.* 2014, Gueta *et al.* 2014).

DNA is sloughed off into the environment (eDNA) by many organisms from bacteria to multicellular organisms. Modern technologies often allow the monitoring of eDNA with great sensitivity, low cost, non-invasive sampling, and in environments that have been difficult to sample specimens directly (Kelly *et al.* 2014). In some cases, eDNA monitoring can be directed at a single species, such as the monitoring of the presence or absence of an invasive species (Rees *et al.* 2014). However, eDNA monitoring can often screen multiple, even thousands, of species from a single sample, and this improves our ability to explore species community diversity and ecosystem-level processes (Bohmann *et al.* 2014). eDNA monitoring has revealed whole communities and ecosystems that we were largely ignorant of before such monitoring was possible. For example, there has been an explosion of knowledge about microbiomes, where eDNA monitoring has revealed the existence of many previously unknown bacterial species and how they define communities and ecosystems that interact with individual multicellular organisms, such as humans (Costello *et al.* 2009, Yatsunenko *et al.* 2012, Le Chatelier *et al.* 2013) and even entire macroscopic communities, such as the native tallgrass prairie in North America (Fierer *et al.* 2013).

In many cases setting restoration goals for a conservation program for the present can be aided by knowledge of the past. These new, highly sensitive DNA protocols can also be applied to surveying ancient DNA from fossils or sediments, thereby providing direct genetic information about the past. For example, Boessenkool *et al.* (2014) surveyed ancient DNA found in sediment cores from two high-altitude volcanic craters in tropical Eastern Africa as part of a study on afro-alpine floral communities. They also examined pollen records found in the same cores. Together, these two sources allowed a detailed reconstruction of past afro-alpine communities that extended into the past well beyond the onset of substantial human-induced effects, with the sedimentary DNA reflecting local vegetation and the pollen the plants from a wider area that spanned several microclimatic zones. Similarly, Wilmshurst *et al.* (2014) surveyed ancient DNA and pollen samples from soil cores taken from some of the mammal-free offshore islands of New Zealand to reconstruct a 2,000-year record of vegetation. These islands were settled by humans around 800 years ago, but then were abandoned for 180 years. The current forest is an angiosperm-dominated forest of native species that was thought to represent the climax forest. However, the ancient DNA record showed that prior to human settlement the forest was podocarp-dominated – information that is critical if restoration replanting is desired.

Overview

Genetics played little role in conservation until the latter half of the twentieth century. Since then, the role of genetics and genomics has increased dramatically. First, there has been the realization that genetic diversity is the form of biodiversity that provides the raw material for evolution and thereby influences all other types of biodiversity. This resulted in a major philosophical shift in many conservation programs because it introduced evolutionary thinking and the idea that conservation is not just about the maintenance of the status quo, but rather the maintenance of the potential for dynamic change. Conservation now is less concerned about preserving what remains as a living museum and more concerned about preserving processes that dynamically shape populations, species, communities, landscapes, and ecosystems. This shift towards preserving and restoring dynamic processes is more important than ever as climate change is occurring at an increasing pace and the future global environment is uncertain.

Second, genetics and genomics serve as assessment and monitoring tools of ever increasing sensitivity and practical applicability from the population level up through the ecosystem level. This practical aspect of genetics in conservation strongly interacts with the evolutionary thinking that genetics has brought into conservation biology. Major philosophical issues within evolutionary biology, such as "What is a species?," have become major practical problems in implementing conservation plans. Genetic methodologies have helped solve many problems in conservation biology, but have also raised many questions. As genetic and genomic tools become even more refined, our ability to answer some of these questions may be augmented, but certainly additional questions will be raised.

Acknowledgments

I thank Anya Plutynski and Sahotra Sarkar for their excellent suggestions on an earlier draft of this chapter.

References

Boessenkool, S., G. McGlynn, L. S. Epp, D. Taylor, M. Pimentel, A. Gizaw, S. Nemomissa, C. Brochmann, and M. Popp. 2014. "Use of Ancient Sedimentary DNA as a Novel Conservation Tool for High-Altitude Tropical Biodiversity." *Conservation Biology* 28: 446–455.

Bohmann, K., A. Evans, M. T. P. Gilbert, G. R. Carvalho, S. Creer, M. Knapp, D. W. Yu, and M. De Bruyn. 2014. "Environmental DNA for Wildlife Biology and Biodiversity Monitoring." *Trends in Ecology & Evolution* 29: 358–367.

Braude, S., and A. R. Templeton. 2009. "Understanding the Multiple Meanings of 'Inbreeding' and 'Effective Size' for Genetic Management of African Rhinoceros Populations." *African Journal of Ecology* 47: 546–555.

Chakraborty, S., D. Boominathan, A. Desai, and T. N. C. Vidya. 2014. "Using Genetic Analysis to Estimate Population Size, Sex Ratio, and Social Organization in an Asian Elephant Population in Conflict with Humans in Alur, Southern India." *Conservation Genetics* 15: 897–907.

Chesser, R. K., O. E. Rhodes, D. W. Sugg, and A. Schnabel. 1993. "Effective Sizes for Subdivided Populations." *Genetics* 135: 1221–1232.

Chesser, R. K., M. H. Smith, and Brisbin, J. 1980. "Management and Maintenance of Genetic Variability in Endangered Species." *International Zoo Yearbook* 20: 146–154.

Clee, P. R. S., E. E. Abwe, R. D. Ambahe, N. M. Anthony, R. Fotso, S. Locatelli, F. Maisels, M. W. Mitchell, B. J. Morgan, A. A. Pokempner, and M. K. Gonder. 2015. "Chimpanzee Population Structure in Cameroon and Nigeria is Associated with Habitat Variation that May Be Lost Under Climate Change." *BMC Evolutionary Biology* 15.

Costello, E. K., C. L. Lauber, M. Hamady, N. Fierer, J. I. Gordon, and R. Knight. 2009. "Bacterial Community Variation in Human Body Habitats Across Space and Time." *Science* 326: 1694–1697.

Crandall, K. A., O. R. P. Binida-Emonds, G. M. Mace, and R. K. Wayne. 2000. "Considering Evolutionary Processes in Conservation Biology." *Trends In Evolution and Ecology* 15: 290–295.

Crow, J. F., and M. Kimura. 1970. *An Introduction to Population Genetic Theory.* New York: Harper & Row.

Debruyne, R. 2005. "A Case Study of Apparent Conflict between Molecular Phylogenies: The Interrelationships of African Elephants." *Cladistics* 21: 31–50.

Eggert, L. S., R. Buij, M. E. Lee, P. Campbell, F. Dallmeier, R. C. Fleischer, A. Alonso, and J. E. Maldonado. 2014. "Using Genetic Profiles of African Forest Elephants to Infer Population Structure, Movements, and Habitat Use in a Conservation and Development Landscape in Gabon." *Conservation Biology* 28: 107–118.

Feist, S., J. Briggler, J. Koppelman, and L. Eggert. 2014. "Within-river Gene Flow in the Hellbender (*Cryptobranchus alleganiensis*) and Implications for Restorative Release." *Conservation Genetics* 15: 953–966.

Fierer, N., J. Ladau, J. C. Clemente, J. W. Leff, S. M. Owens, K. S. Pollard, R. Knight, J. A. Gilbert, and R. L. McCulley. 2013. "Reconstructing the Microbial Diversity and Function of Pre-Agricultural Tallgrass Prairie Soils in the United States." *Science* 342: 621–624.

Frankham, R., C. J. A. Bradshaw, and B. W. Brook. 2014. "50/500 Rules Need Upward Revision to 100/1000: Response to Franklin *et al.*" *Biological Conservation* 176: 286.

Franklin, I. R. 1980. "Evolutionary Change in Small Populations." In *Conservation Biology*, ed. M. E. Soulé and B. A. Wilcox. Sunderland, MA: Sinauer Associates.

Franklin, I. R., F. W. Allendorf, and I. G. Jamieson. 2014. "The 50/500 Rule is Still Valid – Reply to Frankham *et al.*" *Biological Conservation* 176: 284–285.

Funk, W. C., J. K. McKay, P. A. Hohenlohe, and F. W. Allendorf. 2012. "Harnessing Genomics for Delineating Conservation Units." *Trends in Ecology & Evolution* 27: 489–496.

Gonder, M. K., S. Locatelli, L. Ghobrial, M. W. Mitchell, J. T. Kujawski, F. J. Lankester, C.-B. Stewart, and S. A. Tishkoff. 2011. "Evidence from Cameroon Reveals Differences in the Genetic Structure and Histories of Chimpanzee Populations." *Proceedings of the National Academy of Sciences* 108: 4766–4771.

Gray, T. E., T. N. C. Vidya, S. Potdar, D. K. Bharti, and P. Sovanna. 2014. "Population Size Estimation of an Asian Elephant Population in Eastern Cambodia through Non-invasive Mark–Recapture Sampling." *Conservation Genetics* 15: 803–810.

Greenbaum, G., A. R. Templeton, Y. Zarmi, and S. Bar-David. 2014. "Allelic Richness Following Population Founding Events – A Stochastic Modeling Framework Incorporating Gene Flow and Genetic Drift." *PLoS ONE* 9: e115203.

Griswold, C. K. 2015. "Additive Genetic Variation and Evolvability of a Multivariate Trait Can Be Increased by Epistatic Gene Action." *Journal of Theoretical Biology* 387: 241–257.

Gueta, T., A. R. Templeton, and S. Bar-David. 2014. "Development of Genetic Structure in a Heterogeneous Landscape Over a Short Time Frame: The Reintroduced Asiatic Wild Ass." *Conservation Genetics* 15: 1231–1242.

Holsinger, K. E. 1991. "Inbreeding Depression and the Evolution of Plant Mating Systems." *Trends in Evolution and Ecology* 6: 307–308.

Hu, T., W. Banzhaf, and J. H. Moore. 2014. "The Effects of Recombination on Phenotypic Exploration and Robustness in Evolution." *Artificial Life*, 1–14.

Hutchison, D. W., and A. R. Templeton. 1999. "Correlation of Pairwise Genetic and Geographic Distance Measures: Inferring the Relative Influences of Gene Flow and Drift on the Distribution of Genetic Variability." *Evolution* 53: 1898–1914.

Kekkonen, M., and P. D. N. Hebert. 2014. "DNA Barcode-based Delineation of Putative Species: Efficient Start for Taxonomic Workflows." *Molecular Ecology Resources* 14: 706–715.

Kelly, R. P., J. A. Port, K. M. Yamahara, R. G. Martone, N. Lowell, P. F. Thomsen, M. E. Mach, M. Bennett, E. Prahler, M. R. Caldwell, and L. B. Crowder. 2014. "Harnessing DNA to Improve Environmental Management." *Science* 344: 1455–1456.

Lande, R. 1976. "The Maintenance of Genetic Variability by Mutation in a Polygenic Character with Linked Loci." *Genetical Research* 26: 221–235.

Le Chatelier, E., T. Nielsen, J. Qin, E. Prifti, F. Hildebrand, G. Falony, M. Almeida, M. Arumugam, J.-M. Batto, S. Kennedy, P. Leonard, J. Li, K. Burgdorf, N. Grarup, T. Jorgensen, I. Brandslund, H. B. Nielsen, A. S. Juncker, M. Bertalan, F. Levenez, N. Pons, S. Rasmussen, S. Sunagawa, J. Tap, S. Tims, E. G. Zoetendal, S. Brunak, K. Clement, J. Dore, M. Kleerebezem, K. Kristiansen, P. Renault, T. Sicheritz-Ponten, W. M. De Vos, J.-D. Zucker, J. Raes, T. Hansen, H. I. Meta, P. Bork, J. Wang, S. D. Ehrlich, O. Pedersen, and H. Meta. 2013. "Richness of Human Gut Microbiome Correlates with Metabolic Markers." *Nature* 500: 541–546.

Mayr, E. 1970. *Populations, Species, and Evolution.* Cambridge, MA: The Belknap Press of Harvard University Press.

McMahon, B. J., E. C. Teeling, and J. Höglund. 2014. "How and Why Should We Implement Genomics into Conservation?" *Evolutionary Applications* 7: 999–1007.

Neuwald, J. L., and A. R. Templeton. 2013. "Genetic Restoration in the Eastern Collared Lizard under Prescribed Woodland Burning." *Molecular Ecology* 22: 3666–3679.

Payne, J. L., and A. Wagner. 2014. "The Robustness and Evolvability of Transcription Factor Binding Sites." *Science* 343: 875–877.

Ralls, K., J. D. Ballou, and A. R. Templeton. 1988. "Estimates of Lethal Equivalents and the Cost of Inbreeding in Mammals." *Conservation Biology* 2: 185–193.

Reece, J. S., B. W. Bowen, K. Joshi, V. Goz, and A. Larson. 2010. "Phylogeography of Two Moray Eels Indicates High Dispersal Throughout the Indo-Pacific." *Journal of Heredity* 101: 391–402.

Rees, H. C., B. C. Maddison, D. J. Middleditch, J. R. M. Patmore, and K. C. Gough. 2014. "Review: The Detection of Aquatic Animal Species Using Environmental DNA – A Review of eDNA as a Survey Tool in Ecology." *Journal of Applied Ecology* 51: 1450–1459.

Roca, A. L., N. Georgiadis, and S. J. O'Brien. 2007. "Cyto-nuclear Genomic Dissociation and the African Elephant Species Question." *Quaternary International* 169–170: 4–16.

Roca, A. L., N. Georgiadis, J. Pecon-Slattery, and S. J. O'Brien. 2001. "Genetic Evidence for Two Species of Elephant in Africa." *Science* 293: 1473–1477.

Smith, H. M., D. Chiszar, and R. R. Montanucci. 1997. "Subspecies and Classification." *Herpetological Review* 28: 13–16.

Templeton, A. R. 1986. "Coadaptation and Outbreeding Depression." In *Conservation Biology*, ed. M. Soulé. Sunderland, MA: Sinauer.

Templeton, A. R. 1989. "The Meaning of Species and Speciation: A Genetic Perspective." In *Speciation and its Consequences*, ed. D. Otte and J. A. Endler. Sunderland, MA: Sinauer.

Templeton, A. R. 1994. "Biodiversity at the Molecular Genetic Level: Experiences from Disparate Macroorganisms." *Philosophical Transactions of the Royal Society of London Series B – Biological Sciences* 345: 59–64.

Templeton, A. R. 2001. "Using Phylogeographic Analyses of Gene Trees to Test Species Status and Processes." *Molecular Ecology* 10: 779–791.

Templeton, A. R. 2006. *Population Genetics and Microevolutionary Theory*. Hoboken, NJ: John Wiley & Sons.

Templeton, A. R. 2009. "Statistical Hypothesis Testing in Intraspecific Phylogeography: Nested Clade Phylogeographical Analysis vs. Approximate Bayesian Computation." *Molecular Ecology* 18: 319–331.

Templeton, A. R. 2013. "Biological Races in Humans." *Studies in History and Philosophy of Science Part C: Studies in History and Philosophy of Biological and Biomedical Sciences* 44: 262–271.

Templeton, A. R., and B. Read. 1983. "The Elimination of Inbreeding Depression in a Captive Herd of Speke's Gazelle." In *Genetics and Conservation: A Reference for Managing Wild Animal and Plant Populations*, ed. C. M. Schonewald-Cox, S. M. Chambers, B. Macbryde, and L. Thomas. Reading, MA: Addison–Wesley.

Templeton, A. R., and B. Read. 1984. "Factors Eliminating Inbreeding Depression in a Captive Herd of Speke's Gazelle (*Gazella spekei*)." *Zoo Biology* 3: 177–199.

Templeton, A. R., H. Brazeal, and J. L. Neuwald. 2011. "The Transition from Isolated Patches to a Metapopulation in the Eastern Collared Lizard in Response to Prescribed Fires." *Ecology* 92: 1736–1747.

Templeton, A. R., J. L. Neuwald, H. Brazeal, and R. J. Robertson. 2007. "Restoring Demographic Processes in Translocated Populations: The Case of Collared Lizards in the Missouri Ozarks Using Prescribed Forest Fires." *Israel Journal of Ecology and Evolution* 53: 179–196.

Torstrom, S. M., K. Pangle, and B. J. Swanson. 2014. "Shedding Subspecies: The Influence of Genetics on Reptile Subspecies Taxonomy." *Molecular Phylogenetics and Evolution* 76: 134–143.

Wagner, A. 2007. "Robustness and Evolvability: A Paradox Resolved." *Proceedings of the Royal Society B: Biological Sciences* 275: 91–100.

Wang, J. L. 2004. "Application of the One-migrant-per-generation Rule to Conservation and Management." *Conservation Biology* 18: 332–343.

Waugh, J. 2007. "DNA Barcoding in Animal Species: Progress, Potential and Pitfalls." *Bioessays* 29: 188–197.

Weitschek, E., G. Fiscon, and G. Felici. 2014. "Supervised DNA Barcodes Species Classification: Analysis, Comparisons and Results." *BioData Mining* 7(4), DOI: 10.1186/1756-0381-7-4.

Wilmshurst, J. M., N. T. Moar, J. R. Wood, P. J. Bellingham, A. M. Findlater, J. J. Robinson, and C. Stone. 2014. "Use of Pollen and Ancient DNA as Conservation Baselines for Offshore Islands in New Zealand." *Conservation Biology* 28: 202–212.

Wright, S. 1931. "Evolution in Mendelian Populations." *Genetics* 16: 97–159.

Wright, S. 1943. "Isolation by distance." *Genetics* 28: 114–138.

Yatsunenko, T., F. E. Rey, M. J. Manary, I. Trehan, M. G. Dominguez-Bello, M. Contreras, M. Magris, G. Hidalgo, R. N. Baldassano, A. P. Anokhin, A. C. Heath, B. Warner, J. Reeder, J. Kuczynski, J. G. Caporaso, C. A. Lozupone, C. Lauber, J. C. Clemente, D. Knights, R. Knight, and J. I. Gordon. 2012. "Human Gut Microbiome Viewed across Age and Geography." *Nature* 486: 222–227.

19

ESTIMATING BIODIVERSITY LOSS

Yrjö Haila

Introduction

What is known today as the biodiversity crisis is a relative newcomer among environmental concerns. The concept originated in the 1980s, and has since then succeeded in claiming a prominent place on the agenda of global environmental problems. How and why the concept achieved a successful reception is told and retold in a vast literature; I think Lawler *et al.* (2002: 295) offer a good concise explanation by writing: "Biodiversity loss has become a rallying point for conservation activists because it provides a scorecard for how greatly humans are making an impact on nature."

"Loss" is a keyword in this statement. Moreover, loss interpreted as a "scorecard" implies quantitative assessment, and this requires criteria that can be quantified at least on an ordinal scale. The need for criteria and quantification is supported by ambitious international goals concerning the discontinuation of loss of biodiversity, as these have been formulated since the enactment of the Convention on Biological Diversity (CBD). A precondition of reaching such goals is that somebody be able to assess whether loss of biodiversity still goes on or has, actually, been curbed.

My aim in this essay is to chart the approaches adopted to estimate the magnitude of biodiversity loss. Coming up with reliable criteria has turned out to be a hard nut to crack. The all-encompassing character of the concept of "biodiversity" is a source of confusion. Reliable indicators that would join together generality (global significance) and specificity (what is observed locally) are difficult to come by. An even more confusing issue is that the assessment has to be fixed within a temporal horizon from a significant past over the present to a foreseeable future. Mere recording of change in nature is inconsequential; criteria are needed on what kind of change represents loss.

For estimating biodiversity loss, the first necessary step is to specify under what conditions and for which particular purposes this is possible at all. I adopt a pragmatic perspective: it is the need to formulate effective policies and rules of management that require that the characteristics and driving forces of biodiversity loss ought to be understood as precisely as possible. In the next section I explore conceptual difficulties in clarifying the goal by specifying ambiguities inherent in efforts to operationalize biodiversity. In the following sections I review approaches adopted thus far. Finally, I get back to specifying the preconditions of reasonable assessment.

Ambiguities in "biodiversity"

The understanding of the meaning and significance of biodiversity has changed during the short time it has been in focus. To begin with, the way biodiversity loss was originally framed as a crisis deserves attention. The talks given at the foundational event, the National Forum on BioDiversity (Washington, DC, 1986), were published in an edited volume two years later (Wilson and Peters 1988). In his "Editor's foreword" to the volume, Edward O. Wilson named two "more or less independent developments" that supported the crisis perception: "The first was the accumulation of enough data on deforestation, species extinction, and tropical biology to bring global problems into sharper focus … The second development was the growing awareness of the close linkage between the conservation of biodiversity and economic development."

There is an internal tension between Wilson's two points. His first point refers directly to "accumulation of data" and is, in principle, amenable to an assessment by observations. A precondition is, of course, an agreement on what kind of observations are relevant. The second point, by contrast, is primarily about a novel understanding of the value of biodiversity for human sustenance. Wilson's reference to "economic development" is loose, to say the least. It is far from obvious how a link between biodiversity and economic development could be made specific enough to be quantifiable.

The duality inherent in Wilson's characterization shows that analytic and normative appraisal of the problem cannot be torn apart. Specifically, the analytic question "What is lost?" and the normative question "What does the loss mean?" have to be evaluated together. This unavoidable tension implies that the critical question for estimating biodiversity loss is what to estimate, and for what specific purpose. In other words, assessment has to be problem-driven. A critical step is problem definition such that we can define tasks that can be meaningfully taken.

A useful step toward realistic problem definitions is to acknowledge a set of ambiguities that follow from the dual nature of the issue. Such ambiguities concerning problem framing should be understood as concretely as we ever dare, in analogy with famous perceptional ambiguities such as the duck–rabbit duality. When staring at the duck–rabbit image, we can detect either a duck or a rabbit, but not both at the same time. By taking turns in looking at one or the other of the images, we can gain understanding on how the duality works. Similarly, conceptual ambiguities can create clarity of the conceptual field in question if we approach the alternatives in turns.

I take up three ambiguities, partially leaning on Maclaurin and Sterelny (2008). One of the ambiguities they identify (p. 2) is the tension between biodiversity regarded as the object of protection in itself, and biodiversity as providing instrumental benefits for humans. This formulation reproduces E. O. Wilson's two factors in a different phrasing.

Another ambiguity Maclaurin and Sterelny take up, the "units-and-differences problem" (p. 21), is directly relevant for assessment. To succeed in assessment, we have to identify as bookkeeping elements such units of biodiversity that may be lost, but in addition, we have to assess the significance of differences recorded between the units. Taxonomic units such as species provide good examples. Are two taxonomically closely related species "equally different" from each other as two taxonomically distant species? If not – as many authors have argued with good grounds – the significance of the difference has to be included in the assessment.

But there is also a third type of ambiguity, not as commonly recognized: what, precisely, is the "past" that offers a standard of comparison with present and future? A prevalent tendency is to use the "natural state" of nature as a standard. This, however, gives rise to a further question: how to reckon with human environmental modification? Has past human influence always and unexceptionally been bad for biodiversity? Change and loss have to be distinguished from one another. Throughout history, humans have learned to modify biodiversity to better answer to their needs – for instance, by clearing fields and establishing gardens. To assume that "untouched nature" is a gold standard for biodiversity reflects an ideological bias stemming from a dualistic view of humanity versus nature (see Haila 1999a, 1999b).

I refer to these three ambiguities below to bring into focus points of confusion in defining criteria for assessing biodiversity loss. Clearly articulated ambiguities can play a positive role both in scientific thinking and in political discussion and thought (Majone 1989, Rein 2006), but if not taken into account, ambiguities give rise to muddled thinking. Ambiguities can also be used in a purely rhetorical fashion to play down arguments of assumed opponents. In this specific sense the brief history of the biodiversity concern has given rise to interpretative problems. The concern was originally painted as a crisis using the strongest and darkest expressions the proponents were able to come up with. The crisis framing elevated the problem to an enormous social and political scale (Haila 2004). I believe this has been an important and misleading part of the legacy of the concern.

Any assessment of biodiversity loss must be in a close relationship with a clearly articulated perspective on what biodiversity is and what it means. Hence, in what follows I have adopted a dual strategy to structure my argument. I use as my material on the one hand results of investigative practice – the real stuff of scientific work (Dyke 1988) – and on the other hand shifting views on what is the most fertile perspective of framing the biodiversity issue as a problem. Contrast space is a useful conceptual device for getting hold of the framing of arguments, especially in situations involving ambiguity (Garfinkel 1981, Dyke 1988). A contrast space articulates the basic alternatives between which an explanation ought to provide a resolution. Then, if the context is ambiguous, different alternatives can be clarified by constructing several contrast spaces and comparing the conclusions they give rise to. For instance, the units-and-differences ambiguity mentioned above can be clarified by constructing alternative contrast spaces: recording differences in species composition versus assessing the significance of the differences on taxonomic grounds require different contrast spaces.

Quantifying the rate of extinctions

Nature conservation has its roots in broad cultural consciousness concerning proper human behavior toward non-human nature, but extinction was already a major concern at the end of the nineteenth century (Adams 2004). The early focus was on large emblematic animals decimated by hunting, but in the course of the twentieth-century conservation goals became more comprehensive. New investigative practices were developed in particular to survey the geographic range and abundance variation of both single species and specific ecological communities. These efforts gave rise to a rich tradition of practical sampling methods, accompanied with procedures developed for correcting methodological biases to obtain "true" values from samples collected in the field (Haila 1992, Kohler 2002). Articles in Magurran and McGill (2011) provide a rich overview of sampling methods focused on biodiversity. By and large, it seems that the investigative practice of field surveys fed upon itself – more confidence was felt about figures of species richness than is warranted on realistic ecological grounds. Gaston (1996) and Gotelli and Colwell (2011) present overviews of both methodological and conceptual problems of measuring species richness.

Biodiversity brought about a new dimension into conservation concerns: the threat of an imminent biotic impoverishment, driven by an extinction wave. Statistics played a prominent role in efforts to assess the seriousness of the threat: first, as statistics on extinctions, and later on, as statistics on a range of indirect indicators such as the area proportion of protected areas of various political and biogeographic units.

It is very difficult to construct reliable extinction statistics for several interconnected reasons. First of all, direct observations do not offer a place to start. Unambiguous documentation of extinction is well-nigh impossible except for large and visible creatures such as the famous cases from earlier times such as the dodo, the great auk, the passenger pigeon, and so on. A list of such species does not add up to a reasonable estimate of extinction rate. More comprehensive population-level conclusions can only be drawn when a specific valuable and unique habitat has been destroyed; Gentry (1986) gives examples from the Amazon. However, in such cases representativeness remains a problem – that is, it is uncertain whether such observation can be extrapolated to other areas and larger scales.

Extinction statistics are faced with the problem of specificity versus generality: all well-known cases of extinction are unique (just take a look at a classic such as Halliday 1978), but meaningful statistics require generality. Habitat area was adopted as a surrogate for extinction probability in the 1970s. The inspiration came from the species–area relationship, brought into the domain of nature conservation by a short note in the *Theory of Island Biogeography* by MacArthur and Wilson (1967), with a reference to woodlots in Cadiz Township, Wisconsin, leaning on an original data set published by plant ecologist John T. Curtis (1956). Frank Preston (1962) had earlier made a similar argument more extensively. For a while, habitat destruction assessed by the reduction in habitat area was adopted as a primary surrogate for estimates of species extinctions. Wilson (1988) was really explicit on this: "The area-species curves of island systems, that is, quantitative relationship between the area of islands and the number of species that can persist on the islands provide minimal estimates of the reduction of species diversity that will eventually occur in the rain forests."

However, the use of area as a surrogate is problematic on several counts, as has been pointed out almost innumerable times. The analogy with island biogeography assumes that the remaining patches of habitat are surrounded by "biological desert" after the reduction in total area, but this is never the case. Furthermore, if the surrounding areas comprise forests, secondary growth starts on the surrounding areas, and sooner or later some of the forest species find suitable habitats there. The use of area as a surrogate builds upon a flawed contrast space. Ariel Lugo has defended this qualification from the very beginning of the debate, backed by his empirical experience from Puerto Rico (Lugo 1988). I find it very strange that Lugo's work has not found a foothold in the estimates of extinction rate.

Furthermore, statistics based on habitat area imply a highly problematic step from the present to the future. The assumption is that the quality of the habitats both in the preserves and in the surroundings remain the same, but this is never the case. This problem is inherent in the notion of "extinction debt," that is, that populations inhabiting an assumedly too small habitat space are "doomed" to extinction, being merely "living dead" at present (Tilman *et al.* 1994). "Extinction debt" elevates an assumed future to a criterion of the present. While the reality of such dynamics is, indeed, quite plausible, the lack of generality hits back. The lower boundary of area that triggers extinction debt is difficult to pin down, and it certainly varies across taxa, and probably also across biogeographical regions.

Overall, the situation presents an inherent paradox. There is no doubt whatever that human activities have changed and continuously change environments negatively, compared with the previous state. However, precisely documenting and quantifying this change is far from easy. In my view, it is necessary to use some background variables as indicators ("proxies") for estimating change on an ensemble level, no matter that proxies necessarily bring additional interpretative ambiguity into the picture (Haila and Henle 2014).

Lindenmayer and Likens (2011) disagree. They promote, instead, what they call "direct measurement" of the entities of concern. However, the requirements they demand of the "right" entities are very stringent. If the emphasis is on single species or some very well defined habitat types, this approach would certainly work out, but the problem of generality remains. When larger suites of species or environments are of concern, surrogates are necessary, whether we like it or not.

In search of reliable indicators

In any case, the indicator chosen has to be valid, that is, the variation of indicator values must correlate reliably with the variation of what it is supposed to indicate. The theory of island biogeography got a critical role by proposing that area be an adequate surrogate. This interest had a cycle of vigorous life as a concern over habitat fragmentation, that is, the perception that human influence reduces previously continuous habitat areas to sets of "island-fragments" isolated from each other. A merely declarative reference to "habitat fragmentation" leans on the misleading contrast space of "intact habitat" versus "ecological desert on the outside" (see Haila 2002, Lindenmayer and Fischer 2006).

Lists of endangered species offered a remedy to overtly general statistics covering all species and types of environments. The lists were also adopted to indicate extinction risk more generally. Fulfilling this aim became plausible with increasing capacity of computers that could be utilized to develop algorithms for identifying reserve networks such that some specific criteria be fulfilled – for instance, the inclusion of rare and endangered species in the network. Australian ecologists were active in developing this line of inquiry (see Margules and Austin 1991). The total area of preserves and other protected areas has been another indicator. These measures are very coarse, however; as has been emphasized many times, the success or failure of biodiversity preservation takes place outside preserves.

At the opposite end of the range of scales are efforts to compile composite indices that give an overall view of the status of biodiversity in larger and internally varied geographical regions. This approach has good potential when used with care. One example of such an approach is the "biodiversity intactness index" suggested by Scholes and Biggs (2005) for their target region of southern Africa. The index that they propose is compiled in such a way that it would meet the criteria defined by the CBD. The index puts together estimates of the population sizes of vertebrates and major groups of plants, as well as main habitat types. The index requires a reference; Scholes and Biggs (2005) suggest "that which occurred in the landscape before alteration by modern industrial society." We cannot go into the detail of the algorithm they use, but the result is interesting: "Overall, we estimate that $84 \pm 7\%$ of the pre-colonial number of wild organisms persist in present-day southern Africa." As they note, 99 percent of the species persist, which indicates that indices based on extinctions are less sensitive.

Another promising example, methodologically a close relative of the intactness index of Scholes and Biggs (2005), is the so-called Norwegian Nature Index. The idea and its practical implementation are introduced in a set of articles published in *Norsk Geografisk Tidsskrift – Norwegian Journal of Geography* 66(5) (2012); for an overview, see Nybø *et al.* (2012). The Norwegian index was compiled by making detailed interviews with a large number of specialists in different fields of biology and management of natural resources. The process was initiated in 2005, inspired by the goal set within the confines of the CBD to halt the loss of biodiversity by 2010 (later, the focal year was extended to 2020). Similar to the southern African experience, the reference state was chosen in a very pragmatic manner, as the natural state of the environmental types, with baseline year at 1950. The point was merely to enable a convergent perception among the specialists who participated in the exercise, without any deeper meta-physical commitments as to what "primeval" nature of Norway might have looked like. In the case of Norway, such pragmatism makes perfect sense as the whole country was covered by continental ice as recently as 20,000 years ago, save, perhaps, a few glacial refugia in the mountains or along the Atlantic coast.

The southern African and Norwegian experiences are encouraging by suggesting that data aggregated from a large number of heterogeneous sources is usable in assessing the status of biodiversity. Clearly, a basic requirement is that adequate data are available. Norway is certainly a special case, as a country with a strong tradition of field naturalism and relatively small total area. On the other hand, the environmental types found in Norway are very diverse, and hence, the success in compiling the Norwegian Nature Index is very promising indeed.

On an intermediate scale between habitat-specific area calculations and composite indices are comparisons based on the distinction between alpha, beta, and gamma diversities. These notions derive from the interest in patterns of variation in species richness across heterogeneous landscapes that took off in the 1970s. A distinction was drawn between within-site ("alpha"), between-site ("beta") and regional ("gamma") diversity. When ecological realism is given due care, meaningful comparisons are possible; several articles in Magurran and McGill (2011) describe the background and assess the present status of this line of research.

Considerable controversy has been created by the question: if (when) species disappear from local and regional species assemblages, is the effect observable as a decline in local (alpha) diversity or as a change in species identities ("turnover")? Dornelas *et al.* (2014a) conducted an extensive meta-analysis on the level of local communities and concluded "that local assemblages are undergoing biodiversity change but not systematic biodiversity loss" (p. 299). This result is interesting, but it brings up the ambiguity concerning the equality of species: are the species that make up local communities in modified environments "equally valuable" as in the original conditions? An exchange by Cardinale (2014) and Dornelas *et al.* (2014b) focuses on this issue.

Beta diversity, or variation in biodiversity across localities on a larger spatial scale, is another facet of the intermediate scale. Gaston *et al.* (2007) present a thoughtful review of the relevant issues which are firmly embedded in pre-biodiversity investigative practice. Gaston *et al.* (2007) conclude that there is not much by way of general theory, let alone predictive theory in this regard. Hence, it seems variation in beta diversity does not qualify as an indicator of biodiversity loss.

Species turnover in time does not fare much better, although it is difficult to decide as the database is much scarcer than in the case of spatial turnover. The reason is simple: exploring temporal variation requires that the same targets be monitored for extensive periods of time, but field ecologists have only occasionally got that chance. A few remarkable exceptions are well known, though, most prominent among them the long time series of both plants and insects collected at the Rothamstead Agricultural Experimental Station in England. The insect data gave invaluable material for exploring variation in species abundances in the early and mid-twentieth century (Fisher *et al.* 1943, Williams 1964). The contribution of the Rothamstead experiment to ecology was reviewed by Silvertown *et al.* (2006) who point out several insights the experiment allows on factors influencing local variation in plant diversity across the experimental plots through time.

Another important aspect is that in every local assemblage studied over a number of years, species turnover has been observed between the years. This is a natural part of the dynamics of local assemblages – comparable to population "kinetics." Breeding birds, for instance, shift locations of their nests from year to year even when the population level remains stable (Haila *et al*. 1996). The pattern was characterized as "equality of space and time" by Frank Preston (1960).

The instrumental dimension: the functional significance of biodiversity

Quite early on some conservation biologists began to feel that the value of biodiversity in itself does not provide sufficiently convincing grounds for preservation. What is additionally needed according to this perspective is a view on what biodiversity does such that the basis for human sustenance and welfare is strengthened. Edward Wilson made this point in the 1986 Forum by referring to economic benefits, as we saw above.

An emphasis on the role of biodiversity for various ecosystem functions took off in the early 1990s; a collection of essays edited by Schulze and Mooney (1993) is often mentioned as the first effort to give the functional role of biodiversity a systematic presentation. In the foreword to the volume, Paul Ehrlich (1993) makes the following statement: "Of special interest to humanity is the relationship of biodiversity to the variety of services provided by ecosystems and, in particular, to the stability of the flow of those services, such as the maintenance of the gaseous composition of the atmosphere, preservation of soils, recycling of nutrients, and provision of food from the sea." In effect, Ehrlich in this quote gave a precise formulation to Wilson's loose talk of economic benefits: the benefits are ecosystem services, produced by normal functioning of healthy ecosystems.

Exploration of the functional significance of biodiversity has proceeded along two complementary lines which, in a sense, reproduce the division of arguments into analytic and normative questions: is the functional role real?; and which functions are important?, respectively. On the analytic branch, experimental studies were started on the correlation between species number and specified functional aspects of ecosystems. Some of this work predated the biodiversity concern and was primarily inspired by previous interest in the relationship between diversity and stability in local communities. Several essays in Schulze and Mooney (1993) summarize the results. Grassland plants have dominated this line of research, for understandable logistical reasons. On the normative side, specifying what the critical functional features of ecosystems are, and how their role can be demonstrated, has turned out to pose important conceptual and practical problems. Such problems highlight the ambiguity of biodiversity per se versus instrumental value of biodiversity: a precondition for presenting specific instrumental benefits of biodiversity as support for preservation is that those benefits can be named and demonstrated.

Results of the experimental studies have been summarized several times. While the general positive correlation between biodiversity and ecosystem functions is uncontroversial particularly when the numbers of species in the experiments are relatively small, the generalizability of the results remains a tricky problem. Ecosystem ecology and community ecology have partially different pedigrees, and one of the problems in finding a common ground has been the difficulty of mapping structural elements (individuals and populations of different species in communities) to functional elements of ecosystems in which they reside. In the ecosystem tradition, the notion of functional type has been in use for some time (e.g. O'Neill *et al.* 1986), but the ambiguity concerning the "units-and-differences problem" hits back at this point: which types are similar, which are different? A weakness in much of the work on ecosystem functions is that the critical role of microbes is often not recognized at all, but we lack the space to get any deeper into this problem; Meyer (1993) offers an early overview; Øvreås and Curtis (2011) discuss the question albeit mainly from an inventory perspective.

In other words, specifying what the functional types actually are in specific ecosystems is tricky (see also the thoughtful review by Weiher 2011). Conceptual problems concerning functional types feed back to empirical research. It is difficult to get investigative practices on ecosystem function to stabilize. Cardinale *et al.* (2011) draw this conclusion in their comprehensive review. More or less convincing experiments on the statistical relationship between species number and one or another functional feature measurable in a particular ecological community are piling up, but the mechanistic bases of what is observed are mainly unknown. Even if functional groups could be identified with some degree of confidence, assessing their relative quantitative importance poses additional problems, as pointed out by Hooper *et al.* (2005).

Some empirically grounded distinctions have been tried, which are in themselves interesting. Duffy *et al.* (2007) draw a distinction between diversity within trophic levels (horizontal diversity) and across trophic levels (vertical diversity). Another interesting distinction studied by Srivastava *et al.* (2009) is between the effect of bottom-up versus top-down diversity on ecosystem function. They made a meta-analysis of studies on decomposition efficiency and concluded that in such "brown" food webs the diversity of detritivorous organisms has a clear effect on decomposition whereas resource diversity (types of detritus) seems not to have any effect on consumption. As the authors note, their conclusion conflicts with the established view that in "green" foodwebs resource diversity has a major influence on consumption efficiency (see Cardinale *et al.* 2011).

A methodologically relevant side effect of drawing refined distinctions is that the dimensionality of the problem increases. As the relationships among the dimensions are most certainly non-linear, the prospect of drawing firm generalizations fades away. For instance, Cardinale *et al.* (2011) conclude that the positive relationship between producer diversity and ecosystem functioning is well established, but as regards assumptions about possible mechanisms, the literature "is a 'reader beware' field. Less than half of all claims made in the abstracts or discussions of papers are backed by any direct statistical test" (p. 580). As a final conclusion, they present a set of tasks for the future, calling for ambitious long-term experiments focusing on multifunctional and non-linear effects as well as "studies to embrace and try to explain natural variation rather than experimentally control it" (p. 589). In effect, they emphasize problems of interpretation and extrapolation inherent in the relationship between the constructed specificity of controlled experiments and the broad spectrum of natural conditions.

Ecosystem services is a 1990s addition to the discussion of ecosystem functions that are beneficial to humans. I cited Ehrlich above, but in his programmatic statement the services are identified on a very general level: "the maintenance of the gaseous composition of the atmosphere, preservation of soils, recycling of nutrients, and provision of food from the sea." Ehrlich's statement rings true: the global characteristics of the current Earth are, indeed, produced and maintained by life. A critical problem is, however, that the claims are too general to offer sites for drawing distinctions. Everything alive supports the services on the level Ehrlich refers to them – the statement has no diagnostic power.

Another problem that has created quite a lot of controversy is the question of functional redundancy, that is, whether the roles of different species in some ecosystems can be similar enough that they can compensate for the loss of one another. Lawton and Brown (1993) present an early overview of the problem of redundancy. By and large, they support the view that functional redundancy is a real phenomenon. This, of course, makes assessment of biodiversity loss murky: how do we know that extinctions necessarily reduce the functional strength of ecosystems if some species are redundant?

Against assumptions of functional redundancy some authors have argued that high diversity is always needed to maintain ecosystem services (e.g. Isbell *et al.* 2011), claiming support for the precautionary principle that "all species should be conserved because we cannot be certain which species actually produce ecosystem services." As a statement about high principles this sounds convincing, but when looking at the existing ecology of the Earth we come across the ambiguity about the nature of human influence: what is the significance of species inhabiting environments intensively modified by human actions? What about novel ecosystems on sites heavily disturbed by humans? Isbell *et al.* (2015) give additional emphasis to the principle by invoking what they call "ecosystem service debt." From the point of assessment this idea shares with extinction debt the problem that an assumed but basically unknown future is elevated to a criterion to assess the present.

Quite expectedly, the broad use of ecosystem services as an argument for biodiversity protection has also raised criticism. For the perspective of estimating loss, a major criticism draws upon the ambiguity of "units-and-differences." Ridder (2008), for instance, points out that resilient ecosystem services are unlikely to depend on rare species, which are often the focus of conservationists. A distinction between resilient and sensitive ecosystem services emphasizes the context specificity of instrumental arguments.

A policy perspective: assess causes instead of symptoms

A strong motivation to present precise estimates of biodiversity loss has been to support views on the urgency of the problem. But as documentation of the amount and significance of biodiversity loss has turned out to be unexpectedly difficult, a promising alternative, in my view, would be to put more emphasis on such human practices that are known to be harmful. I base this view on a simple argument: wanton destruction of ecological resources should be tackled irrespective of whether its global significance in terms of loss of biodiversity can be quantified. In a sense, this would be an updated version of the age-old concern that biological resources should be used in a sustainable fashion.

Oceanic fisheries provide good examples: there is no doubt that all over the world's oceans, criminally defective and exploitative fishing methods cause destruction of biodiversity. Examples of collapses of fish stocks have accumulated since the cases of the Peruvian anchovy, the Californian sardine and the cod off the northeastern Canadian coast. Worries concerning industrial-scale overfishing in basically all parts of the world's oceans are certainly founded.

The fisheries example raises some problems that have to be acknowledged. The past hardly offers standards, as reliable quantitative data on fish stocks are quite recent (e.g. Caddy and Gulland 1983). Instead, the attention has to be focused on the perspective from the present to the future – in this, of course, qualitative knowledge of old fishing possibilities can be used for advice. I think this is feasible when the focus is restricted enough – on the fishing grounds off the coast of Newfoundland, say. Also, the sustenance of local human communities, dependent on exploiting local fish stocks, has to be woven in. The challenge is to link social and ecological systems in a harmonious way. As a precondition, strong vested interests should be held at an arm's length from management operations. For instance, the fisheries policy of the European Union is too strongly influenced by strong fisheries interests.

Overall, the idea that different types of disturbance are important in ecology, broadly accepted since at least the 1950s, offers one starting point. Dornelas *et al.* (2011) present a thorough overview of different forms of disturbance and also point out that forms of disturbance due to natural and to human-induced causes grade into one another. In general terms, getting human-induced effects on ecological processes to resemble as closely as possible forms of natural disturbance is a reasonable piece of advice, easily stated, but more complicated to follow (Haila and Levins 1992). The "resilience" school of ecological research, building upon the work of C. S. Holling and his collaborators (Gunderson and Holling 2001), has elaborated this approach. The idea of resilience is difficult to quantify on large scales, let alone globally, but it has analytic strength in local, specified contexts for efforts to prevent the destruction of biological resources.

A general problem is that the causes of the destruction of biodiversity may be named using conceptual frames that are way too general to be analytically useful. The frame adopted by the European Environment Agency dubbed "Driving forces – Pressures – States – Impacts – Responses" (DPSIR) is an example. A critical problem of the scheme is that it is presented as a causal chain but each one of the elements is too general to be of any specific analytic use. Maxim *et al.* (2009) argue that the scheme is analytically useless but can, nevertheless, be used for communication between scientists and users of environmental information. That may be a valid argument (which I tend to doubt), but only under the condition of explicitly defining the context.

Pragmatics: problem framings, investigative practices, and critical timeframes

Biodiversity is a rich and multifaceted perspective on the conditions of life on Earth, with deep roots in ecological and evolutionary understanding. Maclaurin and Sterelny (2008) offer an excellent guide to this richness. It is due to this richness that straightforward quantification of the status of biodiversity has been and is difficult. My view is that this is in the nature of the idea itself: there is no single and simple way to gain understanding or to plan policy or management.

Butchart *et al.* (2010) looked at thirty-one general indicators on whether the target of halting biodiversity loss by the year 2010 had been reached by the time of their analysis (2009, apparently). They concluded that some local successes had been achieved, but nevertheless, the general decline of biodiversity had not been halted. The result is not unexpected, taking into account the generality of the criteria that were defined for assessing the target. The indicators are so grossly aggregated and all correlate strongly with overall trends in social development and change that any other conclusion is hardly possible.

I suggest the preconditions of adopting a pragmatic perspective include, at a minimum, three elements: (1) framing research problems such that they illuminate the real-life problems at hand; (2) making sure that we have, or are able to develop, investigative practices which allow solving the problems; (3) considering the temporal perspective within which successful addressing of the problem is possible at all, also taking into account how successfully solving one, perhaps apparently minor problem paves the way for successfully addressing other, deeper problems. Such a "chaining" or "coupling together" of problems is a characteristic feature of any successful policy in the incomplete world of human actors (Majone 1989).

Honest assessment of the situation is a major requirement on every step. In practice, moving along the chain proceeds iteratively, not linearly. A source-book of methods such as the volume edited by Magurran and McGill (2011) offers invaluable help. Note, too, that the assessment always includes both analytic and normative dimensions.

Where does this leave us as regards estimating biodiversity loss? To be honest, my view is that estimating biodiversity loss is no general priority. Biodiversity is an enormously important phenomenon, and doing something about the human infringement on biodiversity is an enormously important task, but "estimating biodiversity loss" on a general level does not help this endeavor in any way. McGill *et al.* (2015) specify fifteen forms of biodiversity trends that conservation ecologists ought to pay attention to. In effect, they also step off from the goal of a unified assessment.

As regards critical temporal frames, a basic task is to draw distinctions between different types of threats to biodiversity; McGill *et al.* (2015) give a timely reminder on this need. Some of the factors bring about risks that can be quantified, however coarsely. This is different from genuine uncertainty that cannot be quantified. The latter situation approaches the setting of "post-normal science," described by Funtowicz and Ravetz (1993); for specifications concerning biodiversity research, see Haila and Henle (2014). A basic question that brings genuine uncertainty into the concern over biodiversity is: what are the critical timeframes for avoiding the possible collapse of key mechanisms that maintain a livable biosphere for a rich biota that includes humans?

References

Adams, W. M. 2004. *Against Extinction. The Story of Conservation.* London: Earthscan.

Butchart, S. H. M., M. Wallpole, B. Collen, A. van Strien, J. P. W. Scharlemann, *et al.* 2010. "Global Biodiversity: Indicators of Recent Declines." *Science* 328: 1164–1168.

Caddy, J. F., and J. A. Gulland. 1983. "Historical Patterns of Fish Stocks." *Marine Policy* 7: 267–278.

Cardinale, B. J. 2014. "Overlooked Local Biodiversity Loss." *Science* 344: 1098.

Cardinale, B. J., K. L. Matulich, D. U. Hooper, J. E. Byrnes, E. Duffy, L. Gamfelt, P. Balvanera, M. I. O'Connor, and A. Gonzalez. 2011. "The Functional Role of Producer Diversity in Ecosystems." *American Journal of Botany* 98: 572–592.

Curtis, J. T. 1956. "The Modification of Mid-latitude Grasslands and Forests by Man." In *Man's Role in Changing the Face of the Earth*, ed. W. L. Thomas Jr, 721–736. Chicago, IL: University of Chicago Press.

Dornelas, M., C. U. Soykan, and K. I. Ugland. 2011. "Biodiversity and Disturbance." In *Biological Diversity. Frontiers in Measurement and Assessment*, ed. A. E. Magurran and B. J. MacGill, 237–251. Oxford: Oxford University Press.

Dornelas, M., N. J. Gotelli, B. McGill, H. Shimadzu, F. Moyes, C. Sievers, and A. E. Magurran. 2014a. "Assemblage Time Series Reveal Biodiversity Change but Not Systematic Loss." *Science* 344: 296–299.

Dornelas, M., N. J. Gotelli, B. McGill, and A. E. Magurran. 2014b. "Response" [to Cardinale, 2014], *Science* 344: 1098–1099.

Duffy, J. E., B. J. Cardinale, K. E. France, P. B. McIntyre, E. Thébault, and M. Loreau. 2007. "The Functional Role of Biodiversity in Ecosystems: Incorporating Trophic Complexity." *Ecology Letters* 10: 522–538.

Dyke, C. 1988. *The Evolutionary Dynamics of Complex Systems. A Study in Biosocial Complexity.* New York: Oxford University Press.

Ehrlich, P. R. 1993. "Foreword. Biodiversity and Ecosystem Function: Need We Know More?" In *Biodiversity and Ecosystem Function*, ed. E.-D. Schulze and H. A. Mooney, vii–xi. Berlin: Springer.

Fischer, R. A., A. S. Corbet, and C. B. Williams. 1943. "The Relation between the Number of Species and the Number of Individuals in a Random Sample of an Animal Population." *Journal of Animal Ecology* 12: 42–58.

Funtowicz, S. and J. R. Ravetz. 1993. "Science for the Postnormal Age." *Futures* 25: 735–755.

Garfinkel, A. 1981. *Forms of Explanation. Rethinking the Questions in Social Theory.* New Haven, CT: Yale University Press.

Gaston, K. J. 1996. "Species Richness: Measure and Measurement." In *Biodiversity. A Biology of Numbers and Difference*, ed. K. J. Gaston, 77–113. Oxford: Blackwell.

Gaston, K. J., K. L. Evans, and J. J. Lennon. 2007. "The Scaling of Spatial Turnover: Pruning the Thicket." In *Scaling Biodiversity*, ed. D. Storch, P. A. Marquet and J. H. Brown, 181–222. Cambridge: Cambridge University Press.

Gentry, A. H. 1986. "Endemism in Tropical versus Temperate Plant Communities." In *Conservation Biology. The Science of Scarcity and Diversity*, ed. M. Soulé, 153–181. Sunderland, MA: Sinauer.

Gotelli, N. J., and R. K. Colwell. 2011. "Estimating Species Richness." In *Biological Diversity. Frontiers in Measurement and Assessment*, ed. A. E. Magurran and B. J. MacGill, 39–54. Oxford: Oxford University Press.

Gunderson, L. H., and C. S. Holling, eds. 2001. *Panarchy: Understanding Transformations in Human and Natural Systems.* Washington, DC: Island Press.

Haila, Y. 1992. "Measuring Nature – Quantitative Data in Field Biology." In *The Right Tools for the Job: At Work in Twentieth-century Life Sciences*, ed. A. E. Clarke and J. H. Fujimura, 233–253. Princeton, NJ: Princeton University Press.

Haila, Y. 1999a. "Biodiversity and the Divide between Culture and Nature." *Biodiversity and Conservation* 8: 165–181.

Haila, Y. 1999b. "Socioecologies." *Ecography* 22: 337–348.

Haila, Y. 2002. "A Conceptual Genealogy of Fragmentation Research: From Island Biogeography to Landscape Ecology." *Ecological Applications* 12: 321–334.

Haila, Y. 2004. "Making Sense of the Biodiversity Crisis: A Process Perspective." In *Philosophy of Biodiversity*, ed. M. Oksanen and J. Pietarinen, 54–82. Cambridge: Cambridge University Press.

Haila, Y., and K. Henle. 2014. "Uncertainty in Biodiversity Science, Policy and Management: A Conceptual Overview." *Nature Conservation* 8: 27–43.

Haila, Y., and R. Levins. 1992. *Humanity and Nature. Ecology, Science and Society*. London: Pluto Press.

Haila, Y., A. O. Nicholls, I. K. Hanski, and S. Raivio. 1996. "Stochasticity in Bird Habitat Selection: Year-to-year Changes in Territory Locations in a Boreal Forest Bird Assemblage." *Oikos* 76: 536–552.

Halliday, T. 1978. *Vanishing Birds: Their Natural History and Conservation*. New York: Holt.

Hooper D. U., S. Chapin III, J. J. Ewel, A. Hector, P. Inchausti, S. Lavorel, *et al.* 2005. "Effects of Biodiversity on Ecosystem Functioning: a Consensus of Current Knowledge." *Ecological Monographs* 75: 3–35.

Isbell, F., P. B. Reich, D. Tilman, S. E. Hobbie, S. Polasky, and S. Binder. 2011. "High Plant Biodiversity is Needed to Maintain Ecosystem Services." *Proceedings of the National Academy of Sciences USA* 110: 11911–11916.

Isbell, F., D. Tilman, S. Polasky, and M. Loreau. 2015. "The Biodiversity-dependent Ecosystem Service Debt." *Ecology Letters* 18: 119–134.

Kohler, R. 2002. *Landscapes and Labscapes: Exploring the Lab–Field Border in Biology*. Chicago, IL: University of Chicago Press.

Lawler, S. P., J. J. Armesto, and P. Kareiva. 2002. "How Relevant to Conservation are Studies Linking Biodiversity and Ecosystem Functioning?" In *The Functional Consequences of Biodiversity. Empirical Progress and Theoretical Extensions*, ed. A. P. Kinzig, S. W. Pacala, and D. Tilmand, 294–313. Princeton, NJ: Princeton University Press.

Lawton, J. H., and V. K. Brown. 1993. "Redundancy in Ecosystems." In *Biodiversity and Ecosystem Function*, ed. E.-D. Schulze and H. A. Mooney, 255–270. Berlin: Springer.

Lindenmayer, D. B., and J. Fischer. 2006. *Habitat Fragmentation and Landscape Change. An Ecological and Conservation Synthesis*. Washington, DC: Island Press.

Lindenmayer, D. B., and G. E. Likens. 2011. "Direct Measurement versus Surrogate Indicator Species for Evaluating Environmental Change and Biodiversity Loss." *Ecosystems* 14: 47–59.

Lugo, A. E. 1988. "Estimating Reductions in the Diversity of Tropical Forest Species." In *Biodiversity*, ed. E. O. Wilson and F. M. Peter, 58–70. Washington, DC: National Academy Press.

MacArthur, R. H., and E. O. Wilson. 1967. *The Theory of Island Biogeography*. Princeton, NJ: Princeton University Press.

Maclaurin, J., and K. Sterelny. 2008. *What is Biodiversity?* Chicago, IL: Chicago University Press.

Magurran, A. E., and B. J. MacGill, eds. 2011. *Biological Diversity. Frontiers in Measurement and Assessment*. Oxford: Oxford University Press.

Majone, G. 1989. *Evidence, Argument and Persuasion in the Policy Process*. New Haven, CT: Yale University Press.

Margules, C. R., and M. P. Austin, eds. 1991. *Nature Conservation: Cost Effective Biological Surveys and Data Analysis*. Canberra: CSIRO, Australia.

Maxim, L., J. H. Spangenberg, and M. O'Connor. 2009. "An Analysis of Risks for Biodiversity under the DPSIR Framework." *Ecological Economics* 69: 12–23.

McGill, B. J., M. Dornelas, N. J. Gotelli, and A. E. Magurran. 2015. "Fifteen Forms of Biodiversity Trend in the Anthropocene." *Trends in Ecology and Evolution* 30: 104–113.

Meyer, O. 1993. "Functional Groups of Microorganisms." In *Biodiversity and Ecosystem Function*, ed. E.-D. Schulze and H. A. Mooney, 67–96. Berlin: Springer.

Nybø, S., G. Certan, and O. Sparpaas. 2012. "The Norwegian Nature Index – State and Trends of Biodiversity in Norway." *Norsk Geografisk Tidsskrift – Norwegian Journal of Geography* 66: 241–249.

O'Neill, R. V., D. L. DeAngelis, J. B. Waide, and T. F. H. Allen. 1986. *A Hierarchical Concept of Ecosystems*. Princeton, NJ: Princeton University Press.

Øvreås, L., and T. P. Curtis. 2011. "Microbial Diversity and Ecology." In *Biological Diversity. Frontiers in Measurement and Assessment*, ed. A. E. Magurran and B. J. MacGill, ed. 221–236. Oxford: Oxford University Press.

Preston, F. W. 1960. "Time and Space and the Variation of Species." *Ecology* 41: 611–627.

Preston, F. W. 1962. "The Canonical Distribution of Commonness and Rarity. Part I." *Ecology* 43: 185–215; "Part II." *Ecology* 43: 410–432.

Rein, M. 2006. "Reframing Problematic Policies." In *Oxford Handbook of Public Policy*, ed. M. Moran, M. Rein, and R. E. Goodin, 389–405. Oxford: Oxford University Press.

Ridder, B. 2008. "Questioning the Ecosystem Services Argument for Biodiversity Conservation." *Biodiversity and Conservation* 17: 781–790.

Scholes, R. J., and R. Biggs. 2005. "A Biodiversity Intactness Index." *Nature* 434: 45–49.

Schulze, E.-D., and H. A. Mooney, eds. 1993. *Biodiversity and Ecosystem Function*. Berlin: Springer.

Silvertown, J., P. Poulton, E. Johnston, G. Edwards, M. Heard, and P. M. Bliss. 2006. "The Park Grass Experiment 1856–2006: Its Contribution to Ecology." *Journal of Ecology* 94: 801–814.

Srivastava, D., B. J. Cardinale, A. L. Downing, J. E. Duffy, C. Jouseau, M. Sankaran, and J. P. Wright. 2009. "Diversity Has Stronger Top-Down than Bottom-Up Effects on Decomposition." *Ecology* 90: 1073–1083.

Tilman, D., R. M. May, C. L. Lehman, and M. A. Nowak. 1994. "Habitat Destruction and the Extinction Debt." *Nature* 371: 65–66.

Weiher, E. 2011. "A Primer of Trait and Functional Diversity." In *Biological Diversity. Frontiers in Measurement and Assessment*, ed. A. E. Magurran and B. J. MacGill, 175–193. Oxford: Oxford University Press.

Williams, C. B. 1964. *Patterns in the Balance of Nature and Related Problems of Quantitative Ecology*. New York: Academic Press.

Wilson, E. O. 1988. "The Current State of Biological Diversity." In *Biodiversity*, ed. E. O. Wilson and F. M. Peter, 3–18. Washington, DC: National Academy Press.

Wilson, E. O., and F. M. Peter, eds. 1988. *Biodiversity*. Washington, DC: National Academy Press.

PART V

Social contexts and global justice

20

PUTTING BIODIVERSITY CONSERVATION INTO PRACTICE

The importance of local culture, economy, governance, and community values

Anya Plutynski and Yayoi Fujita-Lagerqvist[1]

Biodiversity conservation as a practical discipline has been significantly transformed over the past twenty years. Given the extent to which humans influence not only biodiversity loss, but also geographical distribution, and ecological dynamics, there has been a shift in the study of conservation as a scientific discipline from a concern strictly with ecological and biological diversity measures to an interdisciplinary field, drawing upon the human sciences. What has now been called "conservation science" – as opposed to "conservation biology" (Kareiva and Marvier 2012) – is currently more interdisciplinary in character, for two reasons: one pragmatic, and another normative (Callicott *et al.* 1999, Sarkar 2012, Norton 2015). First, there is a growing realization that effective biodiversity conservation must draw upon research in psychology, economics, politics, geography, and anthropology. Second, there is a growing realization that conservation and international justice issues are significantly intertwined (Figueroa and Mills 2001, Dowie 2009, Sarkar 2012).

Turning to the pragmatic issue, the human sciences provide important insights into collective decision-making, economic forces governing institutions and individual behaviors, methods and aims of sustainable practices, and cultural norms and their roles in active resistance to, or commitment to, conservation (Sarkar 2008). Without evidence-based assessment of the role of human actors and institutions in conservation planning and policy, no amount of biological understanding is sufficient to put conservation policy effectively into practice. Such evidence is now being gathered, yielding a variety of practical prescriptions for how to make conservation work, though of course, such prescriptions should be context-sensitive and adapted to various locations. We will briefly review some of the more general positive recommendations for how below, and consider two case studies in illustration.

A second major motivation of the transformation in conservation science is concern with justice and fair treatment of local peoples affected by conservation policy. Until relatively recently, conservation meant forcibly removing people from their ancestral homes – places that they lived in, or traveled through, on seasonal migrations in search of food and shelter (Jacoby 2003). The "fortress" conservation plans of the 1960s and 1970s in sub-Saharan Africa

281

established national parks for the purposes of tourism and hunting under the auspices of wilderness protection, essentially evicting and disenfranchising indigenous populations (Anderson and Grove 1989, Guha 1989, Adams and McShane 1992, Carruthers 1995, Guha 1997, MacKenzie 1997, Neumann 1998, Sarkar 1998, 1999, Schroeder 1999, Woods 2001). In light of deeply unjust practices of forced relocation of local populations from protected areas (Duffy 2000, Dowie 2009), the "fortress" model of conservation has been challenged. In its place is a new focus on participatory practices that involve including all relevant stakeholders in conservation planning, alongside efforts to simultaneously reduce poverty and address biodiversity loss. Community-based conservation (CBC) in the 1980s was the first wave of a movement toward including local stakeholders in conservation planning, particularly in developing countries. However, outcomes of the CBC approach have been mixed (Kellert *et al.* 2000).

International efforts to promote balance between biodiversity conservation and the socio-economic needs of the developing world heightened during the 1990s, particularly after the Rio Environment Summit in 1992. This prompted global conservation agencies to move away from the "fence and fine" approach that excludes local people from accessing resources, to an integrated conservation and development approach that recognizes local inhabitants' socio-economic needs to maintain their lives in proximity to the natural resources. A plethora of Integrated Conservation and Development Programs (ICDPs) were introduced by international conservation agencies especially in Africa and Asia during the 1990s (Wells and Brandon 1992, Enters and Anderson 1999). While activities vary from program to program, ICDPs often include delineation of buffer zones, participatory resource inventory and land use planning, environmental education of community members and professionals, as well as programs that encourage diversification of local livelihoods.

Currently, there is an active literature and institutional support for more systematic analysis of how to improve upon conservation policies that include stakeholder participation (Waylen *et al.* 2010). A variety of important insights have thus been gained about how local cultural context has considerable influence on conservation outcomes, as well as when and how providing education can be effective at promoting conservation, how to sustain positive relationships between stakeholders with competing interests, and how to integrate local with larger state and even international political and economic interests. We review some of these insights below.

Turning to the normative issue, at the same time as this shift has been underway, a growing community of environmental philosophers has begun to acknowledge that practical decisions about what, when, and how to conserve biodiversity require a more pragmatic and "applied" approach to environmental problems (see e.g. Minteer 2011, Sarkar 2012, Norton 2015). Yet, the continued debates over the nature of value of species, ecosystems, and their properties, which seems to systematically leave offscreen the people who live in and surrounding diverse places. This erasure of people from the picture is a symptom of a larger historical tendency by Western environmentalists to ignore the significant role that people play in these diverse places. Biodiversity is at the intersection of a host of political and economic conflicts over land, resources, and power. The most diverse remaining places on Earth – biological hotspots – are places where regulatory frameworks are often weak; yet these places are also frontiers of commercially valuable resources, for example, minerals, water, land. They are also home to some of the most poor and powerless, as well as the last remaining indigenous peoples, whose intimate knowledge of nature and practices of maintaining biodiversity is arguably another aspect of biodiversity worth considering. Thus, conservation planning may require that we compare what some philosophers take to be incommensurable values, and make political and economic compromises.

We argue here that in order for conservation plans to be both effective and sustainable, we need to ensure that legitimate stakeholders' interests be considered from the beginning. This is required not only so that there is agreement with the plan, but also so that the many ways in which biodiversity is valued are respected. Of course, not everyone who presents themselves as a stakeholder is legitimate. And, negotiation over how to prioritize different stakeholders' knowledge and values in conservation planning is bound to be ongoing and involve many compromises (see also Forsyth and Walker 2008). Addressing such issues is a question of justice, but also simply a matter of pragmatics; it is often precisely in those cases where local interests and traditions were not respected or considered that conservation plans have failed (see e.g. Adams and McShane 1992, Peluso 1992, Roy and Jackson 1993, Duffy 2000, Peluso and Watts 2001, Peluso 2003).

Not only are there often conflicts between nation-states for power and authority in conservation planning, but also between international organizations, such as development corporations, non-profits, and even national governments, versus more regional, or local peoples and political or economic organizations. Competing interests both within regions and at the borders of different nations are particularly contentious. Issues of justice arise with respect to both long- and short-term interests; it's surely the case that many generations will be affected by decisions made by current generations, so there is the matter of how to represent the interests of future generations. Apart from the question of who should be at the negotiating table, there is the matter of power inequity. Different parties bring different economic and political power to the negotiating table, and the power dynamics between the parties are also subject to changes over time.

While historical cases illustrate serious inequities in power and control over decision-making, currently the difficulty is far subtler, concerning such issues as stakeholder legitimacy, and how best to include different stakeholders in conservation planning (Sarkar and Illoldi-Rangel 2010). Some voices may be excluded simply due to poor communication. Addressing this kind of subtle failure of justice will require detailed attention to local context (Reed 2008). Perhaps equally important is to understand the complexities and dynamic processes by which powers intersect and influence decisions to manage natural resources. Below, we look at two case studies in more or less successful systematic organization planning as an illustration.

It is important to emphasize that attention to anthropocentric interests is not necessarily in opposition to conservation for other reasons, as some have claimed (Doak *et al.* 2014). Whatever one's "ultimate" reasons for conservation, if we wish conservation efforts to be successful, we cannot ignore the interests of local people, as well as states and other stakeholders, alongside the interests of future generations, in planning and policy-making. Active engagement with the wide spectrum of stakeholders in conservation planning makes for more effective policy.

In keeping with this broader perspective, a more inclusive range of options for biodiversity conservation is required. This might include considering issues of mixed sustainable use, rather than strictly protected areas, captive breeding programs, and more active interventions in service of recovery and restoration. Managed ecosystems now dominate the planet; there are few if any places remaining that are completely unaffected by human influence. In light of this basic fact, the permissibility of sustainable use seems less problematic than it did twenty-five years ago. Introduction of endangered species, moving species to deal with climate change, or active reduction of invasive populations, are now part and parcel of conservation policy. There is also greater optimism with respect to recovery processes.

The same practices will not work everywhere – generalizations about effective biodiversity management will be limited. Consideration of contingent facts about culture, history, politics, economics, and of the environment itself, are all relevant to conservation planning. This is why "systematic conservation planning" is so important (Margules and Pressey 2000, Margules and

Sarkar 2007). Human–environmental interactions are complex systems. Only by an iterated process of adaptive management in light of new information, changing circumstances, and evolving interests can we protect diversity sustainably. Below we discuss two case studies, and then consider some larger generalizations that have been gathered from the empirical literature.

Case study 1: Mekong

The Mekong River Basin, stretching across mainland Southeast Asia including parts of southwest China (Yunnan Province), Myanmar, Laos, Thailand, Cambodia, and Vietnam, is globally known for its vast forest and wildlife species, and comprises a highly productive river system. Its unique biota from the Himalayas to the humid tropics of the Indomalayan region include species that were only discovered in the twenty-first century, and the abundance and diversity of freshwater aquatic resources provide crucial sustenance for the local population (Turner *et al.* 2012, Sunderland *et al.* 2012, Ziv *et al.* 2012, WWF 2013). It is home to an ethnically diverse population of approximately 70 million, whose lives are dependent on natural resources (Poffenberger 1990, Fox *et al.* 2000, Rerkasem *et al.* 2009).

It is only since the 1990s when the protracted political struggles in the region subsided that the uniqueness and the importance of the region's ecosystem came under scrutiny by Western scientists and conservation organizations as well as international donors. One of the key initiatives that reshaped forest administration in the region was the global initiative known as the Tropical Forestry Action Plan (TFAP), led by the Food and Agriculture Organization of the United Nations and supported by the World Bank and the World Resources Institute. TFAP urged countries in the tropics to strengthen their capacities to protect forest resources by classifying forests, delineating resource boundaries, and prescribing resource management practices. Efforts by which state authorities consolidate control over territory are referred to as state territorialization of forest (Peluso *et al.* 1995, Vandergeest and Peluso 1995).

In Laos, a landlocked country located in the lower Mekong River Basin, a forest inventory was carried out for the first time in the 1980s when the socialist state adopted an open-door policy and began to work with international conservation organizations. Following the first national forest reconnaissance survey, which was financed and technically assisted by the Swedish International Development Agency (SIDA), Laos developed its own tropical forest action plan in 1990 with the support of international donors and conservation organizations including the International Union for Conservation of Nature (IUCN), Wildlife Conservation Society (WCS), and the World Wildlife Fund (WWF). The action plan focused on strengthening the forestry administration to identify conservation areas across the country and to legislate land and forest management practices. Throughout the 1990s, international conservation organizations further supported the Laos government to promote conservation programs that were founded on the principle of integrated conservation and development, which comprised resource boundary demarcation and zoning, local engagement in conservation activities, and promotion of alternative income sources for households living in and near the conservation areas.

Similarly in Cambodia and Vietnam, forest conservation was re-institutionalized during the 1990s with the help of international donors that provided financial resources and technical input. In Cambodia, a prolonged period of civil war in the country had disrupted the forest management system, which was initially set up by the French colonial government during the early twentieth century. In 1993, the country introduced the National Protected Areas System, which followed a classification system introduced by IUCN, and thereafter issued a series of legislative actions promoting forest conservation. In Vietnam, the government reviewed national forest policy during the 1990s. Like its neighbours Laos and Cambodia, foreign donors and international conservation organizations contributed in the development of legislation on forest conservation.

Sunderland *et al.* (2012) claim that areas of forest under national parks and protected area systems that were founded on the basis of integrated conservation and development model in Cambodia, Laos, and Vietnam until recently remained relatively intact. However, they also point out that growing demand on natural resources for hydropower, mineral deposits, and expansion of tree and agricultural plantations are accelerating encroachment into forest areas:

> There are now nineteen dams on the Mekong River and countless more on its tributaries – Laos alone has seventy-seven active dam projects. Mines are springing up everywhere – both large industrial mines and small artisanal ones. Plantations of oil palm, rubber, fiber trees, and numerous other crops are expanding rapidly. A region that until recently retained vast tracts of relatively undisturbed natural rain forests is rapidly being sliced up by expanding networks of roads.
>
> (Sunderland *et al.* 2012: 3)

The building pressure on natural resources in the region is particularly highlighted in the report produced by the World Wildlife Fund (2013) that examined the patterns of regional forest cover between 1973 and 2009. Not only does the report highlight the rapid loss of forest areas, but also increasing fragmentation of forests since 2000, suggesting further degradation of forests in the region. Other studies suggest widespread logging for precious timber, for example, rosewood, and poaching of wildlife (Meyfroidt and Lambin 2009, EIA 2012, Singh 2012) are contributing factors that accelerate forest degradation in the region.

Experiences of forest management in the Lower Mekong countries suggest the dynamic relationship between key stakeholders that influence resource management practice. While certain entities may be authorized to control resources at one point in time, such power is not entirely set in stone. Take for example, in Laos where forest administration under the Ministry of Forestry and Agriculture spearheaded efforts to consolidate forest control with external support during the period between the 1980s and 1990s. This effort, which included delineation of conservation forests, has been overwritten by other government agencies during the last two decades. The Ministry of Energy and Mines and the Ministry of Investment and Planning gained authority to determine where development activities such as hydropower, mining, and agricultural plantation can take place. As the government promotes economic growth, led by private-sector investment in natural resources, new ministries such as the Ministry of Energy and Mines, and the Ministry of Investment and Planning have gained powers to authorize concessions rights to land to private-sector investors. The messy overlap of land use in Laos today is a testament to the contested nature of resource control and management (Schönweger *et al.* 2012). It also demonstrates the constant negotiation for control of resources between stakeholders and the tenuous nature of legislation and the regulatory framework to protect and conserve critical natural resources given the changing power dynamics between stakeholders.

Experiences from the Mekong Region also highlight the need to understand dynamic aspects of local lives. Although conservation programs in countries such as Laos were founded on principles of integrated conservation and development models that encouraged local communities and their members to continue to reside in and nearby conservation areas, it did not fully anticipate the speed by which local community members were capable of adapting their livelihood activities to the market economy. As subsistence and smallholder[2] farmers became increasingly integrated into the market economy since the 1990s, they began to radically transform their farming system (Padoch *et al.* 2007). In northern Laos, the transition was particularly rapid in the uplands, where subsistence farmers engaged in swidden[3] cultivation introduced tree crops such as rubber (Ziegler *et al.* 2009). Introduction of rubber in the upland landscape in northern Laos was not primarily driven by the state or development agencies, but more so by smallholders themselves that had trans-border connections (Fox *et al.* 2000, Sturgeon *et al.* 2013).

The rapid integration of smallholder farmers to the regional market economy has had both positive and negative impacts on forest conservation on the ground. For instance, in some of the ethnic communities located on the borders of Laos and China, which traditionally managed extensive land areas in the uplands centered on swidden cultivation and fallow forests, smallholders began to encroach into communal forests that were traditionally reserved for ancestral spirits and for watershed protection. While such an example highlights the negative effect of market integration on community-based forest conservation, other examples from the region suggest that local communities are also capable of adapting their institutional arrangements to exclude outsider encroachment into forests and effectively manage their resources while intensifying their farming practices and seeking alternative livelihoods and thereby reducing pressure on natural resources (Cramb *et al.* 2009, Rerkasem *et al.* 2009). Examples from the rural communities across the Mekong Region suggest the importance of understanding the local context of resource use and practices, and the need to recognize the changing cultural values of communities living proximate to resources and factors enabling (or disabling) conservation practices.

In reviewing fifteen cases of protected areas management in Cambodia, Laos, and Vietnam, Sunderland *et al.* (2012) conclude that there is no silver bullet approach, but rather an adaptive process based on long-term planning that takes into consideration the local context. Such an approach requires continued efforts to coordinate across scale and between multiple stakeholders to discuss the various trade-offs involved. It also requires working across different levels of government, and with public and private sectors, as well as with local communities to build relationships founded on trust (Fisher *et al.* 2008, Sunderland *et al.* 2012). Sayer *et al.* (2012) particularly suggest the need for shared long-term vision among the stakeholders and improving the quality of engagement for various stakeholders. The latter aspect is particularly important for stakeholders who have limited voice and political representation. What is certain from the experiences of the Mekong Region is that culturally appropriate conditions for conservation need to be debated over time, not just among the scientists and policy-makers but also with meaningful engagement of local community members. This can be facilitated through shared experiences in field conservation and on-site training that not only involve scientists and professionals but community members. These activities also need to be designed to instil the long-term value of maintaining a healthy ecosystem (Sunderland *et al.* 2012, Rao *et al.* 2014).

Case study 2: Peru

In a recent paper, Sarkar and Montoya (2011) argue on behalf of a social ecology model for natural habitat and resource management, according to which, "human societies ought to be treated as irreducibly integrated with the natural systems in which they are embedded" (p. 979). It follows from this view that conservation decisions should attend to local cultural values as well as strictly scientific concerns. Sarkar and Montoya contrast their view with two alternative views about conservation – at one extreme, the "Biosphere Reserve" (BR) model emphasized promotion of sustainable development alongside conservation. At the other extreme, a call for a return to the "fortress" model of conservation argues that failure of ICDPs (Integrated Conservation and Development Programs) confounds the ideal of meeting development needs and conservation needs simultaneously. Sarkar and Montoya argue for a social ecology perspective, according to which trade-offs between these goals must be acknowledged and planned for in conservation policy. Nonetheless, they argue that one ought to prioritize the interests of those living in and on the land, as they have a vested interest in its preservation. They use their case study of the Kandozi as a case in point of how one may both acknowledge the inevitability of trade-offs and compromise and negotiation between local, national, and international stakeholders, alongside the privileging of local agents and their interests.

The Kandozi are an indigenous group of fishing communities in the "Pataza fan" or Abanico del Pastaza region of Peru, the largest tropical alluvial fan in the world, situated on the eastern catchment of the Amazon River basin. It is a site of extremely high biodiversity, threatened (much like Laos) with development, particularly petroleum extraction. The Peruvian government both conducted its own extraction, leading to serious environmental and health impacts in the region, and granted access to the Occidental Petroleum Corporation (OXY), without consent from the local community. As a traditional fisheries community, the Kandozi had long been dependent on the ecological health of their local ecosystem, and had a long tradition of managing fisheries in Lake Rimachi.

In 1945, the Peruvian government set up a Fishery Reserve in Lake Rimachi, but permitted commercial fishermen in the area, thus compromising the stock. Kandozi fishermen protested by taking control of the Reserve office, and eventually of the fishery itself, succeeding, with the help of both local and international organizations such as UNICEF and the National Indigenous Association in halting commercial fishing, and returning the stock to their previous levels. However, this process took a great deal of negotiation, education, and communication, both on the part of the Kandozi, who needed to convince WWF not to create a fully protected area in the region, and the local and international non-governmental organizations (NGOs) and government bodies, who needed to develop a management plan that both protected the fishery and local control of their resources by the Kandozi. There is ongoing conflict in the region over further petroleum development. Nonetheless, with more transparent rights and responsibilities, the Kandozi now have a greater voice in such negotiations. This case thus serves as an exemplar of integrated landscape management in the sense of respecting people in place.

Sadly, recent developments in Peru have suggested that continued oil development and failure of adequate monitoring of oil companies has led to serious compromises to the health and well-being of local indigenous people, as well as the environment. Petroperu has recently been responsible for at least 3,000 barrels of crude oil being spilled in an Amazonian region after leaks from Peru's main oil pipeline (BBC News 2016).

Insights from these case studies

Both the case of Laos and Peru are illustrative of the following challenges facing conservation post-2000. First, especially in areas rich in natural resources, competing interests in development and conservation mean that trade-offs are inevitable; while in principle supporting sustainable development and conservation simultaneously is an admirable ideal, it may not always be achieved. However, this does not necessarily mean that a "fortress" model of conservation is the only option. A variety of models of conservation planning occupy a middle ground between the "fortress" or "parks and preserves" model and the development model: the "landscape" approach, the "adaptive management" approach, and the "social ecology" approach are just a few. These have different emphases on different goals, but all base their strategies on case studies, in the struggles facing conservation planning in negotiating trade-offs between national and international stakeholders, and local or indigenous communities. According to Waylen *et al.*'s (2010) and

Reed's (2008) reviews of the literature on the role of local context on success of community-based conservation, there is a variety of factors that make for effective conservation planning:

- Supportive formal and non-formal institutions, for example, land tenure and local participation that are tailored to the local context.
- Attitudes and community participation, control, and education are essential.
- Stakeholder participation needs to be enforced and underpinned by a philosophical "buy in": participants need to believe that they are empowered, and have equal standing and trust in their co-planners.
- Stakeholders should be considered and included as early as possible and throughout the process. Any inequities in power or representation should be addressed.
- Relevant stakeholders should be analyzed and their interests and concerns represented, systematically. Benefits should include both market-based and protected area use, as well as strict conservation.
- Clear objectives need to be agreed upon by all stakeholders.
- Methods should be selected and tailored to the particular cultural context and appropriate to the education and engagement levels of participants.
- Local knowledge should be integrated.
- Supportive attitudes of stakeholders towards community participation.

Similarly, the "landscape approach" to reconciling agricultural development, conservation, and other land uses emphasizes the following principles (Sayer *et al.* 2012):

- Continual learning and adaptive management.
- Common concern entry point.
- Multiple scales.
- Multifunctionality.
- Multiple stakeholders.
- Negotiated and transparent change logic.
- Clarification of rights and responsibilities.
- Resilience.
- Strengthened stakeholder capacity.

Systematic conservation planning, in contrast, emphasizes a set of stages in conservation planning (Margules and Sarkar 2007):

- Delimit boundaries of study area.
- Identify stakeholders.
- Compile, assess, and refine natural feature and sociocultural information for the region.
- Identify constituents of biodiversity.
- Establish conservation and social goals and targets.
- Identify surrogates for biodiversity.
- Establish measures for sociocultural goals.
- Review existing conservation areas.
- Prioritize areas for conservation action.

However, Sarkar and Illoldi-Rangel (2010) have more recently modfied systematic conservation planning in a way that also emphasizes stakeholders (see also Sarkar 2014).

What all these approaches share is a commitment to close examination of the particular history and current practices of land use and current stakeholders, or cultural, economic, historical, and social factors at work in a particular region, and not only biological diversity. For, as the burgeoning literature on the history and practice of conservation in these developing regions amply illustrates, we ignore such factors at our peril. Another lesson is the significance of transparency: both with respect to the interests of stakeholders, and with our measures of biodiversity and of outcomes of relevance to stakeholders – whether economic, social, cultural, or otherwise – we cannot otherwise measure outcomes or set clear goals. As Sunderland *et al.* (2012) argue, and as the battle between those endorsing the BR model and fortress model illustrates, one of the greatest challenges facing conservation efforts is some measure of progress and providing concrete evidence of success. A third is respect for and honoring of legitimate stakeholders; though, of course, it is granted that identifying legitimate stakeholders is not simple or straightforward. A fourth is attention to multifunctionality, a necessary consequence of working toward cooperation and compromise among stakeholders. Fifth, there needs to be ongoing negotiation; this is what "adaptive management" means. In other words, evolving economies, land uses, and interests of stakeholders can and must lead to updated conservation agreements. The rules for these updates need to be as transparent as possible. Last but not least, it is also important to protect the vested interests of locals, and stakeholders with the least power or authority.

Conclusion

Though this part of our history is not emphasized in Ken Burns's majestic documentaries, our national parks – Yellowstone and the Grand Canyon – were once the home of Native American peoples, who hunted, gathered, and actively managed the diversity of these places for their own survival. These "wild" places were not empty, but used for hunting and gathering, as well as serving as locations for the creation of cultural and religious identity of native peoples. Controlled burns were used for purposes of hunting, and food plants were cultivated (Cronon 1983). We in the US are accustomed to thinking of these places as "wilderness" – unaffected by human influence. But, as anthropologists and environmental historians have documented, Native Americans lived in and actively influenced the ecosystem in the places we think of as "wild" before Europeans arrived (Borgerhoff-Mulder and Coppolillo 2004).

After we fenced off and indeed invited the military to patrol and protect our national parks from so-called "squatters and poachers," these places came to represent our ideal of wilderness (Jacoby 2003). This ideal of conservation as the preservation of "wild" places has been exported around the world by Western environmentalists. The "fortress conservation" model was put in place in Africa, South and Central America, and India for much of the early history of conservation. People were not part of the picture; the history of human occupation of the "plains of Africa" was erased in favor of viewing these places as "wilderness." Sometimes this was in service of protecting these places for the exclusive use and enjoyment of European hunters or colonialists. However, well into the twentieth century, native peoples were forcibly removed or evacuated from many of these so-called "wild" places (Dowie 2005).

The fortress conservation model has been challenged by both deliberate acts of sabotage on the part of poachers, and by groups acting on behalf of native peoples, who questioned what they saw as Western biologists' and environmentalists' "imperialist" appropriations of their native hunting grounds or homes (Guha 1997). Critics of "fortress conservation" characterized biologists' instructions to native peoples that they could not or should not continue to use resources, hunt, or live in these places as yet another instance of Western colonialism.

More recently, as we have seen, there have been many attempts by international conservation organizations to work with people living in and surrounding diverse places to jointly support conservation and promote development, so that the well-being of people living in and near biological hotspots may be improved upon simultaneously with meeting the goals of conservation. Critics of this approach, however, argue that meeting the goals of conservation and development together is at best a balancing act, and at worst, threatens our last precious remaining places of endemic diversity with deforestation, overharvesting, and destruction (Terbourgh 1999, Oates 1999). Practically speaking, Sunderland *et al.* (2012) and others studying conservation on the ground argue that particularly in areas of severe poverty like the Lower Mekong, (a) there are often severe trade-offs between conservation and development, and (b) simple measures of the effectiveness of conservation and development performance of projects are hard to come by, given the (c) complex real world, "where even obtaining clarity on shared goals among such diverse stakeholders is difficult" (p. 4).

It is clear that self-governance can play a role in sound management of natural resources in these diverse places (Sarkar and Montoya 2011). Or, local people can at least in some cases sustainably manage natural resources (Ostrom 2009); in these cases, there are often several distinctive features that make for successful management. First, locals must have secure rights to the management of their land, but also need to be in open communication with conservation biologists and NGOs, local governments, and establish unambiguous rules concerning harvesting and punishment of those who violate agreements to harvest resources sustainably. Such indigenous groups also, ideally, should have recourse to outside policing when needed to enforce the rules. These general observations have been noted by social ecologists for some time, though until recently, there has been very little uptake in the context of environmental ethics (Guha 1997). However, it is simply common sense that if a community is either living on the edge of subsistence, or has insecure rights to the resources on which their livelihood depends, they will have no ability to cultivate a practice of conservation.

Common property regimes can lead to resource degradation, for instance, when land tenure is insecure, when there is increased access to extractive technology, immigration, and new markets for resources (Borgerhoff-Mulder and Coppolillo 2004). However, investment in education and an alternative source of continued income to extraction may enable these communities to survive without submitting to the temptation to sell, or over-harvest resources out of sheer desperation.

Philosophers (and indeed, the larger public) often see the "enemy" of biodiversity protection as people (Rolston 1996). Environmental ethicists have argued that the key to changing our behavior is adopting an environmental ethic that "widens" the scope of our moral concern. The metaphor of widening the moral circle has a wonderful rhetorical appeal – it suggests inclusiveness, broader moral consideration, and more advanced development of moral judgment. Indeed, some claim that it is not only a moral ideal, but also an evolutionary inevitability – our concern for the diversity of life is, some argue, a natural and perhaps necessary transformation of ethical concern (Callicott 1989).

Some contend that the enemy of conservation is people, because a tragedy of the commons is inevitable; people cannot control their tendency to overconsumption. However, Ostrom (1999) challenged the inevitability of the tragedy of the commons. Specifically, Ostrom discovered that communities impose costs to themselves to sustainably manage resources when the benefits of such management are transparent and the potential for cheating is sufficiently reduced. The implications of these findings for conservation are profound. Instead of relying on national governments to impose restrictions, or on the endless involvement of NGOs, sustainable conservation can in some cases be achieved by empowering local people to make decisions for themselves, working in cooperation with local governments, national and international organizations.

Arguably, our failure to protect biodiversity and live sustainably stem not from a lack of moral development in our leaders or individual citizens, or an inevitable tragedy of the commons, but instead, institutionalized poverty and inequity. One solution is to grant rights to use and management to local peoples. Attention to these rights is often as, if not more, important to the preservation of these places, as the expansion of moral concern for the protection of biodiversity. Institutionalized inequity, maintained by economic and political forces, especially in resource-rich but developmentally less advanced countries, can compromise efforts in conservation on the ground in surprising and unexpected ways. In sum, it may turn out that in order to value biodiversity, we need to first place greater value on humanity than we do at present.[4]

Notes

1 We are very grateful to Sahotra Sarkar and Justin Garson for comments.
2 A "smallhold" is a piece of land and its adjacent living quarters for the smallholder and stabling for farm animals. It is usually smaller than a farm but larger than an allotment, usually under 50 acres.
3 Swidden agriculture refers to a technique of rotational farming in which land is cleared for cultivation (normally by fire) and then left to regenerate after a few years. This form of agriculture is often pejoratively called "slash-and-burn," due to a mistaken belief that it is a driver of deforestation.
4 Similar points have been made by Sarkar (2005, 2012); though, this debate over the role of "parks versus people" has been a longstanding one in the conservation literature. For a review of the state of the field, see a special issue of *Biological Conservation* (144), edited by Minteer and Miller (2011).

References

Adams, J. S., and T. O. McShane. 1992. *The Myth of Wild Africa*. New York: W.W. Norton and Co.
Anderson, D., and R. H. Grove. 1989. *Conservation in Africa: Peoples, Policies and Practice*. Cambridge: Cambridge University Press.
BBC News. 2016. "Peru Oil Spill Pollutes Amazon Rivers Used by Indigenous Group." February, available online at www.bbc.com/news/world-latin-america-35636738.
Borgerhoff-Mulder, M., and P. Coppolillo. 2004. *Conservation: Linking Ecology, Economics and Culture*. Princeton, NJ: Princeton University Press.
Callicott, J. Baird, Larry B. Crowder, and Karen Mumford. 1999. "Current Normative Concepts in Conservation." *Conservation Biology* 13(1): 22–35.
Callicott, B. 1989. *In Defense of the Land Ethic: Essays in Environmental Philosophy*. SUNY Series in Philosophy and Biology. New York: SUNY Press.
Carruthers, J. 1995. *The Kruger National Park: A Social and Political History*. Durban: University of Natal Press.
Cramb, R. A., C. J. P. Colfer, W. Dressler, P. Laungaramsri, T. L. Quang, E. Mulyoutami, N. L. Peluso, and R. L. Wadley. 2009. "Swidden Transformations and Rural Livelihoods in Southeast Asia." *Human Ecology* 37: 323–346.
Cronon, W. 1983. *Changes in the Land: Indians, Colonists, and the Ecology of New England*. New York: Hill and Wang.
Doak, D. F., V. J. Bakker, B. E. Goldstein, and B. Hale. 2014. "Moving Forward with Effective Goals and Methods for Conservation: A Reply to Marvier and Kareiva." *Trends in Ecology & Evolution* 29(3): 132–133.
Dowie, M. 2005. "Conservation Refugees: When Protecting Nature Means Kicking People Out." *Orion Magazine*.
Dowie, M. 2009. *Conservation Refugees*. Cambridge, MA: MIT Press.
Duffy, R. 2000. *Killing for Conservation: Wildlife Policy in Zimbabwe*. Oxford: James Curry.
Enters, T., and J. Anderson. 1999. "Rethinking the Decentralization and Devolution of Biodiversity Conservation." *UNASYLVA* 199.

Environment Intelligence Agency (EIA). 2012. *Checkpoints: How Powerful Interest Groups Continue to Undermine Forest Governance in Laos*. London: EIA.

Figueroa, R., and C. Mills. 2001. "Environmental Justice." In *A Companion to Environmental Philosophy*, ed. D. Jamieson, 426–436. Malden, MA: Blackwell.

Fisher, R., S. Maginnis, W. Jackson, E. Barrow, and S. Jeanrenaud. 2008. *Linking Conservation and Poverty Reduction: Landscapes, People and Power*. London: Earthscan.

Forsyth, T., and A. Walker. 2008. *Forest Guardians, Forest Destroyers: The Politics of Environmental Knowledge in Northern Thailand*. Chiang Mai: Silkworm Books.

Fox, J., D. M. Truong, A. T. Rambo, N. P. Tuyen, L. T. Cuc, and S. Leisz. 2000. "Shifting Cultivation: A New Old Paradigm for Managing Tropical Forests." *Bioscience* 50(6): 521–528.

Guha, R. 1989. "Radical American Environmentalism and Wilderness Preservation: A Third World Critique." *Environmental Ethics* 11: 71–83.

Guha, R. 1997. "The Authoritarian Biologist and the Arrogance of Anti-humanism: Wildlife Conservation in the Third World." *The Ecologist* 27: 14–20.

Jacoby, K. 2003. *Crimes Against Nature: Squatters, Poachers, Thieves, and the Hidden History of Conservation*. Berkeley: University of California Press.

Kareiva, P., and M. Marvier. "What is Conservation Science?" *Biosciences* 62(11): 962–969.

Kellert, S. R., J. N. Mehta, S. A. Ebbin, and L. L. Lichtenfeld. 2000. "Community Natural Resource Management: Promise, Rhetoric, and Reality." *Society & Natural Resources* 13(8): 705–715.

MacKenzie, John M. 1997. *The Empire of Nature: Hunting, Conservation and British Imperialism*. Manchester: Manchester University Press.

Margules, C. R., and R. L. Pressey. 2000. "Systematic Conservation Planning." *Nature* 405: 249–253.

Margules, C. R., and S. Sarkar. 2007. *Systematic Conservation Planning*. Cambridge: Cambridge University Press.

Meyfroidt, P., and E. F. Lambin. 2009. "Forest Transition in Vietnam and Displacement of Deforestation Abroad." *Progress in National Academics of Science* 106(38): 16139–16144.

Minteer, B. 2011. *Refounding Environmental Ethics: Pragmatism, Principle, and Practice*. Philadelphia, PA: Temple University Press.

Minteer, B., and T. R. Miller. 2011. "The New Conservation Debate: Ethical Foundations, Strategic Trade-offs, and Policy Opportunities." *Biological Conservation* 144: 945–947.

Neumann, R. P. 1998. *Imposing Wilderness: Struggles Over Livelihood and Nature Preservation in Africa*, vol. 4. Berkeley: University of California Press.

Norton, B. G. 2015. *Sustainable Values, Sustainable Change: A Guide to Environmental Decision Making*. Chicago, IL: University of Chicago Press.

Oates, J. 1999. *Myth and Reality in the Rainforest: How Conservation Strategies are Failing in West Africa*. Berkeley: University of California Press.

Ostrom, E. 1999. "Coping with Tragedies of the Commons." *Annual Review of Political Science* 2(1): 493–535.

Padoch, C., K. Coffey, O. Mertz, S. J. Leisz, J. Fox, and R. L. Wadley. 2007. "The Demise of Swidden in Southeast Asia? Local Realities and Regional Ambiguities." *Geografisk Tidsskrift: Danish Journal of Geography* 107(1): 29–41.

Peluso, Nancy Lee. 1992. *Rich Forests, Poor People: Resource Control and Resistance in Java*. Berkeley: University of California Press.

Peluso, N. L., P. Vandergeest, and L. Potter. 1995. "Social Aspects of Forestry in Southeast Asia: A Review of Postwar Trends in the Scholarly Literature." *Journal of Southeast Asian Studies* 26(1): 196–218.

Peluso, Nancy Lee, and Michael Watts. 2001. *Violent Environments*. Ithaca, NY: Cornell University Press.

Peluso, N. L. 2003. "A Look at Environmental Discourses and Politics in Indonesia." In *Nature in the Global South: Environmental Projects in South and Southeast Asia*, ed. P. Greenough and A. L. Tsing, 231–252. Durham, NC: Duke University Press.

Poffenberger, M. 1990. *Keepers of the Forest, Land Management Alternatives in Southeast Asia*. Manila: Ateneo de Manila University Press.

Rao, M., A. Johnson, K. Spence, A. Sypasong, N. Bynum, E. Sterling, T. Phimminith, and B. Praxaysombath. 2014. "Building Capacity for Protected Area Management in Lao PDR." *Environmental Management* 53(4): 715–727.

Reed, M. 2008. "Stakeholder Participation for Environmental Management: A Literature Review." *Biological Conservation* 141(10): 2417–2431.

Rerkasem, K., N. Yimyam, and B. Rerkasem. 2009. "Land Use Transformation in the Mountainous Mainland Southeast Asia Region and the Role of Indigenous Knowledge and Skills in Forest Management." *Forest Ecology and Management* 257(10): 2035–2043.

Rolston, H. 1996. "Feeding People versus Saving Nature?" In *World Hunger and Morality*, 2nd edn, ed. William Aiken and Hugh LaFollette, 248–267. Englewood Cliffs, NJ: Prentice-Hall.

Roy, S. S., and P. Jackson. 1993. "Mayhem in Manas. The Threats to India's Wildlife Reserves." In *Indigenous Peoples and Protected Areas*, ed. E. Kemf. London: Earthscan.

Sarkar, S. 1998. "Restoring Wilderness or Reclaiming Forests?" *Terra Nova* 3(3): 35–52.

Sarkar, S. 1999. "Wilderness Preservation and Biodiversity Conservation – Keeping Divergent Goals Distinct." *BioScience* 49: 405–412.

Sarkar, S. 2005. *Biodiversity and Environmental Philosophy*. Cambridge: Cambridge University Press.

Sarkar, S. 2008. "Norms and the Conservation of Biodiversity." *Resonance* 13: 627–637.

Sarkar, S. 2012. *Environmental Philosophy: From Theory to Practice*. Malden, MA: Wiley-Blackwell.

Sarkar, S. 2014. "Biodiversity and Systematic Conservation Planning for the Twenty-first Century: A Philosophical Perspective." *Conservation Science* 2: 1–11.

Sarkar, S., and P. Illoldi-Rangel. 2010. "Systematic Conservation Planning: An Updated Protocol." *Natureza & Conservação* 8: 19–26.

Sarkar, S., and M. Montoya. 2011. "Beyond Parks and Reserves: The Ethics and Politics of Conservation with a Case Study from Peru." *Biological Conservation* 144: 979–988.

Sayer, J., T. Sunderland, J. Ghazoul, J.-L. Pfund, D. Sheil, E. Meijaard, M. Venter, A. K. Boedhihartono, M. Day, C. Garcia, C. van Oosten and L. E. Buck. 2013. "Ten Principles for a Landscape Approach to Reconciling Agriculture, Conservation, and Other Competing Land Uses." *Proceedings of the National Academy of Sciences* 110(21): 8349–8356.

Schönweger, Olivier, Andreas Heinimann, Michael Epprecht, Juliet Lu, and Palikone Thalongsengchanh. 2012. "Concessions and Leases in the Lao PDR: Taking Stock of Land Investments." Available online at www. lsb. gov. la/decide/MoNRE_Book/Concessions-Leases-LaoPDR_2012. pdf (accessed 18 March 2013).

Schroeder, R. A. 1999. "Geographies of Environmental Intervention in Africa." *Progress in Human Geography* 23(3): 359–378.

Singh, S. 2012. *Natural Potency and Political Power: Forests and State Authority in Post-Socialist Laos*. Honolulu: University of Hawai'i Press.

Sturgeon, Janet C., *et al.* 2013. "Enclosing Ethnic Minorities and Forests in the Golden Economic Quadrangle." *Development and Change* 44(1): 53–79.

Sunderland, T. C., J. Sayer, and M.-H. Hoang. 2012. *Evidence-based Conservation: Lessons from the Lower Mekong*. New York: Routledge.

Terbourgh, J. 1999. *Requiem for Nature*. Washington, DC: Island Press.

Turner, W. R., K. Brandon, T. M. Brooks, C. Gascon, H. K. Gibbs, K. S. Lawrence, R. A. Mittermeier, and E. R. Selig. 2012. "Global Biodiversity Conservation and the Alleviation of Poverty." *BioScience* 62(1): 85–92.

Vandergeest, P., and N. L. Peluso. 1995. "Territorialization and State Power in Thailand." *Theory and Society* 24(3): 385–426.

Waylen, Kerry A., *et al.* 2010. "Effect of Local Cultural Context on the Success of Community-Based Conservation Interventions." *Conservation Biology* 24(4): 1119–1129.

Wells, M., and K. Brandon. 1992. *People and Parks: Linking Protected Area Management with Local Communities*. Washington, DC: The International Bank for Reconstruction and Development.

Woods, M. 2001. "Wilderness." In *A Companion to Environmental Philosophy*, ed. D. Jamieson, 349–361. Oxford: Blackwell.

World Wildlife Fund. 2013. *Ecosystems in the Greater Mekong: Past Trends, Current Status, Possible Futures*. WWF report.

Ziegler, A. D., J. Fox, and J. Xu. 2009. "The Rubber Juggernaut." *Science* 324: 1024–1025.

Ziv, G., E. Baran, S. Nam, I. Rodríguez-Iturbe, and S. A. Levin. 2012. "Trading-off Fish Biodiversity, Food Security, and Hydropower in the Mekong River Basin." *Proceedings of the National Academy of Sciences* 109(15): 5609–5614.

21

SYNERGIES AND TRADE-OFFS

Recognizing the many possible outcomes of community-based conservation

Jeremy Brooks

Introduction

The problem of biodiversity loss has recently begun to resurface as a major environmental issue. A number of scientific papers have highlighted alarming trends in extinction rates, range contractions, and population declines – a process known as defaunation (Dirzo *et al.* 2014). These trends have been identified for both terrestrial (Gatson and Fuller 2008, Dirzo *et al.* 2014, Newbold *et al.* 2015) and marine species (McCauley *et al.* 2015). In addition, a comprehensive literature review has outlined the largely negative effects of these declines on ecosystem functioning and the provision of ecosystem services (Cardinale *et al.* 2012). While estimating extinction rates can be difficult (Haila, this volume) and there is disagreement about how many species are lost annually (He and Hubbell 2011, De Vos *et al.* 2015), recent studies have strengthened the argument that the planet has entered a sixth mass extinction phase (Pimm *et al.* 2014, Ceballos *et al.* 2015). Importantly, this scholarly work has also been supplemented by popular works that have helped put biodiversity loss and species extinctions back on the radar (Kolbert 2014), and have promoted discussion of "re-wilding" to reimagine humanity's place in the natural world and restore key ecosystem processes (MacKinnon 2013).

As biodiversity loss has re-emerged as a core environmental issue, so have debates about how to best address the problem. These debates are largely centered on (i) the degree to which human habitation and activity detracts from, or contributes to, conservation efforts, (ii) the utility of using development as a conservation tool, (iii) the long-term viability of traditional conservation approaches that rely on protected areas (hereafter referred to as PAs), and (iv) whether economic incentives are an appropriate tool for conservation. These debates are not necessarily new. For instance, a recent debate between prominent conservationists in the journal *Conservation Biology* (volumes, 27, issue 5, and volume 28, issue 3, 2014) focuses on a similar set of issues as debates in a special issue of the same journal nearly fifteen years earlier (volume 14, issue 5, 2000). However, these debates remain important because they highlight the trade-offs that are inherent in conservation and different opinions about which trade-offs are acceptable. Sustainable development is supposed to reduce such trade-offs. However, there is now widespread recognition that trade-offs among conservation project outcomes are likely even if they are not readily apparent (McShane *et al.* 2011, Roe *et al.* 2013).

Unfortunately, there has been limited empirical research exploring the commonness of trade-offs, which types of trade-offs are most problematic for conservation projects, how projects can be designed and implemented to reduce the extent or severity of trade-offs, and when and why some projects are able to produce desired synergies and avoid certain kinds of trade-offs. Ultimately, answering these questions about trade-offs and synergies will require systematic data collection for multiple outcomes. Currently, the conservation community does not have sufficient data, collected in a sufficiently rigorous fashion (e.g. Brooks *et al.* 2013, Ferraro and Hannauer 2014, Baylis *et al.*, 2015) to draw conclusions about when synergies are likely or what kinds of conservation approaches are best suited for particular social, ecological, economic, and political contexts.

While better information about the outcomes of conservation projects will not resolve the aforementioned values-driven debates, such information may soften them and can help conservation practitioners and local communities make better decisions about how to design and implement conservation projects. Given the complexity of the social–ecological systems that conservation practitioners must navigate and the need to figure out how to "do conservation better," it is imperative to improve our data collection efforts in order to understand what contributes to the success or failure of conservation projects.

The primary purpose of this chapter is to outline the need for more, and more robust, empirical investigation of the conditions and contexts in which particular conservation strategies are most likely to succeed or fail, with an emphasis on understanding trade-offs and synergies among outcomes. The secondary purpose is to illustrate, via a crude timeline of the evolution of conservation policy, how carrying out such empirical investigations has become more challenging as the objectives associated multi-dimensional conservation approaches have grown and diversified. In the sections that follow, I present a brief summary of the history of conservation practice, discuss the need for (and challenges to) measuring multiple outcomes of conservation projects, and suggest more robust measures of the potential synergies and trade-offs among them. In conclusion, I highlight the need for systematic monitoring of multiple inputs and outcomes of conservation interventions, multiple indicators of success or failure for each outcome, standardizing relevant predictors of outcomes, and careful exploration of synergies and trade-offs.

A brief history of conservation practices and approaches

The history of conservation practice has been nicely described in previous work (Borgerhoff Mulder and Coppolillo 2005, Roe 2008). However, it is important to briefly outline the evolution of conservation approaches and motivations for these approaches because underlying motivations shape the desired outcomes and subsequently the measures used to evaluate project success.

A crude timeline could include three key stages: (1) protection of wildlife and the establishment of scenic national parks for the use and enjoyment of elites; (2) establishment of national parks and other strict PAs for the purpose of biodiversity conservation; and (3) creation of alternatives to PAs that link economic development and poverty alleviation with environmental conservation (see Roe 2008 for a more detailed outline of the evolution of conservation efforts). Broadly speaking, conservation efforts have evolved from strict protectionism towards a more inclusive, human-centered approach, which has had implications for our ability to understand how effective a given conservation strategy has been.

The earliest PAs were wildlife reserves and hunting grounds designed to ensure access to game for royal families. Over time, various types of PAs were established to protect key economic resources for colonial powers or to provide hunting reserves for colonial elites. PAs subsequently took the form of national parks, which were established in areas of scenic beauty and used for recreation and tourism (Borgerhoff Mulder and Coppolillo 2005). Evaluating the success of PAs that have been designed to meet these objectives would have been fairly straightforward and might have involved tracking game populations and habitat quality over time, or measuring the numbers of tourists and their enjoyment of the park.

Evaluating PAs that were designed to conserve biodiversity might require data on species richness or diversity, population changes, vegetation change or forest cover, and/or key ecosystem functions or services. In fact, such measures have been used to argue that PAs can be effective conservation tools and may be necessary for avoiding devastating impacts on species, habitats, and ecosystem functions. PAs have been found to be effective at retaining or recovering natural vegetation (Bruner *et al.* 2001, Armenteras *et al.* 2009), to have less reduction in game populations, less land cleared, and less grazing pressure than surrounding areas (Bruner *et al.* 2001), and to have greater abundance of large mammals than unprotected areas (Caro 1999). Recent studies using quasi-experimental designs have also found levels of deforestation to be lower in PAs than in similar locations that are not under any form of protection (Andam *et al.* 2010, Ferraro *et al.* 2011).

While there is some evidence that PAs can protect habitat and wildlife, they have been criticized on ecological, social, economic, and political grounds (see Table 21.1 for a summary of critiques). From an ecological perspective, even well-funded and well-operated PAs may not be sufficient for protecting all of the species, species assemblages, and ecosystem functions they contain. PAs may be too small and/or disconnected from other patches of suitable habitat and may be ill-suited for species whose ranges will shift with climate change (Staudinger *et al.* 2013). In addition, removal of human activities could have unintended consequences, particularly in ecosystems that have co-evolved with humans and thus depend on some degree of human disturbance. Finally, there is evidence that PAs are disproportionately established in areas of low threat (e.g. in areas with steep slopes and/or high elevations and that are distant from cities and roads; Joppa *et al.* 2009), which limits their contribution to conservation efforts (Andam *et al.* 2010, Ferraro *et al.* 2011).

In addition to concerns about ecological effectiveness, PAs have also been widely criticized for the negative social and economic impacts they can have on local communities including forced eviction, loss of livelihoods and access to resources, and/or increased poverty (Brockington 2002, West *et al.* 2008; but see Ferraro *et al.* 2011 and Andam *et al.* 2010 for evidence that, in some contexts, PAs may alleviate poverty). Even if PAs are effective at conserving biodiversity, the benefits of conservation tend to accrue at national and global scale while the costs fall most heavily on local communities who may be least able to bear the burden (Borgerhoff Mulder and Coppolillo 2005).

The emergence of alternatives to PAs

The aforementioned critiques of PAs, combined with the idea that environmental conservation and economic development could be mutually reinforcing (WCED 1987), led to the development of a suite of alternative strategies for conservation. These strategies have attempted to provide development opportunities and promote more effective conservation both within PAs and in the human-dominated matrix beyond the boundaries of PAs. These alternatives have gone by a variety of names including PA outreach, extractive reserves, eco-tourism, integrated conservation and development projects (ICDPs), co-management, community-based natural resource management (CBNRM), community-based conservation (CBC), payment for ecosystem services, and others. CBC emerged with the hope of making international conservation efforts more just, equitable, and ultimately more effective over the long term, and such projects have proliferated in recent decades. Governments, donors, and non-government organizations

Table 21.1 Critiques of PAs as a conservation strategy (derived from Borgerhoff Mulder and Coppolillo 2005)

Domain	Critiques
Ecological	– Removing human activity can negatively impact important species or ecosystem processes that co-evolved with humans.
	– PAs create islands of habitat that may not be viable for certain species, particularly charismatic megafauna with large home ranges.
	– PAs may cause "leakage" whereby resource extraction simply intensifies in the surrounding matrix, further degrading those ecosystems and isolating the PA (Ewers and Rodrigues 2008).
	– Climate change can shift habitats and home ranges such that the boundaries of PAs are no longer suitable for protecting key species.
	– PAs are sometimes established in areas with low biodiversity because it is politically or economically feasible or because of the aesthetic value of the landscape, rather than for protecting biodiversity.
	– PAs are sometimes established in areas that are under low threat because it is economically and politically feasible. These PAs will have limited value for biodiversity conservation.
Economic	– PAs are often costly to manage effectively. These costs can include physical infrastructure and enforcement of PA boundaries and regulations. The pejorative term "paper parks" refers to PAs that exist on maps and other documents but that lack sufficient financial resources to maintain operations enforcement.
	– PAs can lead to increases in crop and livestock depredation, which has economic implications for local communities and may require funds for compensation if such programs are in place.
Political	– PAs are unlikely to succeed without strong political support, which may be lacking in places where the benefits are not evident, or are perceived to flow mostly to outsiders.
	– Imposing PAs can generate ill-will and lead to enemies of conservation.
Social	– Evictions of local peoples from their homelands for the creating of a PA was not uncommon.
	– PAs can result in a loss of access to vital resources and/or livelihoods, which can exacerbate poverty.
	– PAs can prohibit local communities from accessing culturally important sites.
	– PAs can result in an increase in human–wildlife conflict.

have devoted significant resources to integrative conservation approaches. Between 1980 and 2008, nearly 75 percent of the $18 billion in international biodiversity aid was devoted to such projects (Miller 2014).

The specific details of these approaches differ (Horwich and Lyon 2007) as there is variation in the degree of emphasis on economic benefits, market integration, and transfer of land tenure and decision-making to local communities (Sarkar and Montoya 2011). A more nuanced look would require careful consideration of these differences. However, this class of conservation approaches is broadly similar in that they aim to (i) promote the welfare and cooperation of people living within or adjacent to areas of conservation interest, (ii) actively involve local communities, (iii) provide some form of economic and/or social benefit, and to varying degrees, (iv) devolve rights and responsibilities over natural resources to local communities. As a result of these similarities, I refer to this collection of conservation alternatives as community-based conservation (CBC) throughout the remainder of the chapter.

Evidence suggests that the ecological outcomes of CBC have been mixed, but generally positive. In Colombia, Armenteras *et al.* (2009) found that deforestation was significantly higher in indigenous reserves than in PAs. However, several studies have found the opposite when comparing the effectiveness of various forms of CBC with government-managed PAs. Studies in other parts of Latin America have found that indigenous reserves are as effective as PAs at reducing deforestation and avoiding forest fires (Nepstad *et al.* 2006) and that, while PAs may be more effective at limiting deforestation overall (Nolte *et al.* 2013), indigenous reserves may be more effective than PAs in areas of high threat (Nelson and Chomitz 2011, Nolte *et al.* 2013).

In addition to indigenous reserves, the ecological effectiveness of community forestry programs and other forms of CBC has also been promising. A review of forty PAs and thirty-three community forests found that the latter have lower and less variable deforestation than PAs across the tropics (Porter-Bolland *et al.* 2011). In addition, Hayes *et al.* (2006) found that user-governed systems are just as effective at maintaining vegetation density in forested areas as legally protected forests. In addition, from a global comparative database of CBC projects, Brooks *et al.* (2012) found that 58 percent of the projects that measured ecological outcomes reported ecological success.

However, ecological monitoring of CBC projects has been insufficient, which greatly limits our understanding of their effectiveness (Kremen *et al.* 1994). For instance, Brooks *et al.* (2012) found that only seventy-seven (55%) of the 139 CBC projects they review reported on ecological outcomes. Perhaps more importantly, the range of important outcomes extends beyond changes in forest cover, species richness, habitat quality, and ecosystem functioning.

The importance of measuring multiple outcomes in conservation project evaluations

As a result of the interwoven objectives of environmental conservation, economic development, and social justice, there is a need for conservation projects to serve more diverse sets of stakeholders and to provide multiple benefits at multiple scales. In addition to biodiversity conservation, current conservation programs are also expected to deliver a range of economic and social benefits and link these benefits to environmental outcomes. The diverse set of objectives that results from these expectations requires an equally diverse set of measures across space and time in order for a project to be comprehensively monitored and evaluated.

The five most common measures of CBC effectiveness are attitudinal, behavioral, ecological, economic, and social outcomes (see Fig. 21.1). The latter three are well known from the overlapping circles often used as illustrations of sustainable development. Attitudinal and behavioral outcomes have also been included in evaluations of conservation projects based on the idea that they may be important nodes in the causal connections between the other outcomes (Brooks *et al.* 2006, Waylen *et al.* 2010). For instance, behavioral changes may first require changes in attitudes (or vice versa), which could lead to improved ecological outcomes particularly if the source of the threat is largely local. Economic benefits may incentivize behavioral change and lead to improved attitudes. Positive social outcomes could lead to, or result from, cooperative action that is incentivized by economic benefits, or that emerges as a result of attitudinal change. Importantly, numerous specific measures could be used to operationalize each of these broad outcome domains (see Table 21.2). In addition, recent work has suggested that evaluations of conservation projects also measure a full suite of factors that contribute to human well-being, which could expand the list of important outcomes (Milner-Gulland *et al.* 2014).

Further complicating the evaluation process is the long list of potential predictors of each outcome (Agrawal 2001, Brooks *et al.* 2013). The set of outcomes, specific measures of outcomes, and predictor variables that are measured – and in what way – depends in part on the researcher(s) involved. Conservation biologists, anthropologists, rural sociologists, political ecologists, geographers, and political scientists all study CBC projects. These researchers each bring a disciplinary perspective with a unique set of questions, theoretical backgrounds, and methodological toolkits that can lead them to measure different independent variables and produce different kinds of data.

Thus, researchers from different disciplines can reach different conclusions about the same project because they have measured different outcomes or because they have measured the same outcome in different ways (see Table 21.2). For instance, a social scientist may collect data on economic benefits or a project's impact on social capital or community empowerment. In contrast, a natural scientist may measure rates of deforestation, species richness, or changes in the population of key species. Ideally, these outcomes would be linked such that economic success has led to, or is a result of, positive conservation outcomes. In practice this is not always the case.

In addition, researchers may focus on different dimensions of a particular outcome category. For instance, ecological success may be measured as a forest's ability to sequester carbon (Schwartzman *et al.* 2000a) or as changes in the functional populations of species within the ecosystem (Redford 1992). As such, researchers will disagree over conservation strategies

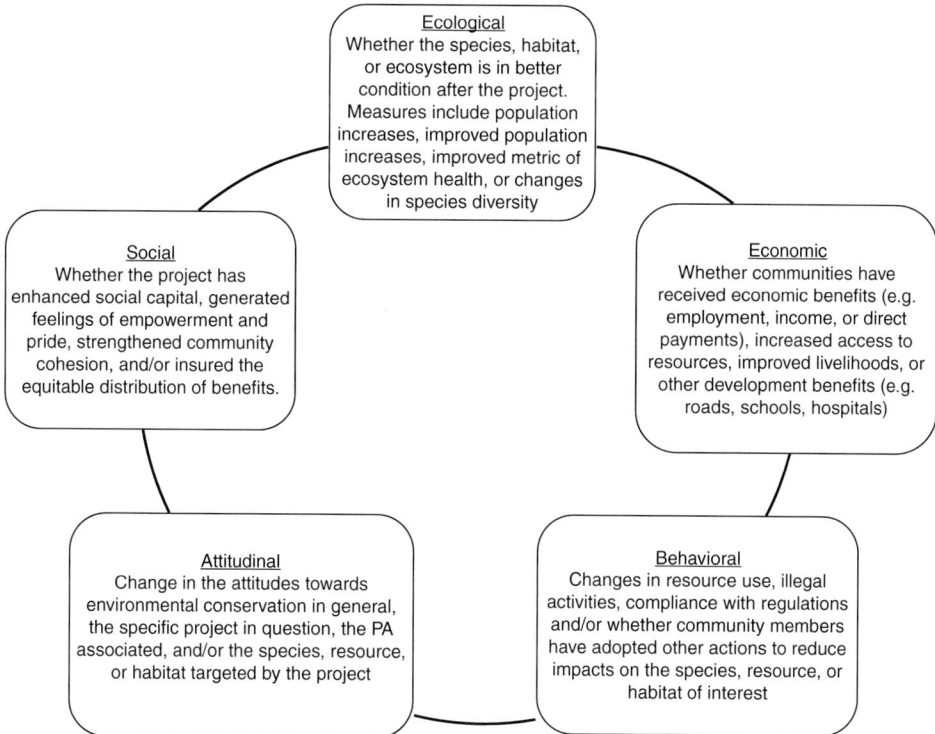

Figure 21.1 Brief descriptions of common outcome domains and potential measures within each domain for CBC projects. Potential relationships between outcomes are not depicted. However, there are theoretical justifications, and in some cases, empirical evidence for causal linkages between and among almost all sets of outcomes.

Table 21.2 Examples of analytical literature reviews or quantitative, comparative analyses exploring factors associated with multiple outcomes of CBC, CBNRM, or related conservation projects. Note that this table is not exhaustive.

Author	Type of analysis	Outcomes measured
Agrawal and Benson (2011)	Analytical literature review	**Economic** (livelihood contributions of commons) **Ecological (**sustainability of the commons) **Social** (equity of benefits allocation)
Agrawal and Redford (2006)	Analytical literature review	**Ecological** (biodiversity conservation, including attitudes, population changes and densities, off-take levels, vegetation cover) **Economic** (poverty alleviation, including income, cultural identity, equity in allocation, participation in decision-making, resource access, infrastructure)
Brooks *et al.* (2006) Brooks *et al.* (2012) Waylen *et al.* (2010)	Analytical literature reviews	**Attitudes** (towards PA, conservation project, or conservation in general) **Behaviors** [levels of resource use, or other behaviors antithetical to conservation addressed by a project (e.g. killing nuisance wildlife)] **Ecological** (condition of habitat or species of interest) **Economic** [economic (e.g. income, direct payments) or other development benefits (e.g. roads, schools, hospitals)]
Chhatre and Agrawal (2009)	Quantitative comparative analysis	**Ecological** (carbon storage) **Economic** [livelihoods index contribution to subsistence needs (firewood, fodder, green biomass for fertilizer, timber for domestic use)]
Cinner *et al.* (2012)	Quantitative comparative analysis	**Economic** (perceived impact on livelihoods) **Behavioral** (compliance with restrictions) **Ecological** [exploitation status of fishery (fish biomass)]
Gutierrez *et al.* (2011)	Analytical literature review	**Ecological (**fishery status, sustainable catches, increase in abundance, add-on conservation benefits) **Economic** (increase in unit prices, increase in catch per unit effort) **Social** (community empowerment, increase in social welfare)
Pagdee *et al.* (2006)	Analytical literature review	**Ecological** (forest conditions, environmental degradation) **Social equity** (equitable sharing of management, equitable distribution of benefits, increased investment in future productivity) **Efficiency** (meet local needs, improve living standards, alleviate poverty, reduce conflicts between locals and authorities, control corruption, resolve management imbalances, reduce misuse of forest)
Persha *et al.* (2011)	Quantitative comparative analysis	**Ecological** (tree species richness) **Economic** [forest contribution to livelihoods-significant reliance on forest products for subsistence or cash income (fuelwood, fodder, charcoal, timber, NTFPs, etc.)]
Persha *et al.* (2010)	Empirical analysis	**Ecological** (tree species richness) **Economic** [livelihoods index (firewood, fodder, timber)]

(Schwartzman *et al.* 2000b) because of (i) the different perspectives of social versus natural scientists, (ii) the multiple outcomes that require attention, (iii) the multiple scales at which outcomes could be evaluated, and (iv) the multiple dimensions of, and measures for, each of those outcomes. Because of these disagreements, and because CBC rests on the idea that multiple interrelated goals must be met to produce long-term conservation success, it is critical that project evaluations include multiple outcome domains and multiple measures within each domain (Baylis *et al.* 2015).

Potential outcome measures can also be collected using a variety of methods and study designs. For instance, much of the research on CBC consists of case studies or cross-sectional studies, with far fewer controlled studies that would allow for a better sense of change over time and project impacts after controlling for pre-existing local social, economic, and environmental conditions (Ferraro and Hannauer 2014, Baylis *et al.* 2015). In fact, all 136 CBC projects that were included in one analytical literature review were categorized as case studies, cross-sectional studies, or case control studies, and 80 percent of these studies were rated as low quality in terms of their risk of bias (Brooks *et al.* 2013). This is not a commentary on the researchers who have conducted these studies, but a reflection of the complexity of the systems in which CBC projects are initiated, and the lack of resources and time that are often available for project evaluation.

While case studies can provide valuable, detailed information about the processes and dynamics specific to a given site, they do not allow for generalization and the illumination of broader patterns, which is a problem that applies to sustainability science more broadly (Waring *et al.* 2015). Of course, local contexts and idiosyncrasies matter, and understanding broad patterns is not a substitute for knowledge of local conditions and cultural nuances. That said, uncovering broad patterns in the factors that tend to be associated with successful outcomes could provide important insights for project design and implementation. For instance, attention to cultural context has been found to be important for CBC success (Waylen *et al.* 2010, Brooks *et al.* 2012). However, what form that attention takes and how local cultural traditions, rituals, or practices are integrated depends on local conditions.

Several recent systematic reviews, meta-analyses and large-*N* analyses have begun to examine local conditions, national contexts, and project design characteristics that tend to be associated with CBC success (Brooks *et al.* 2006, Pagdee *et al.* 2006, Oldekop *et al.* 2010, Tole 2010, Waylen *et al.* 2010, Gutierrez *et al.* 2011, Brookes *et al.* 2012, Cinner *et al.* 2012). While these studies have generated important insights, they often address different outcomes, focus on particular resource domains (e.g. fisheries, forests, pasture), and, in some cases, contain small samples that make generalizations tenuous. Further, just as for ecological outcomes, monitoring of other important outcomes has been lacking. Only 19 percent of projects measured all four outcomes (attitudinal, behavioral, ecological, economic) that were included in one analysis (Brooks *et al.* 2012).

Researchers are still in the early stages of providing strong, evidence-based assessments of when and why conservation projects succeed or fail (Sutherland *et al.* 2004). This endeavor is challenging because conservation approaches have evolved to the point where the number and diversity of objectives has made it difficult for researchers (who often come from a particularly disciplinary perspective) to gain a full understanding of when and why CBC projects succeed. While there has been progress in uncovering the various conditions and project characteristics that tend to correlate with success (see above), much work remains, particularly in regards to understanding relationships among outcomes. Given that the CBC paradigm is based on the assumption that human and ecological well-being are inextricably linked, proper support for this paradigm should demonstrate empirically the interdependence of the multiple measures of success outlined above.

What are synergies and trade-offs and when do they emerge?

For CBC projects to be effective over the long term, generating multiple positive outcomes is crucial. Environmental conservation is expected to be strengthened by economic benefits and provide a foundation for future economic development. Similarly, social equality should *enhance* cooperative efforts that then increase economic rewards and conservation efforts and vice versa. Economic benefits should *result in* better attitudes towards conservation, which should lead to behavioral change that reinforces conservation efforts. Importantly, there is evidence of synergies across outcomes from case studies. For instance, in Ecuador, education programs increased a community's awareness of the ecological importance and economic value of their premontane forests. Awareness of the economic value led to involvement in environmental monitoring efforts, which enhanced community relationships (Becker *et al.* 2005). In another case, a CBC project in Namibia helped create a shared identity among communities (social benefit), which played a role in reducing hunting (behavioral change with ecological implications) (Scanlon and Kull 2009). Generally speaking, however, the degree to which "win–win" ideals have been met has been underwhelming and our understanding of when and how they can be met is still in its infancy.

Before moving ahead with a discussion of synergies and trade-offs, it is important to clarify terminology. The terms "trade-offs" and "synergies" generally refer to negative or positive relationships between outcomes, respectively. However, there is often an element of causality implied with the use of these terms that can be difficult to document in complex conservation projects. For instance, trade-offs can be thought of as a decline in one domain *in order to* achieve a gain in another. In conservation, efforts to protect a species, habitat, or resource might include outright protection or reduced access. These restrictions might lead to a trade-off between ecological outcomes and the income and livelihoods of people affected by the conservation initiative.

Phelps *et al.* (2012: 54) provide an alternative definition of tradeoffs in relation to REDD+ programs (Reduced Emissions from Deforestation and forest Degradation plus conservation). Here, a trade-off requires that one "forego the maximum return of one outcome in exchange for an increase in another outcome." This definition sets a higher bar because an increase in income in communities affected by a conservation program may still be considered a trade-off if the income gains are less than they would have been in the absence of the project. Viewed in this way, the project itself would result in less income compared to the counterfactual condition, even if it has produced an increase in income relative to baseline, pre-project levels. For instance, a community-run ecotourism project could generate income for some households, but this income might be less than that generated by other economic opportunities that could have been pursued.

The term "synergies" also implies causality in the sense that success in one outcome *contributes to* success in another outcome. However, because outcomes may be related through complex feedbacks (Miller *et al.* 2012) and because particular outcomes may be affected by entirely independent factors, it may be best to avoid alluding to causal relationships between outcomes and instead refer to the emergence of multiple positive outcomes as "win–win" scenarios, dual positives, or joint positives (Persha *et al.* 2011, Brooks 2016).

While an objective definition of synergies and trade-offs is useful, perceptions of local residents may be more important. For instance, imagine that a conservation practitioner finds evidence of improved income for a sizeable proportion of households in a community. This researcher might conclude that the project has succeeded from an economic perspective. However, the amount of the increase in local incomes or the proportion of households that benefitted may not have met the expectations of the community members themselves. This is important because the local perception of economic benefits is what will ultimately lead to attitudinal or behavioral change in favor of conservation. A mismatch between objective and subjective measures of success can lead to different interpretations of project effectiveness.

Baird (2014) provides evidence for such a mismatch in Tanzania. He found that communities closer to PAs in Tanzania were the beneficiaries of more development projects than communities that were further from PAs. However, individuals living in communities located near PAs felt that the government and PA authorities had ignored some of their needs and requests and that outsiders benefitted more from economic ventures than they had. Baird (2014) suggests that this dynamic has contributed to negative perceptions of PAs despite the overall positive development impacts in park-adjacent communities relative to communities that were not located near a national park. McShane *et al.* (2011: 968) echo the idea that subjective perceptions are important when they note the need to acknowledge "real, potential, and perceived" losses in their discussion of trade-offs. Brown (2004: 59) also emphasizes the importance of local perceptions when she defines win–win scenarios as those in which "a sufficient number and variety of forest functions can be restored in a landscape to satisfy all stakeholder groups …".

Finally, it is important to recognize that several kinds of trade-offs exist and that they can operate at different spatial and temporal scales (Brown 2004, McShane *et al.* 2011). As an obvious example, there may be a trade-off between particular outcomes like economic development and ecological outcomes, whereby market integration and poverty reduction lead to resource over-exploitation or degradation (Barrett and Arcese 1995, Wunder 2001). There may also be trade-offs between different measures within an outcome category (Persha *et al.* 2011). For example, a project may maintain habitat that benefits one species at the expense of others (Robbins *et al.* 2015) or may enhance the carbon sequestration capabilities of a local forest but at the expense of the local water supply (Cardinale *et al.* 2012). Temporal tradeoffs must also be considered. For instance, initial synergies between ecological and economic success may not persist if harvest rates become unsustainable or market fluctuations reduce the value of a resource. Similarly, a project may purposefully trade off ecological success for economic success in the short term in order to generate enough local support for the project to succeed ecologically over the long term (Miller *et al.* 2012). Finally, there may be trade-offs between and among relevant stakeholder groups. Not all stakeholders, cultural groups, communities, or households have the same preferences, nor will they all be impacted by the project in the same way or to the same degree. While a project may generate higher income on average or improved livelihoods for many communities or households, some communities or households may bear disproportionate costs. Because CBC projects can produce winners and losers, it is important for researchers to consider both average effects and heterogeneity in outcomes to gain an understanding of not just whether a project worked, but for whom (Baylis *et al.* 2015).

Early insights and the challenge of identifying predictors of win–win outcomes and trade-offs

While the idea that "win–win" outcomes are relatively easy to achieve is attractive, the emerging consensus for CBC is that trade-offs are more common than synergies (Tallis *et al.* 2008, Dahlberg and Burlando 2009, Hirsch *et al.* 2011, McShane *et al.* 2011, Persha *et al.* 2011). The conservation community is in the early stages of developing a stronger understanding of how common various kinds of trade-offs are as well as when, why, and how conservation projects can generate success across multiple outcomes. Explicit discussion of synergies and trade-offs has become more prominent in recent years (e.g. Hirsch *et al.* 2011, McShane *et al.* 2011, Roe *et al.* 2013) and a handful of reviews have examined the extent to which tradeoffs occur. For instance, Kusters *et al.* (2006) suggest that synergies between conservation and development are unlikely for commercial NTFP extraction, and Brooks (2016) and Persha *et al.* (2011) found that only about one-third of projects covered in their analyses have win–win outcomes. In contrast, a systematic review by Miller *et al.* (2012) suggests that positive feedback among outcomes (i.e. synergies) is common in the context of conservation and development initiatives.

Careful analysis of the factors associated with joint success is also relatively new. Only a handful of studies have analyzed trade-offs or joint positive outcomes using large-N data sets across national contexts. Two studies on community forest management found that local rule-making autonomy, participation in forest governance, and forest size were associated with win–win outcomes (Chhatre and Agrawal 2009, Persha *et al.* 2011). Another study suggests that capacity building and local participation in project design and implementation can be particularly important for joint ecological and social success and joint ecological and economic success (Brooks 2016). This study also suggests that older projects are more likely to produce joint successes, either because projects improve as they mature or because projects that do not generate multiple positive outcomes fail to secure the funding necessary to persist. These studies represent preliminary efforts to understand the factors associated with "win–win" outcomes but these do not necessarily help us understand complex causal processes and important mechanisms that are likely at play. Instead, additional work is needed to identify causal relationships between and among outcomes and the conditions and characteristics at multiple scales that generate those relationships.

Conclusion

The exploration of trade-offs and joint positive outcomes may be last in a sequence of needs for better knowledge of when and why CBC projects are likely to succeed. Perhaps the most important conclusion to draw from the early analyses noted above is that too few CBC evaluations collect data on multiple outcomes and explicitly consider the relationships between those outcomes (Agrawal and Benson 2011, Miller *et al.* 2012). Thus, we return to a key challenge for CBC. As the objectives of conservation projects have evolved, diversified, and expanded, the need for rigorous, well-designed programs for monitoring and evaluation has grown as has the amount of relevant data that should be collected.

I end this chapter with a short summary of insights for future research that have appeared throughout the chapter. The first is the importance of integrating thorough and systematic monitoring into project design (Stem *et al.* 2003, Sutherland *et al.* 2004, Brooks *et al.* 2006, Waylen *et al.* 2010). This is a call not just for more data, but standardized data. With some effort, academics, NGOs, local communities, and other interested stakeholders could develop a standardized set of predictor variables that is relevant and meaningful across cultural contexts. This will not necessarily be an easy task. Efforts to remedy the lack of standardized data collection and reporting (Cambridge Conservation Forum; Partnership) have produced disagreements about which variables are most critical and how best to collect and organize data related to the inputs, design, and contexts that affect CBC outcomes. However, a systematic and standardized protocol for data collection would provide consistency across project sites, thus increasing the power of analyses to provide insights into key relationships and interactions among variables.

The second suggestion is for projects to measure not just multiple outcomes but to include multiple measures for each outcome domain (see Fig. 21.1). The importance of including multiple outcome measures not only provides a more comprehensive picture of overall project success or failure, but also the ability to uncover trade-offs both within and between outcome domains. The final suggestion is to improve the quality of the studies conducted on CBC projects. Many studies of CBC projects fail to use appropriate control cases or to consider potentially important confounding variables, such as baseline socio-economic conditions, geographic location, or ecological conditions and variation. Emphasis should be given to longitudinal studies that employ quasi-experimental designs in cases where suitable "control" communities can be identified. Such approaches address the problems of endogeneity (such as CBC projects being established in already troubled areas) and the lack of counterfactuals (assuming that changes in outcomes would not have happened in the absence of the intervention). These changes will require large-scale efforts and funding increases, but significant resources are currently being devoted to conservation approaches that often produce mixed results at best.

The ultimate goal of many conservation projects remains the management or protection of ecological systems to stem biodiversity loss. However, the many means by which ecological goals are achieved vary greatly and in many cases involve economic, social, and political components. For better or worse (Soulé 2013, Karieva 2014) conservation is now inextricably linked with a broader suite of issues that directly or indirectly affect local peoples and communities. Given this reality, it is incumbent on conservation practitioners, planners, and scholars to be attentive to these multiple dimensions and to account for them in project design and in project monitoring and evaluation. Without better knowledge of the multiple outcomes of conservation projects and the synergies and trade-offs among these outcomes, we are developing conservation strategy in the dark and may be providing false expectations to the communities that will be impacted.

References

Agrawal, A. 2001. "Common Property Institutions and Sustainable Governance of Resources." *World Development* 29: 1649–1672.

Agrawal, A., and C. Benson. 2011. "Common Property Theory and Resource Governance Institutions: Strengthening Explanations of Multiple Outcomes." *Environmenal Conservation* 38: 199–210.

Agrawal, A., and K. Redford. 2006. "Poverty, Development and Biodiversity Conservation: Shooting in the Dark." Working Paper no. 26. Wildlife Conservation Society.

Andam, K. S., P. J. Ferraro, K. R. E. Sims, A. Healy, and M. B. Holland. 2010. "Protected Areas Reduced Poverty in Costa Rica and Thailand." *Proceedings of the National Academy of Sciences USA* 107: 9996–10001.

Armenteras, D., N. Rodriguez, and J. Retana. 2009. "Are Conservation Strategies Effective in Avoiding the Deforestation of the Colombian Guyana Shield?" *Biological Conservation* 142: 1411–1419.

Baird, T. D. 2014. "Conservation and Unscripted Development: Proximity to Park Associated with Development and Financial Diversity." *E&S Ecology and Society* 19, http://dx.doi.org/10.5751/ES-06184-190104.

Barrett, C. B., and P. Arcese. 1995. "Are Integrated Conservation-Development Projects (ICDPs) Sustainable? On the conservation of large mammals in Sub-Saharan Africa." *World Development* 23: 1073–1084.

Baylis, K., J. Honey-Roses, J. Borner, E. Corbera, D. Ezzine-de-Blas, P. J. Ferraro, R. Lapeyre, U. M. Persson, A. Pfaff, and S. Wunder. 2015. "Mainstreaming Impact Evaluation in Nature Conservation." *Conservation Letters* 9: 58–64.

Becker, C. D., A. Agreda, E. Astudillo, M. Costantino, and P. Torres. 2005. "Community-based Monitoring of Fog Capture and Biodiversity at Loma Alta, Ecuador Enhance Social and Institutional Cooperation." *Biodivers Conservation* 14: 2695–2707.

Borgerhoff Mulder, M., and P. Coppolillo. 2005. *Conservation: Linking Ecology, Economics and Culture.* Princeton, NJ: Princeton University Press.

Brockington, D. 2002. *Fortress Conservation: The Preservation of the Mkomazi Game Reserve, Tanzania.* Bloomington: Indiana University Press.

Brooks, J. S. 2016. "Design Features and Project Age Contribute to Joint Success in Social, Ecological, and Economic Outcomes of Community-Based Conservation Projects." *Conservation Letters*, DOI: 10.1111/conl.12231.

Brooks, J. S., M. A. Franzen, C. M. Holmes, M. Grote, and M. Borgerhoff Mulder. 2006. "Testing Hypotheses for the Success of Different Conservation Strategies." *Conservation Biology* 20: 1528–1538.

Brooks, J. S., K. A. Waylen, and M. Borgerhoff Mulder. 2012. "How National Context, Project Design, and Local Community Characterisics Influence Success in Community-based Conservation Projects." *Proceedings of the National Academy of Sciences USA* 109(52): 21265–21270.

Brooks, J. S., K. A. Waylen, and M. Borgerhoff Mulder. 2013. "Assessing Community-based Conservation Projects: A Systematic Review and Multilevel Analysis of Attitudinal, Behavioral, Ecological, and Economic Outcomes." *Environmental Evidence* 2: 1–34.

Brown, K., 2004. "Addressing Trade-offs in Forest Landscape Restoration." In *Forest Restoration in Landscapes*, ed. S. Mansourian and D. Vallauri. New York: Springer.

Bruner, A. G., R. E. Gullison, R. E. Rice, and G. A. B. da Fonseca. 2001. "Effectiveness of Parks in Protecting Tropical Biodiversity." *Science* 291: 125–128.

Cardinale, B. J., J. E. Duffy, A. Gonzalez, D. U. Hooper, C. Perrings, P. Venail, A. Narwani, G. M. Mace, D. Tilman, D. A. Wardle, A. Kinzig, G. C. Daily, M. Loreau, J. B. Grace, A. Larigauderie, D. S. Srivastava, and S. Naeem. 2012. "Biodiversity Loss and its Impact on Humanity." *Nature* 486: 59–67.

Caro, T. M. 1999. "Abundance and Distribution of Mammals in Katavi National Park, Tanzania." *African Journal of Ecology* 37: 305–313.

Ceballos, G., P. R. Ehrlich, A. D. Barnosky, A. Garcia, R. M. Pringle, and T. M. Palmer. 2015. "Accelerated Modern Human-induced Species Losses: Entering the Sixth Mass Extinction." *Science Advances Science Advances* 1: e1400253.

Cambridge Conservation Forum. Available online at www.cambridgeconservationforum.org.uk/projects/measures/.

Chhatre, A., and A. Agrawal. 2009. "Trade-offs and Synergies between Carbon Storage and Livelihood Benefits from Forest Commons." *Proceedings of the National Academy of Sciences USA* 106: 17667–17670.

Cinner, J. E., T. R. McClanahan, M. A. MacNeil, N. A. J. Graham, T. M. Daw, A. Mukminin, D. A. Feary, A. L. Rabearisoa, A. Wamukota, N. Jiddawi, S. J. Campbell, A. H. Baird, F. A. Januchowski-Hartley, S. Hamed, R. Lahari, T. Morove, and J. Kuange. 2012. "Comanagement of Coral Reef Socio-ecological Systems." *Proceedings of the National Academy of Sciences USA* 109: 5219–5222.

Conservation Measures Partnership. Available online at www.conservationmeasures.org/.

Dahlberg, A. C., and C. Burlando. 2009. "Addressing Trade-offs: Experiences from Conservation and Development Initiatives in the Mkuze Wetlands, South Africa." *Ecology & Society* 14, www.ecologyandsociety.org/vol14/iss2/art37.

De Vos, J. M., L. N. Joppa, J. L. Gittleman, P. R. Stephens, and S. L. Pimm. 2015. "Estimating the Normal Background Rate of Species Extinction." *Conservation Biology* 29: 452–462.

Dirzo, R., H. S. Young, M. Galetti, G. Ceballos, N. J. B. Isaac, and B. Collen. 2014. "Defaunation in the Anthropocene." *Science* 345: 401–406.

Ewers, R. M., and A. S. L. Rodrigues. 2008. "Estimates of Reserve Effectiveness are Confounded by Leakage." *Trends in Ecology and Evolution* 23: 113–116.

Ferraro, P. J., and M. M. Hannauer. 2014. "Advances in Measuring the Environmental and Social Impacts of Environmental Programs." *Annual Review of Environment and Resources* 39: 495–517.

Ferraro, P. J., M. M. Hanauer, and K. R. E. Sims. 2011. "Conditions Associated with Protected Area Success in Conservation and Poverty Alleviation." *Proceedings of the National Academy of Sciences USA* 108: 13913–13918.

Gatson, K. J., and R. A. Fuller. 2008. "Commonness, Population Depletion, and Conservation Biology." *Trends in Ecology and Evolution* 23: 14–19.

Gutierrez, N. L., R. Hilborn, and O. Defeo. 2011. "Leadership, Social Capital and Incentives Promote Successful Fisheries." *Nature* 470: 386–390.

Hayes, T. M. 2006. "Parks People and Forest Protection: An Institutional Assessment of the Effectiveness of Protected Areas." *World Development* 34: 2064–2075.

He, F., and S. P. Hubbell. 2011. "Species Area Relationships Always Overestimate Extinction Rates from Habitat Loss." *Nature Nature* 473: 368–371.

Hirsch, P. D., W. M. Adams, J. P. Brosius, A. Zia, N. Bariola, and J. L. Dammert. 2011. "Acknowledging Conservation Trade-offs and Embracing Complexity." *Conservation Biology: The Journal of the Society for Conservation Biology* 25: 259–264.

Horwich, R. H., and J. Lyon. 2007. "Community Conservation: Practitioners' Answer to Critics." *Oryx* 41: 376–385.

Joppa, L. N., A. Pfaff, and J. Moen. 2009. "High and Far: Biases in the Location of Protected Areas." *PLoS ONE* 4: e8273.

Karieva, P. 2014. "New Conservation: Setting the Record Straight and Finding Common Ground." *Conservation Biology* 28: 634–636.

Kolbert, E. 2014. *The Sixth Extinction: An Unnatural History.* New York: Henry Holt and Company.

Kremen, C., A. M. Merelender, and D. D. Murphy. 1994. "Ecological Monitoring: A Vital Need for Integrated Conservation and Development Programs in the Tropics." *Conservation Biology* 8: 388–397.

Kusters, K., R. Achdiawan, B. Belcher, and M. Ruiz-Perez. 2006. "Balancing Development and Conservation? An Assessment of Livelihood and Environmental Outcomes of Nontimber Forest Product Trade in Asia, Africa, and Latin America." *Ecology & Society* 11: 20–42.

McCauley, D. J., M. L. Pinsky, S. R. Palumbi, J. A. Estes, F. H. Joyce, and R. R. Warner. 2015. "Marine Defaunation: Animal Loss in the Global Ocean." *Science* 347: 1255641–1255646.

MacKinnon, J. B. 2013. *The Once and Future World: Nature As It Was, As It Is, As It Could Be.* New York: Houghton Miifflin Harcourt.

McShane, T., P. Hirsch, T. C. Trung, A. N. Songorwa, A. Kinzig, B. Monteferri, D. Mutekanga, H.V. Thang, J. L. Dammert, M. Pulgar-Vidal, M. Welch-Devine, J. P. Brosius, P. Coppolillo, and S. O'Connor. 2011. "Hard Choices: Making Trade-offs between Biodiversity Conservation and Human Well-being." *Biological Conservation* 144: 966–972.

Miller, B. W., S. C. Caplow, and P. W. Leslie. 2012. "Feedbacks between Conservation and Social–ecological Systems." *Conservation Biology* 26: 218–227.

Miller, D. C. 2014. "Explaining Global Patterns of International Aid for Linked Biodiversity Conservation and Development." *World Development* 59: 341–359.

Milner-Gulland, E. J., J. A. McGregor, M. Agarwala, G. Atkinson, P. Bevan, T. Clements, T. M. Daw, K. Homewood, N. Kumpel, J. Lewis, S. Mourato, B. Palmer Fry, M. Redshaw, J. M. Rowcliffe, S. Suon, G. Wallace, H. Washington, and D. Wilkie. 2014. "Accounting for the Impact of Conservaton on Human Well-being." *Conservation Biology* 28: 1160–1166.

Nelson, A., and K. M. Chomitz. 2011. "Effectiveness of Strict vs. Multiple Use Protected Areas in Reducing Tropical Forest Fires: A Global Analysis Using Matching Methods." *PLoS ONE* 6: e22722–22736.

Nepstad, D., S. Schwartzman, B. Bamberger, M. Santilli, D. Ray, P. Schlesinger, P. Lefebvre, A. Alencar, E. Prinz, G. Fiske, and A. Rolla. 2006. "Inhibition of Amazon Deforestation and Fire by Parks and Indigenous Lands." *Conservation Biology* 20: 65–73.

Newbold, T., L. N. Hudson, S. L. L. Hill, S. Contu, I. Lysenko, R. A. Senior, L. Borger, D. J. Bennett, A. Choimes, B. Collen, J. Day, A. De Palma, S. Dıaz, S. Echeverria-Londono, M. J. Edgar, A. Feldman, M. Garon, M. L. K. Harrison, T. Alhussein, D. J. Ingram, Y. Itescu, J. Kattge, V. Kemp, L. Kirkpatrick, M. Kleyer, D. Correia, C. D. Martin, S. Meiri, M. Novosolov, Y. Pan, H. R. P. Phillips, D. Purves, A. Robinson, J. Simpson, S. L. Tuck, E. Weiher, H. J. White, R. M. Ewers, G. Mace, J. P. W. Scharlemann, and A. Purvis. 2015. "Global Effects of Land Use on Local Terrestrial Biodiversity." *Nature* 520: 45–50.

Nolte, C., A. Agrawal, K. M. Silvius, and B. S. Soares-Filho. 2013. "Governance Regime and Location Influence Avoided Deforestation Success of Protected Areas in the Brazilian Amazon." *Proceedings of the National Academy of Sciences USA* 110: 4956–4961.

Oldekop, J. A., A. J. Bebbington, D. Brockington, and R. F. Prieziosi. 2010. "Understanding the Lessons and Limitations of Conservation and Development." *Conservation Biology* 24: 461–469.

Padgee, A., Y. Kim, and P. J. Daugherty. 2006. "What Makes Community Forestry Management Successful: A Meta-study from Community Forests Throughout the World." *Society & Natural Resources* 19: 33–52.

Persha, L., H. Fischer, A. Chhatre, A. Agarwal, and C. Benson. 2010. "Biodiversity Conservation and Livelihoods in Human-Dominated Landscapes: Forest Commons in South Asia." *Biological Conservation* 143(12): 2918–2925.

Persha, L., A. Agrawal, and A. Chhatre. 2011. "Social and Ecological Synergy: Local Rulemaking, Forest Livelihoods, and Biodiversity Conservation." *Science* 331: 1606–1608.

Phelps, J., D. A. Friess, and E. L. Webb. 2012. "Win Win REDD+ Approaches Belie Carbon Biodiversity Trade-offs." *Biological Conservation* 154: 53–60.

Pimm, S. L., C. N. Jenkins, R. Abell, T. M. Brooks, J. L. Gittleman, L. N. Joppa, P. H. Raven, C. M. Roberts, and J. O. Sexton. 2014. "The Biodiversity of Species and Their Rates of Extinction, Distribution, and Protection." *Science* 344: 987–997.

Porter-Bolland, L., E. A. Ellis, M. R. Guariguata, I. Ruiz-Mallen, S. Negrete-Yankelevich, and V. Reyes-Garcia. 2011. "Community Managed Forests and Forest Protected Areas: An Assessment of their Conservation Effects across the Tropics." *Forest Ecology Management* 268: 6–17.

Redford, K. H. 1992. "The Empty Forest." *Bioscience* 42: 412–422.

Robbins, P., A. Chhatre, and K. Karanth. 2015. "Political Ecology of Commodity Agroforests and Tropical Biodiversity." *Conservation Letters* 8: 77–85.

Roe, D. 2008. "The Origins and Evolution of the Conservation-Poverty Debate: A Review of Key Literature, Events and Policy Processes." *Oryx* 42: 491–503.

Roe, D., J. Elliot, C. Sandbrook, and M. J. Walpole. 2013. *Biodiversity Conservation and Poverty Alleviation: Exploring Evidence for a Link, Conservation Science and Practice*. Hoboken, NJ: Wiley-Blackwell.

Sarkar, S., and M. Montoya. 2011. "Beyond Parks and Reserves: The Ethics and Politics of Conservation with a Case Study from Peru." *Biological Conservation* 144: 979–988.

Scanlon, L. J., and C. A. Kull. 2009. "Untangling the Links between Wildlife Benefits and Community-based Conservation at Torra Conservancy, Namibia." *Development Southern Africa* 26: 75–93.

Schwartzman, S., A. Moreira, and D. Nepstad. 2000a. "Rethinking Tropical Forest Conservation: Perils in Parks." *Conservation Biology* 14: 1351–1357.

Schwartzman, S., D. Nepstad, and A. Moreira. 2000b. "Arguing Tropical Forest Conservation: People versus Parks." *Conservation Biology* 14: 1370–1374.

Soulé, M. 2013. "The 'New Conservation'." *Conservation Biology* 27: 895–897.

Staudinger, M. D., S. L. Carter, M. S. Cross, N. S. Dubois, J. E. Duffy, C. Enquist, R. Griffis, J. J. Hellmann, J. J. Lawler, J. O'Leary, S. A. Morrison, L. Sneddon, B. A. Stein, L. M. Thompson, and W. Turner. 2013. "Biodiversity in a Changing Climate: A Synthesis of Current and Projected Trends in the US." *FEE Frontiers in Ecology and the Environment* 11: 465–473.

Stem, C. J., J. P. Lassoie, D. R. Lee, D. D. Deshler, and J. W. Schelhas. 2003. "Community Participation in Ecotourism Benefits: The Link to Conservation Practices and Perspectives." *Society & Natural Resources* 16(5): 387–413.

Sutherland, W. J., A. S. Pullin, P. M. Dolman, and T. M. Knight. 2004. "The Need for Evidence-based Conservation." *Trends in Ecology and Evolution* 19: 305–308.

Tallis, H., P. Kareiva, M. Marvier, and A. Chang. 2008. "An Ecosystem Services Framework to Support Both Practical Conservation and Economic Development." *Proceedings of the National Academy of Sciences USA* 105: 9457–9464.

Tole, L. 2010. "Reforms from the Ground Up: A Review of Community-based Forest Management in Tropical Developing Countries." *Environmental Management* 45: 1312–1331.

Waring, T., M. A. Kline, J. S. Brooks, S. H. Goff, J. Gowdy, M. A. Janssen, P. Smaldino, and J. Jacquet. 2015. "A Multilevel Evolutionary Framework for Sustainability Analysis." *Ecology and Society* 20: 1–34.

Waylen, K. A., A. Fischer, P. K. McGowan, S. J. Thirgood, and E. J. Milner-Gulland. 2010. "The Effect of Local Cultural Context on Community-based Conservation Interventions: Evaluating Ecological, Economic, Attitudinal, and Behavioural Outcomes." Systematic Review no. 80, Collaboration for Environmental Evidence, Birmingham, UK.

WCED. 1987. *Our Common Future. World Commission on Environment and Development.* New York: Oxford University Press.

West, P., J. Igoe, and D. Brockington. 2008. "Parks and Peoples: The Social Impact of Protected Areas." *Annual Review of Anthropology* 35: 251–277.

Wunder, S. 2001. "Poverty Alleviation and Tropical Forests: What Scope for Synergies?" *World Development* 29: 1817–1833.

PART VI

Biodiversity and other environmental values

22

ECOLOGICAL SUSTAINABILITY

J. Baird Callicott

Introduction

Sustainability is most often associated implicitly with human economic activities or complex systems of human economic activities that are constrained by *environmental* limits – sustainable agriculture, for example. Here I critically examine two classic concepts of sustainability: *sustainable yield* and *sustainable development*. From the perspective of current environmental concerns – and especially concerns about the loss of biodiversity – both of these familiar and essentially economic ways of understanding sustainability are fatally flawed, short of a thorough conceptual overhaul. I introduce and commend a third way of understanding sustainability that I dub *ecological sustainability*. In passing, I also critique a recently developed alternative way of understanding *sustainability* proffered by Bryan G. Norton.

The classic or *received* concepts of sustainable yield and sustainable development and the concept of sustainability à la Norton do not foster the conservation of biodiversity, as I explain in the course of my analysis of them. Putting the received concept of sustainable yield into practice in fishery management, the home turf of that idea, has only succeeded in producing a biodiversity crisis in the world's oceans (Halpern *et al.* 2012). With greater sophistication in its application, however, the concept of sustainable yield might be a useful tool for the conservation of biodiversity (Lewison *et al.* 2015). The received concept of sustainable development has been actually antithetical to the conservation of biodiversity, although in reconceived form, it too might be enlisted in the support of the conservation of biodiversity. Norton's concept of sustainability is more abstract than material, more cultural than biological, and thus is at best only indirectly connected with the desideratum of biodiversity conservation. Ecological sustainability, as I define it, is tailored to the desideratum of biodiversity conservation and related environmental goals.

Sustainable yield

The concept of *sustainable yield* has two permutations: *maximum* sustainable yield and *optimum* (sustainable) yield, the former being the more prominent and having the longer history of application. Classic natural resource management is focused on single species – this kind of game animal, that kind of commercially harvested fish – the goal of management being to attain a *sustainable yield* of the target "resource."

The principal application of the concept of *maximum sustainable yield* has been in marine fishery management (Schaefer 1954). The formulae for calculating maximum sustainable yield are appropriately sophisticated mathematically – involving variables for the species' fertility, growth rate, age of reproductive maturity, and so on (Bousquet *et al.* 2008). They are not, however, at all sophisticated ecologically. The logistic equations for representing the dynamics of species populations are assumed to actually represent the dynamics of species populations – and that assumption is questionable, to say the least (Botkin 2012). Little attention is paid to the feedback loops between those species that strongly interact with the target species and even less attention is paid to those other species that strongly interact with species that strongly interact with the target species (Larkin 1977, Pikitch *et al.* 2014). As a result, the calculation of population growth, stabilization, harvest, and rebound bear little resemblance to the dynamics of actual populations of target species on the ground or in the oceans. In general, the mismatch between classic population models and actual population facts is attributable to the impact of artificially fluctuating populations of the target species on other species in the biotic communities of which the target species is a member. Thus the impact of the fluctuating populations of those species on that of the target species is unaccounted for and may even be chaotic (in the mathematical sense of the word) and thus be unpredictable in principle (Walters and McGuire 1996). This classical way of understanding sustainable yield is further vitiated ethically by the assumption in economics that the natural world is nothing but a "pool" of resources existing to service human wants as well as human needs.

The concept of *optimum* (sustainable) *yield* – the word "sustainable" is usually omitted – is more inclusive of desiderata beyond the "resource" and the industry (mainly commercial fishing) exploiting it. Its locus classicus is the US Magnuson-Stevens Fishery Conservation Act (Public Law 96–265) of 1976 (and amended as recently as 1996) which defines "optimum yield" thusly:

> The term "optimum," with respect to the yield from a fishery, means the amount of fish which (A) will provide the greatest *overall benefit* to the Nation, particularly with respect to food production and recreational opportunities, and taking into account the protection of marine *ecosystems*; (B) is prescribed as such on the basis of maximum sustainable yield of the fishery, *as reduced* by any relevant economic, social, or *ecological factor*; and (C) in the case of an overfished fishery, provides for rebuilding to a level consistent with producing the maximum sustainable yield in such fishery.

To generalize the concept of optimum sustainable yield from this legislative definition, three aspects stand out: (1) benefits to humans collectively other than narrow commercial benefits are taken into account; (2) by definition, the optimum sustainable yield will be lower than the maximum sustainable yield; (3) in calculating the optimum sustainable yield, ecological factors are taken into account, as is the protection of ecosystems. The optimum sustainable yield concept has not been operationalized with the same degree of mathematical precision as has been attempted in the case of the maximum sustainable yield concept.

Personally, I am not convinced that the concepts of maximum and optimum sustainable yield cannot be successfully operationalized. Calculating the former is a technical problem – a daunting one to be sure, but a technical problem nevertheless. Increased mathematical sophistication by, for example, the use of non-linear equations; increased modeling sophistication by, for example, the use of cellular-automata and multi-agent-based models; and the ever increasing computational power of evolving computer hardware and software, all may make it possible for modelers to approximate the behavior of target species populations in a complex nexus of coupled natural and human systems (Filatova *et al.* 2013).

Sustainable development

In response to the politically sensitive tension between economic development for impoverished human populations and the destruction of natural capital that, historically, accompanies human economic development, the concept of *sustainable development* was offered as a resolution. Perhaps it would be possible to eat our environmental cake and have it too – by means of *sustainable* development. Many skeptical environmentalists of a Malthusian bent regard "sustainable development" to be an oxymoron (Bartlett 2012). If poverty is defined as a low standard of living and a low standard of living is equated with low levels of consumption of goods and services, then economic development for impoverished human populations entails increased consumption of goods and services, which in turn entails increased conversion (destruction) of natural capital on a finite, small planet – which is not sustainable – or so the neo-Malthusians grumble (McKee 2012).

Taking the concept of sustainable development seriously might require challenging the rather conventional assumptions that its critics make (Roudsepp-Hearne *et al.* 2010). Might not an impoverished people's standard of living be increased without a corresponding and proportionate increase in its levels of consumption of goods and services? Possibly, I would think. But if not, can increases in the consumption of goods and services be attained without a corresponding and proportionate increase in the conversion of natural capital? Again, possibly, I would think. And here too, the problems are partly technical, but they are also partly psychological, cultural, and political. Better agricultural techniques, for example, might increase food production without converting more land to crops; alternative techniques of generating energy might raise standards of living without adversely affecting the environment. Re-envisioning the good life – from consumerist values to those of association, environmental aesthetics, education, vegetarianism, human and ecological health – might increase standards of living without an inevitable adverse impact on the natural environment (Ybarra 2016).

The very famous definition of "sustainable development" in the Brundtland Report – "development that meets the needs of the present without compromising the ability of future generations to meet their own needs" – reveals a deeper, essentially philosophical flaw in the concept of sustainable development, if so defined (United Nations World Commission on Environment and Development 1987). To be sure, the Brundtland Report's definition of "sustainable development" is laudable to the extent that it implicitly distinguishes between *needs* and *wants*. But, as things presently stand in our consumerist culture, satisfying wants is thought to be a need. So, even the Brundtland Report's laudable focus on needs is less than satisfactory, in my opinion, without explicitly divorcing needs from wants. That it makes no reference whatever to environmental constraints on development is what is really striking about this definition. Consider a possible alternative definition: development that meets the needs of the present and those of future generations without compromising environmental quality – without increasing atmospheric, oceanic, or freshwater pollution; without intensifying species extinction and the erosion of biodiversity; without accelerating deforestation and global climate change …

Doesn't the Brundtland Report's definition of sustainable development at least imply sustainably harvesting renewable natural resources (sustainable yield) and preserving ecological services, without which future generations could not meet their own needs? The Report may not mention environmental conservation explicitly, but it implies the necessity of sustaining environmental quality as a condition for future generations meeting their own needs, doesn't it? No, it doesn't. In neo-classical economics there is the "gross substitution axiom" (Fisher 1972). As a heavily exploited natural resource becomes scarce its price increases, making investment in finding a substitute become increasingly attractive. Thus, there is no need to conserve any particular natural resource (Solow 1993). For example, when we begin to run short of copper for making telephone wires, someone will (as someone did) invent fiber optics and then wireless cell phones. Such accumulated anecdotal evidence suggests that market forces will always stimulate the discovery or invention of substitutes for any natural resource – from petroleum to Madagascar periwinkles. As one species of marine finfish is harvested to commercial extinction, the fishing industry just exploits another; when all marine finfish stocks are depleted, someone will figure out how to make sushi with a fish-flavored and fish-textured soy product. According to this prevailing way of thinking in economics, the present generation can, therefore, meet its own needs – including the presently prevailing need to satisfy wants – by rapidly exploiting existing organic natural resources to commercial if not to biological extinction and bequeathing a legacy of wealth and technology and a culture of business and inventiveness to future generations, by means of which they can meet their own needs (Barry 1989).

As noted in the beginning, sustainability is most often implicitly associated with human economic activities or complex systems of human economic activities that are constrained by *environmental* limits. But as the foregoing considerations demonstrate, the two classic conceptual permutations of the naked concept of sustainability – *sustainable yield* and *sustainable development* – are embedded in the larger conceptual domain of neo-classical economics and share its assumptions. The former regards the natural environment as a pool of resources and the latter, even more radically, regards the natural environment not at all. To arrive at an environmentally sensitive concept of sustainability, we might make a fresh start by examining the naked concept of *sustainability*.

The naked concept of sustainability

Sustainability is a property of an activity or complex system of activities capable of going on and on indefinitely, if not forever. Any human activity, whether economic or otherwise, may or may not be sustainable, nor does sustainability necessarily reference environmental constraints. Gambling losses of a thousand dollars a week, week after week, are not sustainable by a person who earns 50,000 dollars a year. A national health-care system the cost of which increases by 10 percent per year is not sustainable.

Because the property of sustainability is a property of activities – processes – the concept of sustainability implicitly, but obviously, references time. Further, although the temporal scale of sustainability is rarely specified, we unconsciously and often very vaguely scale the temporal parameters of sustainability relative to the human activity or complex system of human activities in question. Continuous running is a human activity that may or may not be sustainable, but the temporal scale of sustained continuous running is calibrated in minutes and hours. I am unable to sustain running for much more than thirty minutes. One might ask of a passing marathon runner, on pace for a three-hour finish, if that pace is sustainable. I would suppose that a national health-care system that is designed to function well for a century would be deemed sustainable. But I would also suppose that an agricultural system would be deemed unsustainable if it too were designed to function well for only a century, after which period it might well collapse. One would probably not, however, withhold the sustainability descriptor from an agricultural system that functioned well for a million years followed by collapse. Where should the temporal parameter of sustainability be drawn for an agricultural system? At a thousand years, two thousand, ten thousand …? I don't know. My point simply is that the temporal scale of sustainability is not infinity and, though often both implicit and vague – as the word "indefinitely" in the first sentence of the previous paragraph is meant to suggest – the temporal scale for assessing sustainability is relative to the human activity or complex system of human activities in question.

Weak sustainability, strong sustainability, and Nortonian sustainability

Another, more recent definition of "sustainability" is proffered by Bryan G. Norton (2005). Norton rejects what is sometimes called "weak sustainability," the economistic understanding of sustainability that the famous Brundtland Report definition of *sustainable development* invites and which Barry (1989) and Solow (1993) explicitly endorse. He also rejects what is sometimes called "strong sustainability" – the view that artificial capital cannot substitute for natural capital – championed by ecological economists and conservation biologists. What they regard as a foundational natural legacy for future generations – biodiversity, for example – Norton demeaningly refers to as "stuff," a term that Norton (2005: 306) attributes to Brian Barry, citing a "personal communication," and a term which Norton warmly embraces as perfectly expressing his own assessment of strong sustainability. Advocates of strong sustainability think it necessary to bequeath some natural "stuff," such as biodiversity, to future generations. Norton (2005: 336) goes on to define "sustainability" in terms of a less tangible legacy: "*sustainable* and *sustainable development* are not themselves general *descriptors* of states of societies or cultures but rather refer to many specific sets of commitments on the part of specific societies, communities, and cultures to perpetuate place-based values and project them into the future."

What those place-based values are is up to specific societies, communities, and cultures to determine democratically. While I certainly agree with Winston Churchill (2013: 574) that "democracy is the worst form of government, except for all the other forms," I am less sanguine than is Norton that place-based values will always project environmental quality into the future. That may happen in some specific societies, communities, and cultures – say Ashland, Oregon and Boulder, Colorado. But it may not happen in others – say Talladega, Alabama and Odessa, Texas, where the place-based values likely to be projected into the future center, respectively, on NASCAR racing and wildcat drilling for the petroleum industry's gold standard – light, sweet Texas crude. Thus even the concept of sustainability proffered by Norton makes no provision for environmental constraints on human economic activities – biodiversity being just a bunch of stuff.

Further, Norton's implicit bioregionalism – a geographical ontology of "place" and a socio-logical ontology of "*specific* societies, communities, and cultures" – is quaint. The twenty-first century is characterized by globality – for better or worse. Our most daunting and urgent environmental challenge is *global* climate change. Specific places, societies, communities, and cultures can no longer be considered in isolation from one another and can no longer plausibly make independent commitments to sustainability. Without international cooperation to abate carbon emissions, no place will look like the same place fifty years from now, no matter what commitments to place-based values its denizens make and project into the future.

Twenty-first-century ecologists have also abandoned the concept of closed, self-regulating, ontologically robust ecosystems (Pickett and Ostfeld 1995). Ecosystems are partly artifacts of ecological hypotheses as Arthur Tansley (1935) long ago noted. Once bounded as such, they are porously open to all sorts of comings and goings, from invasive organisms to minerals from afar blown in on the wind and washed down by the rain. For example, the scant fertility of Amazonian soils depends on African dust blown across the Atlantic Ocean on the prevailing easterlies (Korin *et al.* 2006). If global climate change affects wind patterns across the tropical Atlantic Ocean, the Amazon rainforest could be starved for nutrients, with a huge loss of biodiversity. For another example, the drought cycle, constraining the ecosystems in the American Southwest that Aldo Leopold (1979) long ago noticed and wondered about, is traceable to the Pacific Decadal Oscillation cycle (McCabe *et al.* 2004). Sustainability has an implicit spatial scale, as well as an implicit temporal scale, and it is becoming increasingly clear that that scale is necessarily global, not local. Sustainability is thus not a matter of "place-based" independent commitments on the part of "specific societies, communities, and cultures" as per Norton's definition.

Although problematic in the ways just indicated, there is a kernel of insight in Norton's understanding of sustainability in terms of commitments to values projected into the future. De-dichotomized – by rejecting Norton's zero-sum stuff versus values dichotomy – and scaled up from place to planet, Norton has something valuable to contribute to an understanding of sustainability in the third millennium and its inescapably global dimensions. To that something I return at the conclusion of this chapter.

Ecological sustainability

To fully capture the implicit sense of sustainability ambient in contemporary discourse – human economic activities or complex systems of human economic activities that are constrained by *environmental* limits – I would characterize sustainability ecologically as follows. The human economy is a subset of the economy of nature. Need it be recalled that the words "ecology" and "economy" were coined from the same Greek root, *oikos*, meaning "home"? Less well known, the notion of an "economy of nature" is our oldest proto-ecological metaphor, coined by none other than Carl Linnaeus in the eighteenth century (Linnaeo 1749). Sustainable human economic activity would not disrupt the globally integrated ecological processes and functions of our global home, planet Earth. Obversely put, sustainable development consists in devising artificial human economic systems – call them econo/ecosystems – that are modeled on and symbiotically adapted to the economy of nature (the global ecosystem, the living biosphere). The macro-economy of nature is the model for a sustainable human economic microcosm.

How does the economy of nature work? In other words, what are the fundamental principles of the economy of nature?

That it is solar-powered is its first principle. Green plants convert radiant energy from the sun into potential chemical energy by stripping the carbon atoms from atmospheric carbon dioxide (CO_2) and photosynthesize them with the hydrogen and oxygen atoms in water (H_2O) to form carbohydrates – for example, glucose ($C_6H_{10}O_6$), among the simplest such compounds. Respiring herbivorous animals oxidize (burn, as it were) the hydrocarbons composing plants, converting the chemical energy therein into kinetic energy to power their metabolisms and locomotion, finally scattering the degraded energy in thermal form to their surroundings.

That it is cyclical is the second principle of the economy of nature. While energy flows through the economy of nature on a one-way trip from solar source to the sink of outer space, the materials of nature's economy cycle. In the simplified example of the previous paragraph, green plants dump the excess oxygen atoms – generated as a waste product (O_2) of their photosynthesis of carbohydrates – into the atmosphere. When respiring animals use atmospheric O_2 to burn carbohydrates they, in turn, dump the waste products (CO_2 and H_2O) of oxidization back into the atmosphere, completing a cycle. But the carbon cycle is just one of many materials cycles in the biosphere – others involve nitrogen, phosphorus, calcium, potassium, iron, and other plant "nutrients."

In the simplest terms, therefore, an ecologically sustainable human economy – an econo/ecosystem – will be powered by solar energy and will thoroughly recycle its material components.

A good, explicit example of devising artificial econo/ecosystems that are adapted to and modeled on natural ecosystems is the perennial polyculture envisioned by Wes Jackson (1980) and in process of development at the Land Institute in Salina, Kansas. Jackson's model is the prairie, which, under various names (grassland, savannah, steppe) on various continents (North and South America, Asia, Africa) is a globally distributed biome. According to Jackson (1980), four kinds of plants must constitute a perennial polyculture: cool-season grasses, warm-season grasses, legumes, and sunflowers – all perennialized. Fossil-fuel input is reduced by eliminating the need for annual plowing; and artificial nitrogen input is eliminated by mixing grains with nitrogen-fixing legumes. If successful (or rather, when successful, as progress has been notable), Jackson's would (will) represent a second agricultural revolution, the first, based on annual monocultural grasses, having occurred worldwide about 10,000 or 11,000 years ago (Barker 2009). Another example is the suite of artificial econo/ecosystems going under the rubric of "industrial ecology" – the general idea being that the waste product of one industry is the resource of another (Ehrenfeld 2004). Spent fry oil, to take but one instance, from fast-food restaurants can be (and is in some places) the feedstock for the manufacture of biodiesel fuel for the engines of cars, trucks, and buses (Kemp 2006).

Sustainability ethics

This way of conceptualizing sustainability – as *ecological* sustainability – points to the emergence of a new field in philosophy: sustainability ethics. Sustainability ethics emerges at the interface of environmental ethics and business ethics and is currently being developed at several institutions.[1] Ecological sustainability revolves around the concept of economy – the planetary economy of nature and the globalized human economy. Ecological sustainability is a matter of adapting human economic systems to and modeling them on the economy of nature in which the globalized human economy is embedded and in relation to which it should stand as microcosm to macrocosm. This ecological way of *understanding* sustainability devolves from environmental ethics. But to *actually achieve* sustainability, so understood, would seem to me to be a matter of business practice, constrained by environmental ethics wed to business ethics.

If the concept of economy is what ecological sustainability revolves around, why would actually achieving sustainability not be a matter of applied economics, instead of being a matter of a hybrid environmental-business ethics constraining business practices? Because economics is a descriptive social science, with predictive ambitions based on a number of controversial assumptions, among them the invidious gross substitution axiom. That all values are preferences and that human welfare consists of maximizing "preference satisfaction" is also among those controversial assumptions (Randall 1988). To treat all values as preferences enables economists to quantify the way humans value various things in a monetary metric for purposes of comparison – thus making economics a totalizing discipline, beyond the reach of which there is nothing. But all values are not preferences. I may prefer strawberry ice cream to chocolate, but I do not just prefer say democracy to autocracy. My personal preferences are often frustrated by democratically determined public policies and when they are I sometimes wish the world were run by a philosopher king, such as Plato envisioned in *The Republic*. But such a dark thought is fleeting because autocracy flouts the principles of human autonomy, dignity, and equality – things that moral human beings *value,* not things that we merely prefer. Such "transcendent values" (as we might call them) constrain preferences and things so valued should not be subjected to shadow pricing and benefit–cost analysis (see Callicott, "What Good Is It, Anyway?", this volume). Nor does human welfare consist of maximizing preference satisfaction; to think that it does is to think like a two-year-old.

Business ethics consists precisely in exploring various transcendent values, and recommending that such values negatively constrain and positively inspire and guide human economic activities (Goodpaster *et al.* 2006). At a minimum, business ethics demands that human economic activities be constrained by the aforementioned values of human autonomy, dignity, and equality which would, among other things, prohibit cost-cutting by employing child or prison labor in workplaces that are a hazard to life, lung, liver, or limb. Business ethics should also inspire and guide human economic activities by recommending that businessmen and women set a goal of enhancing the commonwealth as well making a private profit.

At the conjunction of environmental and business ethics, sustainability ethics would demand that transcendent *environmental* values also serve as constraints on and inspirational guides for human economic activity. Such values would range from a concern for animal welfare to clean air and water to biodiversity conservation. Sustainability ethics, as I conceive it, is business ethics informed not only by the extra-economic values of traditional social ethics, but also by those of contemporary environmental ethics, all in service of the goal of achieving ecological sustainability – that is, attuning the human economic microcosm to the macrocosmic economy of nature.

Comparative sustainability ethics and economics

We live in a global environment culturally as well as environmentally. Thus any viable twenty-first-century understanding of sustainability must include representation from non-Western traditions of thought (Callicott 1994). Therefore, comparative philosophy can make a crucial contribution to sustainability ethics in a variety of ways. At the metaphysical level or the level of "first philosophy" in Aristotle's sense of the term, many of the great philosophical traditions of Asia locate human being, human society, and, yes, the human economy in a cosmic setting and urge a harmonization of things human with cosmic goings on. Such a harmonization of "society" with "heaven" (cosmically not religiously understood) is at the heart of Confucianism (Tucker 2013). Daoism, of course, also leaps to mind, especially with its concepts of the *dao, wu wei*, and *feng shui* (Lai 2014). At the moral level, certainly Buddhism draws a sharp distinction between needs and wants and between preferences and values proper, and offers as searing a critique as one could hope to find of a consumerist lifestyle (Kaza 2014).

Comparative environmental philosophy and ethics has by now been very well explored and developed in theory and in the abstract (Callicott and Ames 1989, Callicott and McRae 2013). Thus I can add little of substance here – which has not already been worked out by scholars more able and more expert than I – to that body of scholarship. As it seems to me, less well explored territories for working by comparative environmental philosophers and ethicists are the ecologically sustainable economies that remain in place in various parts of the world.

The dream of Japanese conservationists to restore the iconic crested ibis (*Nipponia nippon*), first on Sado Island and eventually on Honshu and all the other Japanese islands, depends directly on ecologically sustainable paddy-rice cultivation (Nishimiya and Hiyashi 2010). The toki, as the bird is affectionately known in Japan, was driven to near global extinction by hunting, pollution, and habitat loss. It nests in pine trees on wooded uplands and feeds on amphibians, small fish, and small mammals that inhabit nearby wetlands. In the traditional paddy-rice system of farming, uplands are kept forested to protect the watershed and stream flow from which water is diverted to the small rice fields, in which the crested ibis feeds symbiotically with the rice cultivation (Totman 1989). The paddies are, as it were, artificial wetland habitat for the bird. And the toki performs pest-and-vermin control services for the Japanese rice farmer. One might even imagine that the toki and the Japanese rice paddy mutually co-evolved, they are so well adapted to one another.

But what of distinctly philosophical substance does one find in this example of ecological restoration and ecologically sustainable economics? The future of philosophy – mainstream, as well as comparative and environmental – lies in a turning outward to other intellectual disciplines for the purpose of interdisciplinary problem solving. The inward-looking, over-specialized pre-occupations of twentieth-century philosophy are no longer viable as the twenty-first century unfolds. Ornithologists work with the bird. Ecologists work with the bird's habitat. Agronomists work with the rice cultivation. Anthropologists work with the local folklore. Political econo-mists work with Japanese government policy-makers to keep rice cultivation small-scale and traditional. (Japan prohibits the importation of rice, thus helping to preserve its traditional rural landscape and thus helping to restore its emblematic avifauna.) And what can philosophers do? We can synthesize all the knowledge gathered by the other disciplinarians into a coherent whole and locate it in a fitting metaphysical superstructure – in the case of Sado Island, into a syncretic Buddhist–Shinto metaphysical superstructure (Grapard 1984, Toyoda 2013). And, most importantly, we can try to understand how such local efforts to achieve ecological sustainability and ecological restoration can be integrated into a global network of resonant efforts around the world.

Philosophical value added

What unique contributions can the more general discipline of philosophy make toward enhanc-ing our understanding of what ecological sustainability is and how ecologically sustainable goals can be accomplished? The foregoing is a discussion based in the discipline of philosophy pri-marily aimed at enhancing our understanding of what sustainability is and, secondarily, aimed towards how sustainable goals can be accomplished. So let's stand back and look at what is going on.

First, the discipline of philosophy is concerned with conceptual clarification and precision of expression. Sometimes philosophical conceptual clarification and precision of expression can be annoying because it may seem too fastidious and (indeed be as well as seem) trivial. But clarifying the difference between the concepts of sustained yield, sustainable development, and ecological sustainability goes to the substantive heart of the matter at hand.

Second, the discipline of philosophy is concerned with critical thinking. Here I subjected the concept of maximum sustained yield and the definitions of "sustainable development" in the Brundtland Report and "sustainability" in Norton's book to extensive critical examination and clearly exposed the inadequacies of each. I also subjected the assumptions of mainstream economics to a more cursory regimen of critical thought.

Third, the discipline of philosophy is concerned with values. In other disciplines, especially those with scientific pretenses, values are often treated as purely subjective, personal, arbitrary, and irrational. Preferences may be all these things, but, as noted, not all values are preferences. And values are of the utmost importance. Whether we admit it or not, values drive all human activity and all public policy. Recognizing values for what they are, exposing them to people who have been taught to marginalize and trivialize them, and stressing their ultimate impor-tance is a central and unique contribution that the discipline of philosophy can make to any human endeavor, including enhancing our understanding of what sustainability is and how ecological sustainability can be achieved.

Fourth, the discipline of philosophy is concerned with ontology, with what exists and what does not. I criticized Norton's ontological assumptions about geography and society, his hypostatization of place and specific communities. To make more explicit another ontological issue lurking in the foregoing discussion, let me ask: do ecosystems exist robustly – as robustly as say a snake or a monkey? I hinted that they do not. They exist, but their existence is less robust than that of snake or a monkey. That ecosystems have fuzzy boundaries is the least ontologically problematic thing about them (Pickett *et al.* 1992). More problematic ontologically, their (often fuzzy) boundaries are determined by the particular interests of the ecologists who investigate them (Tansley 1935, Allen and Hoekstra 1992). In the instance mentioned in the foregoing discussion, the Amazon rainforest is taken to be an ecosystem sustained by dust blown in from Africa. But by an ecologist interested in energy flows in Amazonian food webs, the boundaries of the ecological object of study would be drawn very much more narrowly. Ecosystems are thus partly scientific artifacts, they come into being, partly, only when interrogated by ecologists.

Fifth and finally, the discipline of philosophy is concerned with epistemology – with what we know, how we know it, and the limits of human knowledge. Scientific knowledge is often privileged epistemologically in comparison with other knowledge claims. At the same time, uncertainty is becoming an ever more prominent epistemological issue in science itself, an epistemological issue that is also ever more prominent at the interface of science and policy and politics (Jamieson 2014). Epistemological issues hardly appear at all in the foregoing discussion, but perhaps they should have taken more of a center-stage position. Who gets to declare what is and what is not a sustainable human activity? Perhaps unconscionably, I disparaged the epistemic credentials of Talladega rednecks and Odessa roughnecks. Perhaps arrogantly, I dismissed without hesitation their various claims to know what is and what is not sustainable. A signal and laudable virtue of Norton's definition of "sustainability" is that it is not so dismissive as is my account of ecological sustainability. But should every claim to knowledge be treated with respect and given an equal hearing with every other? What about the knowledge claims of Scientologists who believe that they arrived on a spaceship from another planet or evangelical Christian fundamentalists who claim to know that the Earth has only been around for about six thousand years? Aren't some extra-scientific knowledge claims just as worthy of contempt as others are of respect? How do we determine which knowledge claims warrant appropriate respect and which warrant appropriate contempt? That is an epistemological problem; and epistemology is a subdiscipline of philosophy.

The future of ecological sustainability inquiry

Looking toward the future, we might ask, what are the most important topics of future inquiry that ecological sustainability theorists need to investigate?

Assuming that ecological sustainability requires a multidisciplinary approach and that ecological sustainability theorists inhabit a wide variety of disciplines, then the topics of future inquiry might be sorted by discipline. In the foregoing discussion I mention two disciplines in which topics of future inquiry are quite technical: perennial polyculture agronomy and biofuels engineering. In the case of other disciplines, economics, for example, the concept of ecological sustainability challenges some very deep assumptions: substitutability, for one example; the reduction of all values to preferences. More basically, the concept of ecological sustainability challenges the economistic understanding (in terms of preference satisfaction) of what it means for human beings to fare well. In addition, the concept of ecological sustainability challenges the practice by economists of discounting future benefits at the current rate of interest. Ecological sustainability is, as noted, an inherently temporal concept, and implicitly references an indefinite future. As long as future benefits are discounted at a significant rate – or, really, at any discount rate at all – economics as it is now practiced will be an impediment to an ecologically sustainable economy, not an analytic aid toward achieving it.

As these reflections indicate, the topics of future inquiry by ecological-sustainability theorists are myriad. I will limit my remaining reflections on topics of further inquiry for ecological-sustainability theorists in the discipline of philosophy.

One topic for future philosophical inquiry is the temporal scale of ecological sustainability – as I have characterized it in terms of attuning the human economy to the economy of nature. The spatial scale, as noted, must be global because of two primary considerations. The first relates to the indistinct ontology of the Earth's many ecosystems in comparison with the robust ontology of the living Earth itself considered as a systemic unit, the biosphere (Margulis 1998). The second relates to the globalization of the human economy, rendering its local and regional subdivisions no more ontologically independent than are its local and regional ecosystems. So what *temporal* scale is fitting and correlative to the global scale of a sustainable human economy? It lies somewhere between a scale calibrated in decades and one calibrated in millions of years, but narrowing that vast range to the proper temporal scale is indeed a topic of future philosophical inquiry.

A second topic for future philosophical inquiry into sustainability is articulating a fitting moral ontology. To whom – or perhaps better, to what – is the present generation obligated by way of our widely recognized obligation to "future generations"? Resolving this problem does not depend on a definitive resolution to the problem of determining a fitting temporal scale for ecological sustainability; nevertheless, that topic for future inquiry puts this one in temporal perspective. Whatever the fitting temporal scale of ecological sustainability, we can be sure that it extends beyond what we might think of as the *extended personal future*.

By "extended personal future," I mean that for members of my generation, future generations already exist; and my age mates and I are personally acquainted with some members of those future generations. At about seventy-five years of age, my own personal future is very limited. My son is about forty-five years old and may well live another forty or fifty years. That forty- or fifty-year temporal horizon is thus part of my *extended* personal future – because I am personally very concerned about the world my son will live in as he approaches my present age. His son, my grandson, is about fifteen years old. He may well live another seventy or eighty years, enlarging my extended personal future – members of the future generations with whom I am personally acquainted and for whose welfare I am personally concerned – almost throughout the whole twenty-first century. If I live to see the birth of my grandson's son or daughter, my extended personal future – as here characterized, my window of personal concern for the future persons I care about – will grow well into the twenty-second century.

The paradox of future-generations ethics

I want to help insure that the world that my son, grandson, and great grandchild will live in remains habitable and pleasant. For that to happen, I and others with similar concerns – that would be pretty much everyone who has an extended personal future – must join together and effect radical changes in the human economy. Creating a sustainable econo/ecosystem (as here characterized) will be necessary to mitigate and to adapt to global climate change. But radical changes in the human economy will entrain radical changes in human lifestyles, which will affect the reproductive chances and choices of the members of my grandson's generation. That leads to the "Parfit paradox," as I call it, or the "non-identity problem" as it is more widely known (Parfit 1984). If business goes on as usual, my grandson and other members of his generation will meet, mate, and have one set of children. But if radical changes in the human economy are initiated, with attendant lifestyle changes, my grandson and other members of his generation will meet, mate, and have a set of children different from the set of children they would have had had business gone on as usual. So it is impossible for me to be concerned *now* about the individual welfare of my unborn great great grandson or -daughter, *qua individual*, and that of his or her unborn cohort, *considered individually*. For if I succeed in helping make the world of the twenty-second century habitable and pleasant, those individuals who would have existed, had business gone on as usual, would not exist at all; rather, different individuals would exist in their stead.

Assuming that the temporal scale of ecological sustainability, as I understand it, extends beyond one century into the future, the ontology of future-generations ethics must also be scaled up from one of individual persons to something more proportional to the temporal scale of ecological sustainability, whatever that turns out to be (Callicott 2013). As just noted, the appropriate *objects* of moral considerability for intergenerational ethics cannot be yet unborn future *individual* human persons – because their very existence or non-existence, as individual persons, will depend on what we presently do or leave undone. Well then, is the appropriate object of moral considerability for intergenerational ethics *Homo sapiens* – the human species? I think not. The natural lifespan of a large mammalian species is about a million years – which, I would think, exceeds the temporal scale of ecological sustainability. Further, our highly adaptive species may well survive an environmental apocalypse and hang on with a much-reduced population living in a scarcely habitable and very unpleasant world – if we can trust the vision of post-apocalypse fiction writers, film-makers, and Gaia theorists (Lovelock 2006, McCarthy 2006, Robertson 2013).

No, I think that the appropriate object(s) of moral considerability for intergenerational ethics is (are) neither individual human persons nor the human species, but global human civilization and biodiversity. In addition to biodiversity and more generally environmental quality, ecological sustainability is about projecting things like the visual arts, music, poetry, literature, science, philosophy, architecture, law, and government into the future. Sustaining biodiversity and biospheric health and integrity are necessary conditions for sustaining these other things that enrich and ennoble human life. One aspect of what I am getting at here is "stuff" *sensu* Norton – the cultural achievements of the past – Plato's *Republic*, the *Bhagavad Gita*, *Hamlet*, the "Mona Lisa," the Taj Mahal, … The other aspect is the visual arts, music, poetry, literature, science, philosophy, architecture, law, and government of the future. That's essentially a commitment to perpetuate cultural values and project them into the future *sensu* Norton. But aren't such future achievements of global human civilization equally subject to the Parfit paradox? Yes, they are – if we think of the future achievements of global human civilization in terms of particular artifacts. Radical changes in the human economy will result in radical changes in the visual arts, music, poetry, literature, science, philosophy, architecture – and, indeed in law and government (because the latter two must also be fully globalized). Making the radical lifestyle, economic, and political changes necessary to avert catastrophic global climate change will surely affect the cultural productions of the future; but if we go forward with business as usual, cultural productions worthy of being called the visual arts, music, poetry, literature, science, and philosophy may not be forthcoming at all in a world of failed states and remnant bands of survivalists led by sociopathic warlords. So it's not a matter of what specific visual arts, music, poetry, literature, science, philosophy, and architecture will be forthcoming than a matter of perpetuating these human *activities* themselves. Thus moral concern for the future achievements of human civilization, though subject to, is not vitiated by the non-identity problem.

Therefore, in some ways, this view of ecological sustainability as adjusting the human economy to the economy of nature aligns with that of Bryan Norton, but in other ways it does not.

This way of understanding sustainability, like Norton's way, projects values into the future (those of global human civilization), but it also projects stuff (the precious past achievements of human civilization) in addition to biodiversity and environmental health and integrity. Indeed, without projecting the stuff as exemplars of the values of global civilization it would be virtually impossible to project into the future the values that they embody and exemplify. Thus to understand sustainability in terms of projecting global human civilization into the future is to project both the accumulated artifacts (stuff) of human civilization – the extant literature, art, music, architecture, science, philosophy that is our precious human heritage – and the culture, the values, and skills which fostered the creation of those artifacts.

And what of sustaining biodiversity? Well, we have long realized that the species concept in biology is an abstraction – and one that remains contested (Agapow *et al.* 2004). Organisms exist concretely and by a process of abstraction we organize them into Linnaean categories. Biodiversity, conceptually comprising not just species diversity, but genetic diversity and various kinds of ecological diversity – for example, β and γ, as well α diversity (*sensu* Whittaker 1960) – is an even more abstract idea. So to project biodiversity into the future by actively conserving it (whatever "it" may be) is not exactly to project "stuff" into the future. However, without projecting the actual organisms that embody biodiversity along with their ecological inter-relationships we cannot project that non-thing that we call biodiversity into the future.

The further difference I have with Norton's understanding of sustainability is partly a matter of scale and partly a matter of governance. He thinks locally; I am suggesting we think globally about sustainability; nor should we view it through a narrow window of time, but in the open vista of thousands of years past and thousands of years future. Moreover, the sustainability of biodiversity, ecological integrity, and human global human civilization should not depend on the vagaries of democratic governance. In the foregoing, I have more than once affirmed the preeminent value of democracy as a form of government. But democratic governance, at least in the case of the United States, is constrained by a constitution, just as it was in ancient Athens. Sustaining global human civilization and biodiversity should be more like an article of the implicit constitution of the social and natural contracts (*sensu* Serres 1992); it should be beyond the vagaries of parliamentary legislation and town-hall decision-making.

That's just my opening gambit concerning the moral ontology of intergenerational ethics. The question has scarcely been asked, and so a definitive answer is impossible to even broach. This along with many others is an important topic of future inquiry in the nascent field of ecological-sustainability philosophy.

Note

1 For example, The Markkula Center of Applied Ethics at Santa Clara University, www.scu.edu/ethics/practicing/focusareas/environmental_ethics/lesson4.html (accessed 12 February 2015), The Rock Ethics Institute of Penn State University, http://rockethics.psu.edu/climate/sustainability-ethics (accessed 12 February 2015), and the Center for Humans and Nature, www.humansandnature.org/ethical-aspects-of-sustainability-article-51.php (accessed 12 February 2015).

References

Agapow, P. M., G. M. Mace, O. R. Biinda-Emons, and K. A. Crandall. 2004. "The Impact of Species Concept on Biodiversity." *Quarterly Review of Biology* 79: 161–179.
Allen, T. F. H., and T. W. Hoekstra. 1992. *Toward a Unified Ecology*. New York: Columbia University Press.
Barker, G. 2009. *The Agricultural Revolution in Prehistiry: Why Did Foragers Become Farmers?* New York: Oxford University Press.
Barry, B. 1989. *Democracy, Power and Justice: Essays in Political Theory*. New York: Oxford University Press.
Bartlett, A. 2012. "Reflections on Sustainability and Population Growth." In *On the Brink: Environmentalists Confront Overpopulation*, ed. P. Cafaro and E. Christ, 29–40. Athens: University of Georgia Press.
Botkin, D. 2012. *The Moon in the Nautilus Shell: Discordant Harmonies Reconsidered*. New York: Oxford University Press.
Bousquet, N., T. Duchesne, and L. P. Rivest. 2008. "Redefining the Maximum Sustainable Yield for the Schaefer Population Model Including Multiplicative Environmental Noise." *Journal of Theoretical Biology* 254: 65–75.

Callicott, J. B. 1994. *Earth's Insights: A Multicultural Survey of Ecological Ethics from the Mediterranean Basin to Australian Outback*. Berkeley: University of California Press.

Callicott, J. B. 2013. *Thinking Like a Planet: The Land Ethic and the Earth Ethic*. New York: Oxford University Press.

Callicott, J. B. and R. T. Ames, eds. 1989. *Nature in Asian Traditions of Thought: Essays in Environmental Philosophy*. Albany: State University of New York Press.

Callicott, J. B. and J. McRae, eds. 2013. *Environmental Philosophy in Asian Traditions of Thought*. Albany: State University of New York Press.

Churchill, W. 2013. *Churchill by Himself: The Definitive Collection of Quotations*. London: Ebury Press.

Ehrenfeld, J. 2004. "Can Industrial Ecology Be the 'Science of Sustainability'?" *Journal of Industrial Ecology* 8: 1–3.

Fisher, F. M. 1972. "Gross Substitutes and the Utility Function." *Journal of Economic Theory* 4: 82–87.

Filatova, T., P. H. Vurburg, D. C. Parker, and C. A. Stanard. 2013. "Spatial Agent-based Models for Socio-ecological Systems." *Environmental Modeling and Software* 45: 1–7.

Goodpaster, K. E., L. L. Nash, and H. C. deBettignies. 2006. *Business Ethics: Policies and Persons*, 4th edn. New York: McGraw Hill Irwin.

Grapard, A. 1984. "Japan's Ignored Cultural Revolution: The Separation of Shinto and Buddhist in Meiji: A Case Study." *History of Religions* 23: 240–265.

Halpern, B S., C Longo, D. Hardy, K. L. McCloud, J. F. Samhouri, S. K. Katona, *et al.* 2012. "An Index to Help Assess the Health and Benefits of the Global Ocean." *Nature* 488: 615–620.

Jackson, W. 1980. *New Roots for Agriculture*. Lincoln: University of Nebraska Press.

Jamieson, D. 2014. *Reason in Dark Time: Why the Struggle Against Climate Change Failed – and What It Means for Our Future*. New York: Oxford University Press.

Kaza, Stephanie. 2014. "Acting with Compassion: Buddhism, Feminism, and the Environmental Crisis." In *Environmental Philosophy in Asian Traditions of Thought*, ed. J. B. Callicott and J. McRae, 71–98. Albany: State University of New York Press.

Kemp, W. 2006. *Biodiesel: Basics and Beyond*. Tamworth, Ont.: Aztext.

Korin, I, Y. J. Kaufman, R. Washington, M. C. Todd, J. V. Matins, and D. Rosenfeld. 2006. "The Bodélé Depression: A Single Spot in the Sahara that Provides Most of the Mineral Dust to the Amazon Forest." *Environmental Research Letters* 1: 01405.

Lai, K. L. 2014. "Conceptual Foundations for Environmental Ethics: A Daoist Perspective." In *Environmental Philosophy in Asian Traditions of Thought*, ed. J. B. Callicott and J. McRae, 173–195. Albany: State University of New York Press.

Larkin, P. A. 1977. "An Epitaph for the Concept of Maximum Sustained Yield." *Transactions of the American Fisheries Society* 106: 1–11.

Leopold, A. 1979. "Some Fundamentals of Conservation in the Southwest." *Environmental Ethics* 1: 131–141.

Lewison, R., A. J. Hobday, S. Maxwell, E. Hazen, J. R. Hartog, and D. C. Dunn. 2015. "Dynamic Ocean Management: Identifying the Critical Ingredients of Dynamic Approaches to Ocean Resource Management." *BioScience*, doi: 10.1093/biosci/biv018.

Linnaeo, Carolo. 1749. *Specimen Academicum de Oeconomi Naturae*. Stockholm: Isacus J. Bieberg.

Lovelock, J. 2006. *The Revenge of Gaia*. New York: Basic Books.

Margulis, L. 1998. *Symbiotic Planet: A New Look at Evolution*. New York: Basic Books.

McCabe, G. J., M. A. Palecki, and J. L. Betancourt. 2004. "Pacific and Atlantic Ocean Influences on Multidecadal Drought Frequency in the United States." *Proceedings of the National Academy of Sciences* 101: 4136–4141.

McCarthy, C. 2006. *The Road*. New York: Alfred A. Knopf.

McKee, L. 2012. "The Human Population Footprint on Global Biodiversity." In *On the Brink: Environmentalists Confront Overpopulation*, ed. P. Cafaro and E. Christ, 91–97. Athens: University of Georgia Press.

Nishimiya, H., and K. Hiyashi. 2010. "Reintroducing the Japanese Crested Ibis in Sado, Japan." TEEBweb (accessed 30 January 2015).

Norton, B. G. 2005. *Sustainability: A Philosophy of Adaptive Ecosystem Management*. Chicago, IL: University of Chicago Press.

Parfit, D. 1984. *Reasons and Persons*. Oxford: The Clarendon Press of Oxford University.

Pickett, S. T. A., and R. S. Ostfeld. 1995. "The Shifting Paradigm in Ecology." In *A New Century for Natural Resource Management*, ed. R. L. Knight and S. F. Bates, 261–279. Washington, DC: Island Press.

Pickett, S. T. A., V. T. Parker, and P. L. Fielder. 1992. "The New Paradigm in Ecology: Implications for Conservation Biology Above the Species Level." In *Conservation Biology*, ed. P. L. Fielder and S. K. Jain, 50–65. New York: Chapman and Hall.

Pikitch, E. K., K. J. Rountas, T. E. Essington, C. Santora, D. Pauly, R. Watson, *et al.* 1992. "The Global Contribution of Forage Fish to Marine Fisheries and Ecosystems." *Fish and Fisheries* 15: 43–74.

Randall, A. 1988. "What Mainstream Economists Have to Say about Biodiversity." In *Biodiversity*, ed. E. O. Wilson, 217–223. Washington, DC: National Academy Press.

Robertson, M. 2013. *Crash: A Post-apocalyptic Tale*. Kindle Edition.

Roudsepp-Hearne, C., G. D. Peterson, M. Tengö, E. M. Bennett, T. Holland, K. Benessaiah, G. K. McDonald, and L. Pfeifer. 2010. "Untangling the Environmentalists' Dilemma: Why is Human Well-being Increasing while Ecosystem Services Degrade?" *BioScience* 60: 576–589.

Schaefer, M. B. 1954. "Some Aspects of the Dynamics of Populations Important to the Management of the Commercial Marine Fisheries." *Bulletin of the Inter-American Tropical Tuna Commission* 1: 25–56.

Serres, M. 1992. *Le Contrat Naturel*. Paris: Éditions Flammarion.

Solow, R. M. 1993. "Sustainability: An Economist's Perspective." In *Economics of the Environment: Selected Readings*, ed. R. Dorfman and N. Dorfman, 131–138. New York: W. W. Norton.

Tansley, A. G. 1935. "The Use and Abuse of Vegetational Concepts and Terms." *Ecology* 16: 284–307.

Totman, C. 1989. *The Green Archipelago: Forestry in Preindustrial Japan*. Berkeley: University of California Press.

Toyoda, M. 2013. "Revitalizing Local Commons." *Environmental Ethics* 35: 279–293.

Tucker, M. E. 2013. "The Relevance of Chinese Neo-Confucianism for the Reverence for Nature." In *Environmental Philosophy in Asian Traditions of Thought*, ed. J. B. Callicott and J. McRae, 133–148. Albany: State University of New York Press.

United Nations World Commission on Environment and Development. 1987. *Our Common Future*. New York: United Nations.

Walters, C., and J. McGuire. 1996. "Lessons for Stock Assessment from the Northern Cod Collapse." *Reviews in Fish Biology and Fisheries* 6: 125–137.

Whittaker, R. H. 1960. "Vegetation of the Siskiyou Mountains, Oregon and California." *Ecological Monographs* 30: 279–338.

Ybarra, P. S. 2016. *The Good Life: Mexican American Writing and the Environment, 1848–2010*. Tucson: University of Arizona Press.

23

ECOLOGICAL RESTORATION AND BIODIVERSITY CONSERVATION

Justin Garson

Ecological restoration is the practice of restoring ecosystems in the aftermath of damage or neglect. It is motivated by the observation that there are fewer and fewer undamaged places in the world to conserve. If we want to achieve conservation goals such as biodiversity, sustainable habitats, and the semblance of wild nature, we must deliberately construct, or reconstruct, ecosystems for that purpose.

Ecological restoration is closely connected with the idea of historical fidelity, that is, the idea that we should be as faithful as possible to an ecosystem's own past when restoring that ecosystem. The demand for historical fidelity raises a series of philosophical questions and challenges. First, can we ever truly restore nature, or are we always just manufacturing artificial imitations of it? Second, is there any non-arbitrary way to select a historical baseline for restoration projects? Third, if we could, in principle, construct an ecosystem that satisfies all of our other conservation goals such as biodiversity and sustainability, but that does not exhibit fidelity to any past, wouldn't historical fidelity be irrelevant?

Another question that should be raised here is, how does ecological restoration relate to biodiversity conservation? Are these two goals tightly interwoven? That is, should we think of ecological restoration as a *means* to biodiversity conservation? Or are these goals independent of one another? Is it possible for these goals to conflict? And when they do conflict, which one should we prefer?

The following has six parts. In the first section, I will explicate the idea of ecological restoration, with emphasis on the idea of historical fidelity. In the second, I will enumerate some of its benefits. In the third, fourth and fifth sections, I will raise, and respond to, the three philosophical challenges noted above. In the concluding section, I describe how ecological restoration relates to other conservation goals, specifically, biodiversity conservation.

What is ecological restoration?

Ecological restoration is, most broadly, the attempt to restore, recreate, or reconstruct ecosystems in the aftermath of damage or neglect. Although many people have attempted to define precisely what "ecological restoration" means, few definitions have been universally acknowledged (see Higgs 2003: 107–110, for an overview of such attempts). In the following, I'll adopt the definition most recently given by the Society for Ecological Restoration (SER) as, "the process of assisting the recovery of an ecosystem that has been degraded, damaged, or destroyed."[1] The practice of ecological restoration is largely motivated by the rapid disappearance of habitats that have been only minimally transformed for human use. Simply put, we are running out of places to conserve.

"Restoration" can refer either to a process or to a product, though the SER definition emphasizes process. As a *process*, it refers to a special kind of activity carried out by groups of people. As a *product*, it can refer to the outcome of that activity, namely, a restored ecosystem. There is also a difference between "restoration ecology" and "ecological restoration." Restoration ecology is a science; more specifically, it is a sub-discipline within environmental management devoted to the practice of ecological restoration. Restorations differ greatly in spatial scale and in terms of the sheer magnitude of intervention they entail. Large-scale restoration efforts may include procedures such as the removal of relics of recent human activity, reintroduction of native species, reconstruction of soils, redirection of waterways, sediment dredging of lakes, controlled burns, and so on.

A theme that runs through most of the proposed definitions of "restoration" is the idea of *historical fidelity* (Higgs 1997, 2003). Historical fidelity is the idea that we ought to make the ecosystem resemble the way it was in the past. We should be faithful to that ecosystem's own history. Moreover, historical fidelity is not about making a given ecosystem resemble just any past ecosystem. Rather, it has to do with being faithful to the past of that very ecosystem or place. (The point is obvious when we think of restoring a house or car.) The practice of re-creating English gardens in colonial-era India was not an instance of historical fidelity in this sense, though it does exhibit fidelity to some historical state.

Achieving the goal of historical fidelity can be very costly, knowledge-intensive, and labor-intensive, depending on how seriously we take it. That is because it may be difficult to assess what exactly the past ecosystem was like. In these cases, we must rely on sources of data such as paleontology, comparative data, and photographs or oral history (Desjardins 2015: 88). This is a cost-intensive and labor-intensive process, so it requires some philosophical justification.

I'll put the point more sharply: suppose (even hypothetically) that we could design an ecosystem that satisfies various goals such as enhancing biodiversity, sustainable land-use, wildlife habitat, educational and other social and cultural benefits, and so on, but that is entirely unprecedented for that region. The phased reconstruction of Governor's Island, a small island just south of Manhattan, comes to mind (see Garson 2014). In the early twentieth century, Governor's Island was a landfill for debris generated by the construction of Manhattan's subway system. In the 1960s it was given to the Coast Guard to serve as a residential base, but by the turn of the century the island served no meaningful conservation or recreation purpose. Recently, the Governor's Island Trust – a city-funded non-profit group – started to implement a massive reconstruction project. The flat, barren landscape will be outfitted with rolling hills. (The demolished remains of the Coast Guard buildings will provide the infrastructure for the hills.) Non-native plants and shrubs will be planted along the perimeter of the island. These plants were chosen partly to enhance marine biodiversity and partly to withstand the effects of climate-change-induced sea-level rise. A network of thin, paved paths will traverse the island for cyclists and joggers. The idea of historical fidelity played no role in the design or implementation of the plan, yet the plan, once fully implemented, will serve a number of valuable goals. Some ecologists are vocal advocates for just this sort of novel ecosystem construction (e.g. Hobbs *et al.* 2009).

Presumably, the advocate of ecological restoration would object to this practice becoming a norm or ideal for environmental management. But why? What *additional* value is bestowed on a place or an ecosystem when it exhibits fidelity to its own past? Is historical fidelity just an expression of collective nostalgia (which need not be a bad thing)? Or can we give it a deeper justification?

Before I move on, I wish to address one definitional wrinkle. Some scientists use the term "restoration" very broadly, to refer to any sort of intensive habitat modification, regardless of whether it achieves fidelity (Hobbs and Cramer 2008: 40, suggest such broad usage). Sarkar (2011: 337, 2012: 139) claims that this broad usage is typical among practitioners. The question of how frequently, or infrequently, scientists use this broad sense of "restoration" is a sociolinguistic claim that I am not prepared to assess. Moreover, I think the attempt to substantiate that claim (that is, that this broad sense of "restoration" is typical in the field) would lead to difficult interpretive problems (Garson 2014). At any rate, in the following I will always use the term "restoration" to designate a project that centers around historical fidelity, and I will use the term "reconstruction" (following Sarkar 2011) for the broader category, that is, to denote any sort of beneficial habitat modification. The more interesting question is: why should we restore ecosystems?

Why is ecological restoration valuable?

There are two kinds of benefits associated with restoration: the benefits associated with the *product* of the restoration (the restored ecosystem), and the benefits associated with the *process* of restoration (the human activity). The benefits of the product can be vast and fairly obvious. They include goals such as promoting wildlife habitat, biodiversity, and the sustainable harvesting of "natural capital" (Aronson *et al.* 2007). This is particularly so if an ecosystem's ability to provide such goods has been crippled by damage or harm. Psychologically and socially, restored ecosystems may provide greater opportunities for education, entertainment, research, and the interaction of people with nature.

The practice of restoring ecosystems can promote values such as community, volunteerism and teamwork, particularly in small-scale, local restorations. It can also provide opportunities for education, and for exposing people to nature. I'll give examples of both sorts of benefit.

First, there are clear benefits associated with the product of restoration, that is, having restored ecosystems. On April 20, 2010, the Deepwater Horizon drilling rig exploded off the Gulf of Mexico. The explosion caused the release of over 200 million gallons of oil into the Gulf Coast, and jeopardized livelihoods of entire communities, particularly along the Mississippi–Louisiana coast.[2] In the wake of that spill, President Barack Obama made several speeches calling for the "restoration and recovery" of the Gulf Coast and calling on oil giant BP (to whom the rig was leased) to pay a large part of it.[3]

Interestingly enough, the president used that opportunity to promote a restoration plan that the White House had been designing even *prior* to the spill.[4] As he emphasized in those speeches, the rig explosion was only the latest in a series of environmental insults to the US Gulf Coast carried out for over a century. Much of the harm to the region has come from anthropogenic modification of the Mississippi River. These modifications include dredging – artificially deepening the river by extracting silt from the riverbed – and the construction of levees that prevent occasional flooding of the river. Dredging and levee construction have the effect of destroying wetlands along the banks of the river.

The restoration plan that the president promoted – the "Roadmap for Restoring Ecosystem Resiliency and Sustainability" – enumerates at least six different benefits of restoration (that is, the benefits of having a restored ecosystem). These include: flood and storm protection for coastal residents; biodiversity conservation; reduction of potential impacts of climate change (particularly sea level rise); commercial fishing; aesthetic and recreational value; and preservation of the cultural legacy of coastal communities. Among the main mitigation practices are the "beneficial use of dredged materials," that is, transporting sediment dredged from the river directly to the wetlands, and diversions, which include breaches that permit the waters of the river to enter adjoining wetlands and replenish them directly.

There are also benefits associated with the practice of restoring ecosystems. These benefits are particularly evident in the context of small-scale, local restoration projects that rely heavily on volunteer help. One example of the community benefits of restoration is "prairie restoration day," a venture supported by the Sierra Club, which gives inner-city adolescents the opportunity to participate in prairie restorations. In 2009, the Illinois chapter of the Sierra Club brought fifteen teenagers to Theodore Stone Forest Preserve and taught them to identify and safely cut down invasive European buckthorn trees. Most of the teenagers were from South Lawndale, a primarily Latino neighborhood plagued by the city's oldest coal-burning power plants and a near-absence of green space for children. In this case, restoration primarily served the value of exposing young people to the natural world and cultivating a sense of responsibility for it.[5] Such examples have led theorists such as Eric Higgs (1997; also see Light and Higgs 1996) to claim that community participation should be considered a *criterion* of a good restoration, in addition to more "technical" criteria such as fidelity and sustainability.

Another example of the benefits of the practice of restoration is the restoration of Tiritiri Matangi Island of New Zealand (see Craig and Vesely 2007). Once forested, Tiritiri Matangi was partially cleared by Maori settlers in the sixteenth century, and then farmed intensively in the mid-nineteenth century when it was taken over by British colonists. In 1971, the government decided to allow the land to regenerate "naturally," but after a decade, few changes took place. In 1984, a restoration plan was implemented which involved reconstruction of forests and reintroduction of select bird species.

From its inception, one of the project's main goals was to encourage public support and to reconnect urban communities to their natural heritage. To that extent, the restoration was wildly successful; the opportunity to plant trees and observe endemic and newly introduced species drew so many volunteers that it became necessary to ration the number of seeds each person could plant. A volunteer organization, Supporters of Tiritiri Matangi (SoTM), started in 1987 and has become the largest supporter of the island. The island now draws an estimated 35,000 visitors annually, primarily from Auckland.

In short, there is little question about whether restorations can sometimes be beneficial. (There is another question here regarding how seriously these projects took the goal of historical fidelity. Certainly, historical fidelity played some role in guiding these projects, even if the projects gave themselves much more creative leeway than historical fidelity, rigidly adhered to, would allow.) To the extent there is a deep philosophical question here, the question has to do with whether comparable benefits could not be achieved in some other way, one that involves less money and time. I will examine this prospect in the last section.

Is ecological restoration a big lie?

Some philosophers think that the very idea of ecological restoration, or "nature restoration," more generally, contains an inherent contradiction. [That is why Katz (1992) called it the "big lie."] They also think that, once we see the contradiction, we will no longer be disposed to attach much value to such projects.

The argument is almost deceptively simple, though there are minor variations on it. I'll rely on Katz's (1992) influential version of the argument. There are actually two parts to the argument. First, what makes something "natural" is the fact that people haven't designed it for anything. This lack of being designed for something is what makes something a piece of nature. It is what distinguishes a rock from a paperweight, or a sheet of Arctic ice from a laptop projector. As soon as people start to modify or manipulate something for their own purposes, it becomes an artifact and is no longer natural. This is true even if their purposes are beneficial. So, "nature restoration" is a contradiction in terms. All we can ever do is produce more artifacts.

The second part of the argument tries to show that artifacts are less valuable than the pieces of nature they replace. A natural system that has been damaged by human activity, say, flooding caused by climate change, is still a piece of nature. It has not been deliberately designed or engineered by people, so it is still better than an artificial ecosystem. In short, since restored ecosystems are artifacts, and artifacts are not as valuable as the natural states they replace, we should generally avoid restoring ecosystems.

Why are natural states so much better than artifacts? Theorists have different views here. In Katz's view, the reason is that nature that has not been designed by humans is autonomous, in the sense that it is free of human domination (Katz 1992: 239; also see Katz 2010). In one expression of this view, he uses the Kantian distinction between an end-in-itself rather than a means to an end. When humans restore nature, they redesign the natural world to serve human purposes (even benevolent ones). They thereby change its moral status from being an end-in-itself to a means to some human end (Katz 1993: 230).

Elliot (1982, 1997) has a different view about the value of nature. In his version of the argument, restored ecosystems are like counterfeit paintings. A counterfeit Van Gogh may have some value or another, but not the same value as the original, even if the original were marred somehow. What gives a Van Gogh painting its inestimable value has to do, in part, with historical facts about it and not just its current-day properties. This point is easy to recognize when we are talking about paintings, but Elliot thinks the same point holds for ecosystems.

He asks us to suppose that we are walking through what we take to be a pristine piece of nature. Suddenly, we discover that a vast corporation constructed the entire ecosystem only years earlier, and even designed the natural feel. (Consider that the nineteenth-century architects of New York's Central Park, Frederick Law Olmsted and Calvert Vaux, designed the Park's North Woods, very deliberately, to convey the impression of wild nature, down to the placement of specific trees and rocks. One encountering the North Woods for the first time, without any knowledge of the history or the mechanics of the park, might think that the architects had simply discovered it that way and had left it alone. There is one stream in particular that runs down a formation of rocks and flows into a pond; engineers can literally turn it on and off with a hidden faucet.) Elliot thinks that we would attach less value to the experience, much as if we discovered that the Van Gogh we were admiring was a fake. Elliot's argument seems plausible to me, though I understand that people have different intuitions about this case.

It might be tempting to dismiss this argument as a philosopher's riddle, one that has little bearing on real conservation practice. But that would be an error. Anti-restorationist arguments have managed to escape from purely academic discourse into the realms of practice and policy. For example, a recent scientific primer on ecological restoration devoted a section to discussing what it calls the "big lie" of restoration (Vidra and Shear 2010: 205). (Although the "lie" that they refer to is not quite what Katz had in mind, the language is obviously borrowed from Katz.) These arguments have also provided a philosophical basis for opposition to specific restoration plans, such as a recent plan to cull lake trout (*Salvelinus namaycush*) from Yellowstone for the purpose of protecting cutthroat trout (*Oncorhynchus clarkii*). According to its critics, given that the supposed "restoration" of the cutthroat habitat requires intensive human management, such a system would no longer be "natural."[6] That shows that worries about the nature/artifact divide run deep in restoration.

Elliot's and Katz's concerns have produced a sizable philosophical literature. There are two main sorts of objections that have been repeatedly raised. The first challenges the sharpness of the distinction between nature and artifact that Katz's argument relies on. After all, aren't humans an evolved feature of the natural world? And don't some non-human animals create artifacts, like beaver dams (e.g. Scherer 1995, Hettinger 2002, Ladkin 2005; see Katz 2002 for a response; Sober 1995 and Ereshefsky 2009 develop similar themes.) Along these lines, some critics think that Katz's argument relies on the sort of dualistic view of the human–nature relationship that we should move beyond, and that we should replace with a more positive, interactionist viewpoint (e.g. Cowell 1993, Jordan 1994, Chapman 2006).

A second response is to accept that the restored ecosystems are artifacts, but to challenge the moral implications of that fact. Artifacts are not always less valuable than the natural states they replace (Lo 1999, Light 2000). Even Katz agrees that "a house is better than a tree or a cave," and "the complex artefactual system we call medical practice is better than letting natural diseases take their course" (Katz 2002: 142). So why can't we say the same about ecosystems? At the very least, one might think that the loss of autonomy is, in part, offset by the kinds benefits of restoration that I enumerated above.

One of the points that Katz's critics have not generally noticed, however, is that Katz's argument rests crucially upon a problematic analysis of the concept of *function* – as in, for example, "the function of the heart is to circulate blood," or, "the function of the pocket watch is to tell time." (Though see Lo 1999, and Siipi 2003, who emphasize the idea of function in their criticism of Katz.) Here, in a nutshell, is his argument. He begins by asking, what is the difference between nature and artifact (p. 234)? What makes one thing natural and another thing artificial? He proposes that the difference between them is that artifacts have *functions* and entities in nature do not. Very specifically, he says that in order for something to have a function, it must have been designed for a purpose (1992: 235). In this restrictive sense of the term, laptops and tables have functions, but trees and organs do not. As he puts it, "natural individuals were not designed for a purpose. They lack intrinsic functions, making them different from human-created artifacts" (1992: 235). He thinks that, even though ecologists sometimes attribute functions to populations, populations do not really have functions because nobody designed them for anything: "[a]lthough we often speak as if natural individuals … have roles to play in ecosystemic well-being … this kind of talk is either *metaphorical or fallacious*. No one created or designed the mountain lion as a regulator of the deer population" (1992: 235).

Katz's restrictive view about function underpins his argument that restored ecosystems can never be anything more than unnatural artifacts. The idea is that as soon as human beings deliberately modify or manipulate something, they impose a function on it. The thing now has the function of doing whatever it was designed for. As a consequence of this (this acquisition of a function), it loses its membership in the category of natural things, and it becomes an artifact. This is precisely why "restored areas will never be natural – they will be anthropocentrically designed human artifacts" (1992: 235).

But I think Katz's argument is mistaken. Katz's view of function seems to have the absurd implication that, say, the heart does not really have the function of pumping blood. Many philosophers of biology believe that parts of individuals, such as internal organs, as well as individuals within an ecosystem, can possess functions, in a literal sense (see Garson 2016, for a critical overview of the functions literature). So I think there is a false premise in his argument. If we do not accept his view of function, then we should not accept the way that he distinguishes nature and artifact. Of course, Katz may have some other way of distinguishing between nature and artifact, one that does not rely on function. If so, that distinction should be spelled out.

Is ecological restoration arbitrary?

When we decide to restore an ecosystem, we have to select a historical baseline. We have to decide what point in time we wish to reset the ecosystem to. Some people believe that the choice of a baseline is arbitrary (see Callicott 2002, Sarkar 2011). Suppose we agree that we should restore the Gulf Coast, and that historical fidelity matters. We still have to decide how far back to go. Ten years ago, shortly before the rig explosion? A hundred years ago, before the era of levee construction and dredging? How about 500 years ago, before the arrival of European settlers? Or over 15,000 years ago, before the arrival of *Homo sapiens* to North America? There is a kind of arbitrariness here. As Callicott (2002: 412) puts it, "The condition that Dane County, Wisconsin was in at the moment European settlers saw it in the 1840s ... is but a snapshot in its ever-changing ecological odyssey. Why seize on that condition as the norm for restoration, rather than its condition at some earlier or, for that matter, later moment?"

The arbitrariness objection is particularly pointed when we reject the idea that there is some sort of equilibrium state that ecosystems naturally return to after they have been disturbed (Callicott 2002). This is an idea that was promoted by the early twentieth-century ecologist Frederic Clements (1916), who compared ecosystems to organisms. Clements argued that just as the fetus naturally moves toward a certain end-point, namely, the mature adult, so too does the ecosystem, once disrupted, move toward a certain end-point or equilibrium which he called a "climax." If that picture of ecosystems were correct, then restoration could be viewed, most favorably, as an attempt to hasten the ecosystem back to this point of equilibrium. In the absence of such guiding paradigm, we are back to the threat of arbitrariness.

In addition to the choice of a baseline, there are other sorts of choices that must be made. Which features should be the target of restoration? If we are to restore the Louisiana Gulf Coast, presumably we do not care very much about restoring the precise shape of the coast as much as we care about, say, its ability to sustain wildlife. What degree of similarity between the current ecosystem and the baseline is sufficient before we determine that the restoration has been satisfactory (e.g. Falk *et al.* 2006)? There does not seem to be any principled way to answer these questions. But I will focus on the choice of a temporal baseline.

It seems to me that this arbitrariness is not a serious problem. It would only be a problem if we thought that historical fidelity was the only value that should ever guide restoration projects. (If I asked my stakeholders, "what sort of ecosystem do you want?" and they replied, "one that exhibits historical fidelity!" then I would be at a loss to help them.) Fortunately, historical fidelity is not the only value that guides environmental planning. Other goals include maximizing biodiversity, creating sustainable habitats, creating a sense of wildness, promoting cultural and scientific opportunities, and so on. There is nothing arbitrary about invoking such principles in order to compare the merits of different restoration plans.

Here is one way of thinking about it. Historical fidelity sets a preliminary filter on the sorts of plans that are worth considering. In other words, when we set about restoring an area, we first contemplate all of the different historical baselines we could plausibly restore to (e.g. one century ago, five centuries ago, several thousand years ago ...). This will give us a large set of potential baselines. Once we have these plans in mind, we then apply other values, such as biodiversity, and so on, to decide which baseline is the best. Restoring the Gulf Coast back to the way it was before the last glacial period would amount to submerging it in the ocean. This would be bad for biodiversity (though perhaps not so bad for marine biodiversity) and bad for human livelihood, so we should reject it.

Callicott attempted to formulate more principled guidelines regarding the choice of a baseline. He alludes to hierarchy theory in ecology and notes that ecological processes can be sorted into different temporal "scales" (Holling 1992), in particular, the microscale, mesoscale, and macroscale. Each scale is marked off by a certain sort of recurring ecological event. Processes such as vegetation growth and soil erosion can be measured over days and years (microscale). Processes such as speciation and the movement of continents are measured in millennia (macroscale). Somewhere in between, there is a third group of processes, the "mesoscale." These processes can be marked off by the regularity of disturbance regimes, which are measured in centuries: "… for coastal environments we might measure ecological time by the periodicity of disturbance by hurricane-force winds; for riparian environments by the periodicity of floods of various magnitudes, from seasonal fluctuation to the hundred-year flood cycle; for upland forests and grasslands, ecological time might be measured by the frequency of fire" (p. 414).

When we think about possible historical baselines, there are different temporal scales we may have in mind. Callicott thinks that we should generally prefer mesoscale-level time scales, and we should think about restoration in terms of centuries, rather than decades or millennia. His argument is that this scale coheres best with conventional restoration planning, and it coheres well with our current way of distinguishing native and non-native species. To my understanding, this is the main argument he gives for preferring the mesoscale level when thinking about restoration, namely that it provides a scientific justification for "the classic norms of ecological restoration" (p. 415), and in particular its preference for native over non-native species when it comes to restoration. This would exclude certain time scales, such as the attempt to recreate late-Pleistocene conditions in North America.

So, there is nothing inherently arbitrary about the choice of a historical baseline. There can be very good reasons for selecting one over another. The deeper question here is simply this: why should we care about historical fidelity at all, if comparable benefits can be achieved without it?

Is historical fidelity worth pursuing?

Sarkar (2011, 2012) develops an argument against historical fidelity, which I call the "replacement" argument. The idea is that, if we reflect on the reason we care about historical fidelity, we will see that it has an overtly instrumental character. For example, one might prefer historical fidelity because one believes that the past ecosystem was more self-sustaining than the present one. If so, then self-sustainability is ultimately what that person is after. If sustainability could be achieved through the construction of a novel ecosystem that bears little resemblance to the past, then the value of fidelity would drop out as irrelevant.

To give another example, one might prefer historical fidelity because one believes that, in the past, people had more meaningful interactions with the natural world. But in that case, what one is ultimately after is a more meaningful interaction with the natural world. If one could achieve this in some other way (for example, by taking up gardening or hiking), then, again, fidelity would become irrelevant. So fidelity is always a kind of placeholder for something else, something that matters to us in a more ultimate way.

Simply put, the replacement argument says that, when one appeals to the value of historical fidelity in the context of conservation planning, one can always appeal to some other goal instead, and doing so would not diminish, at all, the overall (expected) value of the conservation project. Sarkar is not against historical fidelity, per se. But he thinks we should recognize its instrumental character and not "deify" it as if it has some inherent value, independently of the goods it helps us to obtain. That also means that, if we come up with some alternate way of achieving conservation goals for a certain area, one that does not exhibit historical fidelity, we should be open to pursuing that.

How can we respond to this challenge? One argument is to say that, conceptually speaking, historical fidelity has merely instrumental value, but that practically and empirically speaking, it is difficult to achieve those other goals (biodiversity, and so on) without, in fact, recreating the past. I think this would be a difficult argument to sustain. First, we often do not have very much information about what the past ecosystem was like, but, at least sometimes, we achieve our goals (biodiversity, and so on) well enough without it. Moreover, presumably as our knowledge grows and our technology advances, we will have a better ability to construct entirely novel ecosystems that successfully promote a range of conservation goals.

Another sort of argument is a precautionary one (Desjardins 2015: 81). In some cases, such as the Louisiana Gulf Coast, we have good reasons to think that the past ecosystem functioned very well (that is, immediately before the era of mass dredging and levee construction). We cannot be entirely sure that a novel ecosystem, one entirely unprecedented, will work as well, because there are just too many uncertainties. Sometimes our attempts to transform ecosystems in entirely novel ways prove disastrous. It is probably safest just to recreate, as well as possible, the way things were in the past. Note that this precautionary argument, strictly speaking, does not contradict Sarkar's view. Sarkar's view is that historical fidelity merely has an instrumental character (in this case, it is instrumental to promoting a well-functioning ecosystem). But this precautionary argument, if it is correct, provides some justification for the idea that historical fidelity should be a default mode of environmental planning.

I think the precautionary argument is a reasonable one. However, in some cases we should note that the same sort of reasoning would lead to the opposite conclusion, namely, that we should not restore an ecosystem, because we know, given changed ecological conditions, that the past will *not* work. The reconstruction of Governor's Island, noted above, is a good example. In this case, the design team that is leading the reconstruction chose not to replenish the island with historically native vegetation, because that vegetation would not be able to withstand the effects of sea-level rise due to climate change. Instead, they decided to plant non-native but salt-resistant vegetation.

It seems to me that in order to assess Sarkar's argument, we must reflect on our reasons for attaching value to any of the traditional conservation goals, such as biodiversity or wild nature. Why does biodiversity matter? Why does wild nature matter? Sarkar's own view is a broadly human-centered one, which is based on what he calls "transformative power" (Sarkar 2012: 51). This is loosely related to Bryan Norton's (1984) "weak anthropocentrism." The idea is that what confers value on biodiversity is its power to shape, or transform, human values. This is not the claim that biodiversity has value because people value it. It is to say that biodiversity matters because encounters with biodiversity have the power to change the things we value. He thinks that, as a rule, we ought to protect things that have this effect on us.

But if transformative power is the source of biodiversity's value (or the value of wild nature), why should we not say the same thing about historical fidelity? Can't the process of restoring nature have a transformative effect on people? In other words, I do not see the reason for treating its value as secondary to the value of biodiversity, rather than on a par with it. At the very least, the question should not be "is historical fidelity valuable?" The question should be "does historical fidelity have the same sort of value as other conservation goals such as biodiversity or wild nature?" In order to answer this question well, we must think deeply about the source of biodiversity's value itself.

Sarkar (2014) thinks that it is implausible that restored ecosystems, generally, possess transformative power. But I think it is more plausible when we consider that there are two different ways that restoration can have transformative power. First, the product of an ecological restoration can have transformative power. But perhaps more importantly, the process of restoration, that is, the activity of restoring ecosystems, can have transformative power, particularly in the context of small-scale or local restorations. For example, the reason that the restoration of New Zealand's Tiritiri Matangi Island was so successful – judged in social terms – was not just that it gave people an opportunity to interact with nature in some way or another. Rather, the process of restoration gave participants the sense that they were connecting with an important part of their national heritage. That has a distinctive psychological quality, which Higgs (2003) describes as bringing about "narrative continuity." Of course, that argument is most effective in the context of small-scale restoration projects that rely on volunteer help. It would not be as applicable to a massive restoration such as the restoration of the Gulf Coast. In those cases, it is probably better to rely on precautionary arguments in order to justify historical fidelity.

Ecological restoration and biodiversity

I have already implicitly stated my view about the relationship between ecological restoration and other goals, such as biodiversity conservation, but I will take a moment to explicate it. First, there is no conceptual or logical relation between biodiversity and ecological restoration. A restored ecosystem can, in principle, be less diverse than a current one, and a novel ecosystem, one that bears little resemblance to the past, might be more diverse than an existing one.

I see the two values as having comparable status, among other values such as sustainability and wild nature. They should all, jointly, play a role in designing conservation projects. There are rigorous and formal models that allow decision-makers to incorporate multiple values into conservation planning (or "multi-criteria analysis" – see Sarkar and Garson 2004, Moffett and Sarkar 2006). Such models allow decision-makers to consider several different values, or constraints, simultaneously when ranking potential conservation plans, such as biodiversity, wildlife habitat, and socio-cultural opportunities or costs. I think historical fidelity is best understood as one among other such constraints. In some cases, such as the reconstruction of Governor's Island, that constraint may be minimized or trumped by the collective importance of the other criteria. Historical fidelity is a defeasible goal, but not one that should be dispensed with entirely.

Acknowledgments

I am grateful to David Frank, Eric Katz, Jay Odenbaugh, Anya Plutynski, and Sahotra Sarkar for discussion of some of the material contained here.

Notes

1 www.ser.org/docs/default-document-library/ser_primer.pdf (accessed 1 August 2015).
2 www.nytimes.com/2010/08/03/us/03spill.html (accessed 3 August 2015).
3 www.whitehouse.gov/the-press-office/remarks-president-nation-bp-oil-spill (accessed 3 August 2015).
4 www.whitehouse.gov/administration/eop/ceq/initiatives/gulfcoast (accessed 3 August 2015).
5 http://vault.sierraclub.org/sierra/200911/prairie.aspx (accessed 3 August 2015).
6 www.nytimes.com/2011/08/24/us/24trout.html (accessed 4 September 2011).

References

Aronson, J., S. J. Milton, and J. M. Blignaut, eds. 2007. *Restoring Natural Capital: Science, Business, and Practice*. Washington, DC: Island Press.

Callicott, J. B. 2002. "Choosing Appropriate Spatial and Temporal Scales for Ecological Restoration." *Journal of Biosciences* 27: 409–420.

Chapman, R. L. 2006. "Ecological Restoration Restored." *Environmental Values* 15: 463–478.

Clements, F. E. 1916. *Plant Succession: An Analysis of the Development of Vegetation*. Washington, DC: Carnegie Institute of Washington.

Cowell, C. M. 1993. "Ecological Restoration and Environmental Ethics." *Environmental Ethics* 15: 19–32.

Craig, J. and E.-T. Vesely. 2007. "Restoring Natural Capital Reconnects People to Their Natural Heritage: Tiritiri Matangi Island, New Zealand." In *Restoring Natural Capital*, ed. J. Aronson, S. J. Milton, and J. N. Blignaut, 103–111. Washington, DC: Island Press.

Desjardins, E. 2015. "Historicity and Ecological Restoration." *Biology and Philosophy* 30: 77–98.

Elliot, R. 1982. "Faking Nature." *Inquiry* 25: 81–93.

Elliot, R. 1997. *Faking Nature: The Ethics of Environmental Restoration*. London: Routledge.

Ereshefsky, M. 2009. "Defining 'Health' and 'Disease'." *Studies in History and Philosophy of Biological and Biomedical Sciences* 40: 221–227.

Falk, D. A., M. A. Palmer, and J. B. Zedler. 2006. *Foundations of Restoration Ecology*. Washington, DC: Island Press.

Garson, J. 2014. "What is the Value of Historical Fidelity in Restoration?" *Studies in History and Philosophy of Biological and Biomedical Sciences* 45: 97–100.

Garson, J. 2016. *A Critical Overview of Biological Functions*. Dordrecht: Springer.

Hettinger, N. 2002. "Humans in the Natural World." *Ethics and the Environment* 7(1): 109–123.

Higgs, E. 1997. "What is Good Ecological Restoration?" *Conservation Biology* 11: 338–348.

Higgs, E. 2003. *Nature by Design*. Cambridge, MA: The MIT Press.

Hobbs, R. J., and V. A. Cramer. 2008. "Restoration Ecology: Interventionist Approaches for Restoring and Maintaining Ecosystem Function in the Face of Rapid Environmental Change." *Annual Review of Environment and Resources* 33: 39–61.

Hobbs, R. J., E. Higgs, and J. A. Harris. 2009. "Novel Ecosystems: Implications for Conservation and Restoration." *Trends in Ecology and Evolution* 24: 599–605.

Holling, C. S. 1992. "Cross-scale Morphology, Geometry, and Dynamics of Ecosystems." *Ecological Monographs* 62: 447–502.

Jordan, W. R. 1994. "'Sunflower Forest': Ecological Restoration as the Basis for a New Environmental Paradigm." In *Beyond Preservation: Restoring and Inventing Landscapes*, ed. A. D. Baldwin, J. De Luce, and C. Pletsch, 17–34. Minneapolis: University of Minnesota Press.

Katz, E. 1992. "The Big Lie: Human Restoration of Nature." *Research in Philosophy and Technology* 12: 231–241.

Katz, E. 1993. "Artefacts and Functions: A Note on the Value of Nature." *Environmental Values* 2: 223–232.

Katz, E. 2002. "Understanding Moral Limits in the Duality of Artifacts and Nature: A Reply to Critics." *Ethics and the Environment* 7(1): 138–146.

Katz, E. 2010. "Anne Frank's Tree: Thoughts on Domination and the Paradox of Progress." *Ethics, Place, and Environment* 13: 283–293.

Ladkin, D. 2005. "Does 'Restoration' Necessarily Imply the Domination of Nature?" *Environmental Values* 14: 203–219.

Light, A. 2000. "Ecological Restoration and the Culture of Nature." In *Restoring Nature*, ed. P. H. Gobster and R. B. Hull, 49–70. Washington, DC: Island Press.

Light, A., and E. Higgs. 1996. "The Politics of Ecological Restoration." *Environmental Ethics* 18: 227–247.

Lo, Y. S. 1999. "Natural and Artifactual: Restored Nature as Subject." *Environmental Ethics* 21: 247–266.

Moffett, A., and S. Sarkar. 2006. "Incorporating Multiple Criteria into the Design of Conservation Area NETWORKS: A Minireview with Recommendations." *Diversity and Distributions* 12: 125–137.

Norton, B. G. 1984. "Environmental Ethics and Weak Anthropocentrism." *Environmental Ethics* 6: 131–148.

Sarkar, S. 2011. "Habitat Reconstruction: Moving Beyond Historical Fidelity." In *Philosophy of Ecology*, ed. K. de Laplante, B. Brown, and K. Peacock, 327–361. Amsterdam: Elsevier.

Sarkar, S. 2012. *Environmental Philosophy: From Theory to Practice*. Malden, MA: Wiley-Blackwell.

Sarkar, S. 2014. "Environmental Philosophy: Response to Critics." *Studies in History and Philosophy of Biological and Biomedical Sciences* 45: 105–109.

Sarkar, S., and J. Garson. 2004. "Multiple Criterion Synchronization (MCS) for Conservation Area Network Design: The Use of Non-dominated Alternative Sets." *Conservation and Society* 2: 433–448.

Scherer, D. 1995. "Evolution, Human Living and the Practice of Ecological Restoration." *Environmental Ethics* 17: 359–379.

Siipi, H. 2003. "Artefacts and Living Artefacts." *Environmental Values* 12: 413–430.

Sober, E. 1995. "Philosophical Problems for Environmentalism." In *Environmental Ethics*, ed. R. Elliot, 226–247. Oxford: Oxford University Press.

Vidra, R. L., and T. H. Shear. 2010. "Ethical Dimensions in Ecological Restoration." In *Ecological Restoration: A Global Challenge*, ed. F. A. Comin, 100–111. Cambridge: Cambrdge University Press.

INDEX

Page numbers in **bold** refer to a table or figure in the text.

Sarkar, S. and Montoya, M. 286–7
satisfaction, human 206
satisfaction, moral 190
satisfaction, relative 243, 247
scepticism 51, 58–9, 60, 61, 65, 66, 114–16
Schnall, S. 202–3
Scholes, R. J. and Biggs, R. 270
Schulze, E-D. and Mooney, H. A. 272
Scientism 44, 45–8, 52
Scovell, E. L. 14
sea level rise 32, 111, 327, 328, 334
sediment 145, 262, 329
sentience 170, 171, 193
sentimentalism 199–203, 204, 207, 208
Sepkoski, J. John, Jr 35, 36
Serengeti National Park 196
Shannon Evenness Indices 45, 99, 104, **112**
Shelford, Victor 18–19
Sierra Club 329
Silvertown, J. 271
Simberloff, Dan 219, 221
similarity 101; dissimilarity 70, 73, 74, 75; natural/
    unnatural kinds 61–2, 63, 64, 65, 128, 132–3
Simon, Herbert 203
Simpson, E. H. 101
Singer, P. 165n6, 171
single nucleotide polymorphisms (SNPs) 255
slavery 173–4, 176, 177, 180
Smithsonian Institution 28, 43, 97
snakes, Australian 121
Sober, E. 78
social ecology 51, 286, 287
social equality 302, 317, 318
social outcomes 298–301
Society for Conservation Biology (US) 44, 174
Society for Ecological Restoration (SER) 326
soil 113, 205, 244, 327; biodiversity loss 272, 273;
    microbial diversity 141, 142, 144, 146
solar energy 316
Solow, R. M. 315
Soulé, M. 44, 78, 97–8, 155, 174, 248
Southern Oscillation Index 233
Southwood, T. R. E. 56
spatial areas 4
spatio-temporal scale 148, 271, 303, 333;
    contextual pluralism 98, 101; ecological
    hierarchy 116, 119–20; ecological sustainability
    314, 316, 321, 322
species abundance 104, 139, 141, 235
species-area effect 33
species conservation 12, 70, 96, 105, 216,
    217–18, 220–1
species diversity 87, 251–62; contextual pluralism
    98–9, 100, 101, 104, 105; ecological hierarchy 4,
    111, **112**; microbial diversity 142, 145; models
    of biodiversity 60, 66; nature of biodiversity 11,
    12, 13–15, 20, 22

species, endangered 15, 118, 270, 283;
    biodiversity indicators 232, 233, **234**; genetic
    measurement 252–5, 256–7; intrinsic value
    168–9, 176, 177, 180; structured decision-
    making 242, 244, 245
species introductions 125, 130, 131, 132, 136n6
species, invasive 134, 213
species, keystone 130, 189, 234
species loss 2–3, 26–37, 58, 91, 99, 122, 204;
    assisted colonization 212–13, 221, 222;
    de-extinction 130, 131, 132, 136n5, 221;
    measurement and methodology 6, 229, 237n2,
    268–9; microbial diversity 146–7, 148; models
    of biodiversity 69, 76–7; stability 28–31, 32, 33,
    36; unnatural kinds 130, 131, 132, 136n5,
    see also biodiversity loss
species, native 91, 333, 334
species, non-native 91, 218–20, 286, 327, 333, 334
species richness, *see* richness
species, sibling 261
species turnover 271–2
sponges, marine 147
Srivastava, D. 273
ß (beta)-diversity 44, 46
stability 1, 21, 44, 111, 163, 272–3, 332; definition
    of biodiversity 120–1; diversity-stability
    relationship 45, 88, 92–3, 111; ecological
    hierarchy 113–14, 117–19; eliminativism 88,
    92–3; species loss 28–31, 32, 33, 36
Steele, K. 89
Sterelny, K, *see* Maclaurin, J. and Sterelny, K.
sticklebacks 255, 259, 260
'stuff' 315, 322–3
subjectivity 171–4, 180–1, 247
subspecies 53n8, 258, 259, 260, 261
substitution 135, 313, 320
success clause 64, 65, 66
suitability 81, 128, 133–4, 242, 269, 296
Sullivan, Sian 163
Sunderland, T. C. 285, 286, 289, 290
supernatural 127
surrogates 74, 105, 288; biodiversity indicators 233,
    234; biodiversity loss 269, 270; eliminativism 86,
    87–8, 91, 92; nature of biodiversity 46, 49–50,
    53n13, 57, 59–60, 61
surrogates, true 59–60, 87–8, 107n7
Survival Service Commission 15
sustainability 7, 49, 122, 147, 311–23; definition
    of biodiversity 58; local culture 283, 291;
    measurement and methodology 237n1, 274
sustainability ethics 317–19
sustainable development 294; ecological
    sustainability 311, 312–13, 315; local culture
    286, 287
sustainable yield 311–14, 319
Swedish International Development Agency
    (SIDA) 284